# World geomorphology

# World geomorphology

E.M. BRIDGES

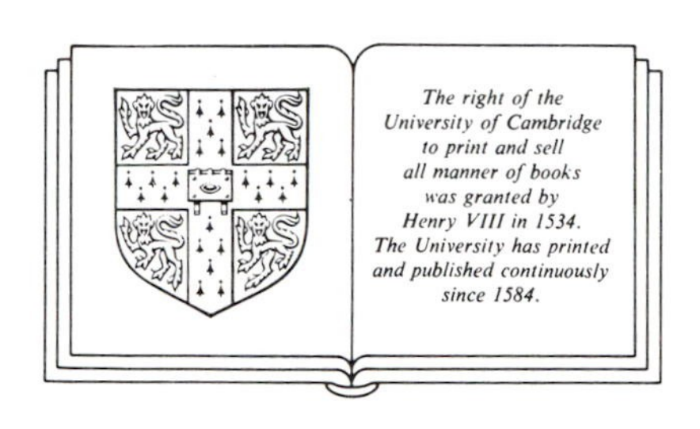


CAMBRIDGE UNIVERSITY PRESS

*Cambridge*
*New York Port Chester*
*Melbourne Sydney*

Published by the Press Syndicate of the University of Cambridge
The Pitt Building, Trumpington Street, Cambridge CB2 1RP
40 West 20th Street, New York NY 10011, USA
10 Stamford Road, Oakleigh, Melbourne 3166, Australia



First published 1990

Printed in Great Britain at the Bath Press, Avon

*British Library cataloguing in publication data*

Bridges, E.M. (Edwin Michael) 1931–
World geomorphology,
1. Geomorphology
I. Title
551.4

*Library of Congress cataloguing in publication data*

Bridges, E.M. (Edwin Michael)
World geomorphology/E.M. Bridges.
p. cm.
Includes bibliographical references.
ISBN 0 521 38343 9. – ISBN 0 521 28965 3 (paperback)
1. Geomorphology. I. Title.
GB401.5.B75 1990
551.4'1–dc20 90-31050 CIP

ISBN 0 521 38343 9 hardback
ISBN 0 521 28965 3 paperback

SE

# *Contents*

# Preface

The study of geomorphology is an academic discipline devoted to the explanation of the earth's surface relief and to an understanding of the processes which create and modify landforms. In recent years, geomorphological textbooks have concentrated almost exclusively upon an examination of the detailed processes which take place in the weathering of rocks and the transport of debris as landforms are created and destroyed. An academic journal, *Earth surface processes* even takes its title from this aspect of geomorphology. The study of processes has enabled great progress to be made in geomorphology as process studies lend themselves admirably to measurement and modelling techniques. However, there is also a broader, regional aspect to geomorphological studies which has been rather neglected during the past two or three decades; the disillusionment which many physical geographers felt in the 1950s about the denudation chronology approach led eventually to the virtual abandonment of wider-scale studies.

A plea for more attention to large-scale geomorphology was made by Gregory in *The nature of physical geography* in 1985, but most geomorphologists, with a few notable exceptions, have avoided consideration of geomorphology on a large scale or on a regional basis. One notable occasion when broader issues were discussed was the symposium held by the British Geomorphological Research Group on 'Mega-geomorphology' and published in 1983. It is this background which provoked the author to provide a text which would cover this field of study and hopefully to interest students of all ages in the broader aspects of geomorphology.

Many geological and geophysical discoveries have taken place in the last two or three decades which are of considerable importance for the study of the large-scale geomorphology of the Earth. A new understanding of the formation and structure of oceanic basins brought a fresh, dynamic approach to the study of geology which must also be reflected in geomorphological investigations and explanations. So, this book begins with the large lithospheric plates, composed of continental and oceanic crust materials. These large fragments of the Earth's crust are then sub-divided into the broad physical regions of the emergent and submergent parts of the Earth's surface. Geomorphology does not stop at the low tide line.

Clearly, an understanding of the location of the continents and oceans on the surface of the earth has great geographical significance in terms of climate, soil and vegetation, but the past history and changing distribution of the lithospheric plates also provides clues for the explanation of many geological, biogeographical and geomorphological features. The theory of plate tectonics provides the most acceptable explanation currently available for the origin, development, distribution and modification of the Earth's major features.

All books are a compromise and this one is no exception. It would have been easy to bring in more examples of different landforms thus adding to the book's length but not its impact. Its aim has been to introduce students of geography, earth sciences, environmental sciences and biologists to the Earth's major landforms and to stress their importance to all persons interested in an academic study or practical assessment of the physical nature of the Earth's surface. Although the emphasis is upon the broad scale, sufficient detail has been included to highlight the differences between the various provinces and sections which comprise the major sub-divisions of the Earth's geomorphology.

Experience is the best teacher of physical geography and the author's good fortune in being able to set foot on all the major landmasses, with the exception of Antarctica, has helped to develop the eye for country, which is at the heart of geomorphological study. I would like to acknowledge the enlightenment I received from Professor D.L. Linton during my undergraduate studies and for the shared experiences on excursions with many friends and colleagues in the fields of geography and soil science. This book, and others which preceded it would have been impossible without the wholehearted support of my wife, who has endured not only my absences but has also helped significantly in the preparation of the manuscript in readiness for publication.

E.M. BRIDGES

# *Acknowledgements*

Preparation of this book has taken place over several years. Since it was first agreed to produce it several CUP commissioning editors have come and gone. The author is grateful to all of them for their patience and for keeping open the possibility of eventual publication, even when little progress was apparent. I am particularly grateful to Miss Lucy Purkiss and Dr Caroline Roberts, who after encouraging me to finish this book have seen it through the publication process. I wish to acknowledge with gratitude the sub-editing work of Howard Farrell whose deft handling of manuscript and author have resulted in a readable text with an attractive layout. The manuscript was prepared for publication on the word processors of 'Words', Word Processing Services, where Mrs G. Bridges and Mrs M. Owen provided typographical and editorial skills which have turned illegible writing into a manuscript. I am grateful to Miss Jane Ward, who extracted relevant material on the geomorphology of Africa and Asia in the early stages of preparation. It is a pleasure to record my thanks to members of the Cartographic Unit of the Department of Geography, University College of Swansea. Mr G.B. Lewis and his assistants A. Lloyd, P. Taylor and Mrs N. Jones have all contributed their skill to the illustrations contained in this book. Mr A. Cutliffe and Mr D.E. Price have assisted with photographic work by preparing prints and line illustrations in readiness for publication. Photographic illustrations are mainly those taken by the author on various pedological or geographical excursions, but grateful acknowledgement is made to Dr R.D.F. Bromley and Mr H.J.R. Henderson for photographs of landscapes in South America and South Africa respectively.

The most accessible references for further reading in this large subject are given at the end of the text in a bibliographic commentary (pp. 253–5) but the author also wishes to express his grateful thanks for the ideas and sources of information derived from the work of many other people not named individually which have come together in this book. *World geomorphology* originated as a companion volume to three other books published by CUP; *World vegetation* by Riley and Young, *World climate* by Riley and Spolton and *World soils* by E.M. Bridges. These four books cover basic aspects of the physical environment of interest to all students of geography and environmental science throughout the World.

# 1

# Introduction

An inspection of the globe immediately reveals an imbalance of areas of land and sea and the almost infinite variation of shape and relief forms which characterise the continental and oceanic areas (Figure 1.1). It is the intention of this book to describe and account for the development of the large-scale features of the Earth's surface. An attempt will be made to indicate why the composition, configuration and structure of the Earth's crust is of fundamental interest to all geographers and natural scientists.

The development of geomorphology can be traced from nineteenth-century physical geography, through a study entitled 'Physiography' which was propounded by T.H. Huxley, to the geomorphology of W.M. Davis at the beginning of the present century. Davis was responsible for introducing the three basic elements of geomorphology: structure, process and stage. Structure embraces the materials of the Earth's surface from which landforms are produced by the active processes of weathering, erosion and sedimentation. The stage, or morphological development of a landform, was related to the factor of time through which the processes had worked. In the last 20 years geomorphologists have tended to concentrate much of their energies unravelling the processes which operate and on how to measure them. Whilst in no way decrying this trend, this account of world geomorphology attempts to refocus attention onto the major physical features of the Earth's surface, emphasising their significance in terms of man's environment and economic well-being.

It would appear that the time is ripe to bring back to to geomorphology a wider perspective which is implicit

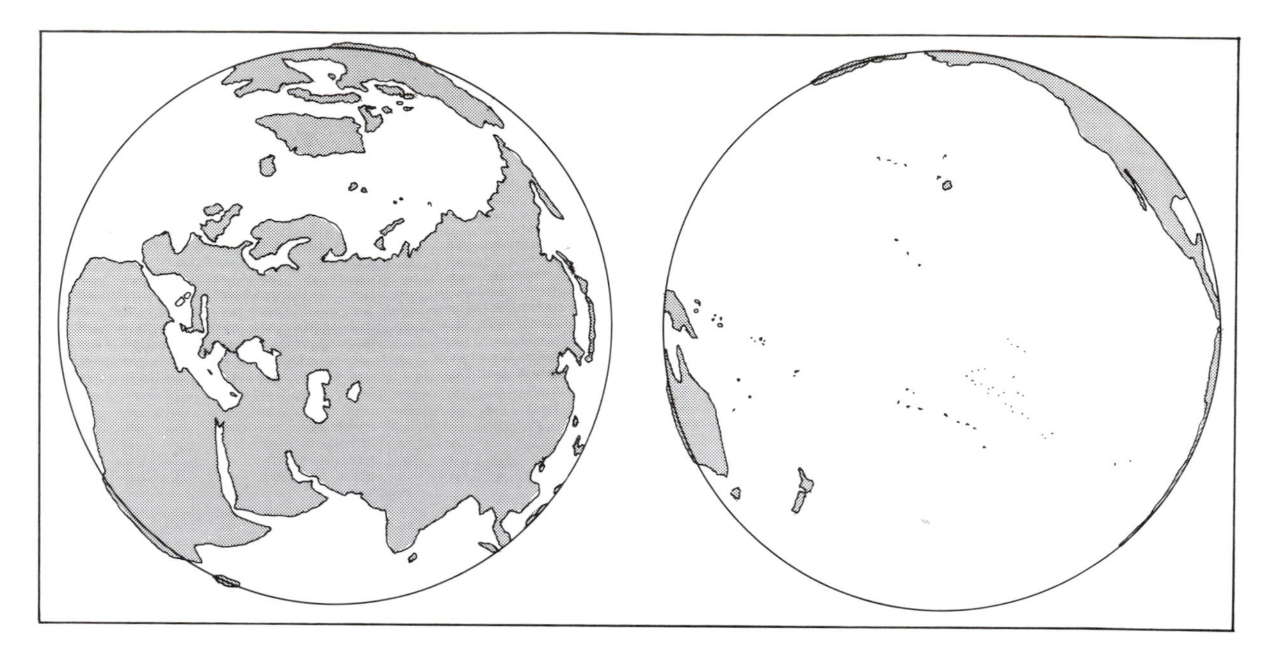

*Figure 1.1.* The unequal distribution of land and sea.

in the title *World geomorphology*. By definition geomorphology is concerned with the configuration of the Earth's surface, with relief and the way relief forms have developed into the landscape of the present day. The underlying significance of landscape, with its natural resources, is that the whole of human economic and social activity takes place upon it, and is inescapably linked to it.

Geographers require the analysis of geomorphology as a basis for an understanding of land form development and regional synthesis. Landforms are used extensively by geologists in mapping the distribution of rock outcrops. Soil scientists find a geomorphological analysis of the landsurface invaluable for mapping the distribution of the different types of material and soil; different sites tend to show different combinations of the soil forming factors. Ecologists and hydrologists also find the scientific description of landforms, as well as an understanding of landform evolution, to be significant in their studies. Land-use planners and civil engineers have found a knowledge of the origin, character and distribution of landforms useful in avoiding unnecessary expense in planning urban development and carrying out large-scale civil engineering projects.

As an explanation of the past and present distribution of continents, the theory of plate tectonics provides the most coherent account available. Besides giving an explanation of the fundamental relief of the Earth and why the continents are in the positions they now occupy, it can also be extended to explain the origin and distribution of mountain ranges, plains, plateaux, deep sea trenches and other major geomorphological features. Secondly, hazards such as earthquakes and volcanicity are not random in their distribution but are related to the margins of lithospheric plates. Thirdly, the occurrence of many economically valuable minerals and fuels has been shown to coincide with the plate margins. Fourthly, the break-up and movement of the continental masses provides some answers to the puzzle of palaeontological and present biogeographical distributions.

Following this introduction in which the geomorphological units of study are described, as well as the geological time-scale, the internal structure of the Earth is considered. This is an essential precursor for the understanding of the shape and disposition of the continents. In Chapter 2 the evolution of the sea floor and the continental masses is considered, setting the scene for discussion of the geomorphology of the major lithospheric plates. The general geological history of each lithospheric plate, including both continental and related oceanic areas, is described and the major divisions introduced in Chapters 3–9. The geomorphology of each province within these major divisions is then

Table 1.1. *The geological time-scale (in millions of years: my)*

| Era | Period | | Epoch | Base |
|---|---|---|---|---|
| Cenozoic era | Quaternary | Holocene | | |
| | | Pleistocene | | 2 my |
| | Tertiary | Neogene | Pliocene | 7 my |
| | | | Miocene | 26 my |
| | | Palaeogene | Oligocene | 38 my |
| | | | Eocene | 63 my |
| | | | Palaeocene | 65 my |
| Mesozoic era | Cretaceous | | | 135 my |
| | Jurassic | | | 190 my |
| | Triassic | | | 225 my |
| Upper Palaeozoic era | Permian | | | 280 my |
| | Carboniferous | | | 345 my |
| | Devonian | | | 395 my |
| Lower Palaeozoic era | Silurian | | | 430 my |
| | Ordovician | | | 500 my |
| | Cambrian | | | 570 my |
| Eozoic or Pre-Cambrian time | Upper Proterozoic | | | 1000 my |
| | Lower Proterozoic | | | 2000 my |
| | Archean | | | 3000 my |
| | Katarchean | | | 4500 my |
| | Origin of the Earth | | | |

presented in so far as information is available. It will be obvious to the reader that the information is uneven in amount, content and reliability. However, within the broad scale employed, it is possible to present an outline account of the many geomorphological regions of the world.

## The geological time-scale

The landscape which we see around us is the culmination of many millions of years of geological activity. The familiar relief of hills and valleys was formed by the operation of processes which, it is assumed, have been working in a similar manner throughout the Earth's history. Essentially, Earth history is deciphered from the evidence of rock strata and particularly the order in which they have been deposited. The sequence of sedimentary rocks was first worked out in the British Isles and the approach taken then has been applied and found satisfactory in other parts of the world. The sequence of rock strata is accompanied by the evolutionary development of their contained fossil fauna thus enabling correlation of rock strata across country from one outcrop to another.

The geographical formations and the evidence of evolving flora and fauna contained as fossils can be conveniently divided into four eras (Table 1.1). The *Eozoic*, a period in which there was little sign of life forms, was followed by the *Palaeozoic* in which there was a considerable variety of invertebrate fauna. The *Mesozoic* is characterised by the dominance of different forms of reptile, in particular the dinosaurs, and finally the *Cenozoic* covers the period of development of the mammals and other recent and present-day life forms. Geological formations of the Palaeozoic and Mesozoic are frequently referred to as the Primary and Secondary rocks respectively but the much shorter duration of the Cenozoic is unequally divided between the Tertiary and Quaternary, the latter referring to only the last two million years of Pleistocene and Holocene formations.

In terms of the evolution of the major continental areas of the world, developments since the Carboniferous formation was laid down are significant as considerable rearrangement of land and sea has occurred since that time. When considering the age of the oceanic floor, it has been found that there are few areas which are older than the Cretaceous. The Quaternary, including the glaciations of the Pleistocene, has great significance for the landforms of many regions of the world where glaciation took place. In many other regions the indirect effects of glaciation, such as those of low sea level, can be observed to have also had profound effects on landscape development.

The Earth is thought to have had its origin between four and five thousand million years ago. Evidence from the oldest parts of the continents indicates an age of over two thousand million years, but the application of normal stratigraphical methods of dating is made difficult in the Pre-Cambrian by the absence of palaeontological remains. A sub-division is made possible based on

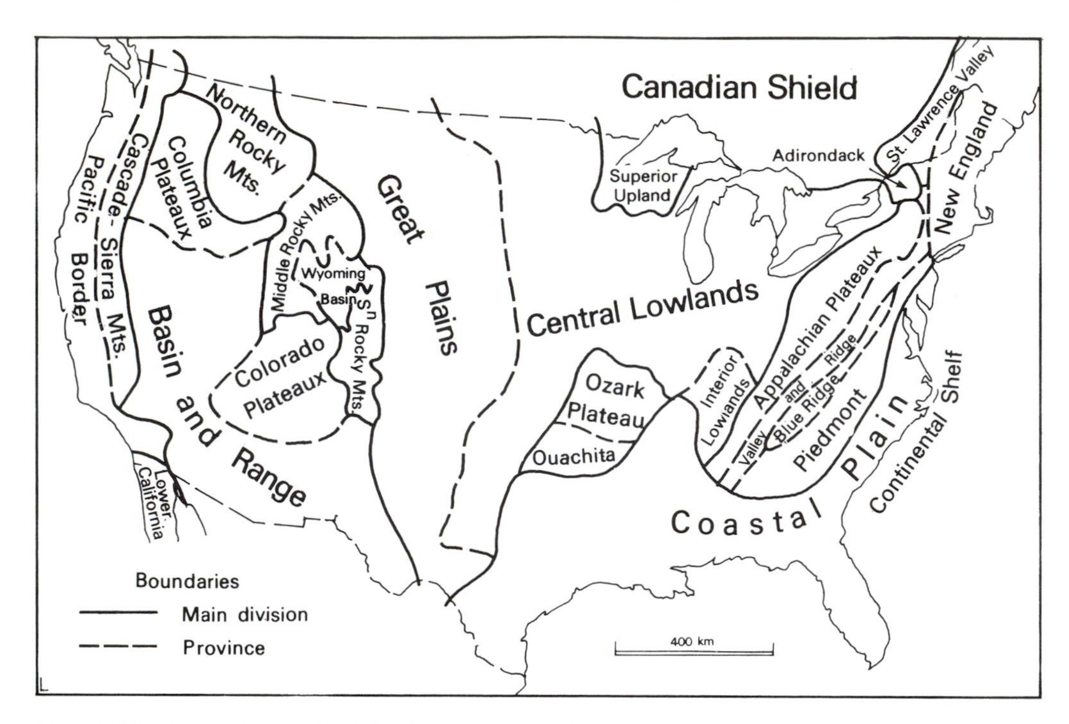


*Figure 1.2.* Physical provinces of the United States as prepared by Fenneman for the U.S. Geological Survey.

lithology and succession, backed by radioactive decay methods of dating. Four sub-divisions of Pre-Cambrian time are recognised: the Katarchean, the Archean and the Lower and Upper Proterozoic. There is evidence of several phases of mountain building throughout the Pre-Cambrian during which the continental nuclei were added to by geosynclinal action around their edges. This process is described in Chapter 2. Mountain-building episodes continued at intervals throughout the Earth's history and in part serve to sub-divide the long period of geological time.

Although the lithology of the rocks is more important than their age in the development of landforms, it is still necessary for the geomorphologist to have a clear picture in his mind of the sequence of events which has produced the materials upon which the processes of erosion act.

## Divisions of relief

When using weights and measures it is convenient to have units of different sizes, and the same principle exists in descriptive geomorphology where landforms of vastly different sizes are being discussed. It is possible therefore to speak of several different 'orders' of relief. After the whole Earth, the first order of relief a geomorphologist must take into account is the fundamental division of the Earth's surface into ocean basins and continents. As Linton has written, 'nature offers us two inescapable morphological unities and two only; at the one extreme, the indivisible flat or slope, at the other the undivided continent'. Between these two extremes it is necessary for the geomorphologist to search for the geomorphological equivalent of the highest common factor which will enable the delimitation of successively lower orders of relief features.

In 1916 a map showing the physiographic regions of the USA was produced by a committee of the Association of American Geographers. N.M. Fenneman who chaired this committee also produced a detailed account of the *Physiography of the United States* in two volumes. This major contribution to physical geography has stood the test of time and is used by Thornbury and others as a basis of their more recent accounts of the physical geography of the USA's part of the North American continent (Figure 1.2). The approach of Soviet geographers is briefly outlined in *The Geographical Magazine* volume 48, number 5 (the issue for

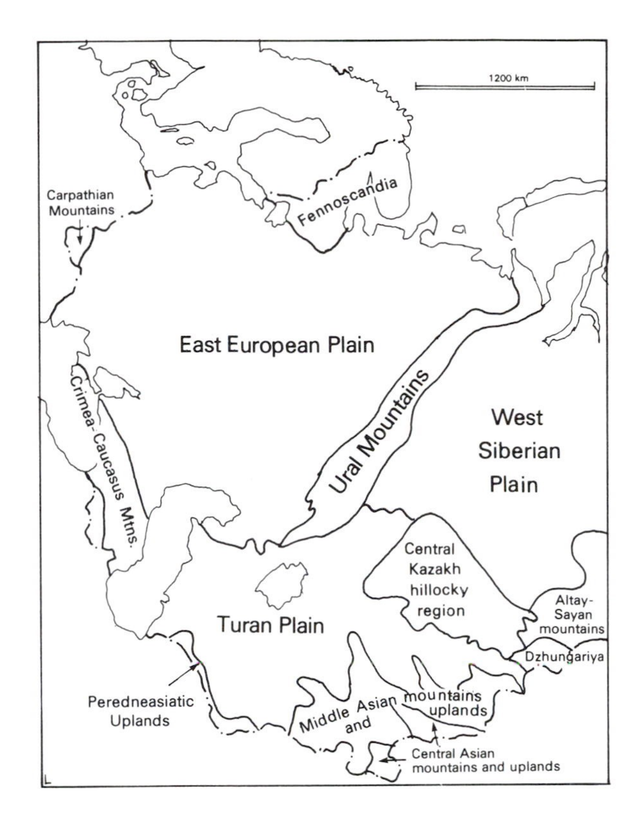


*Figure 1.3.* Natural environment complexes of the western USSR. (After *Geographical Magazine* 1976.)

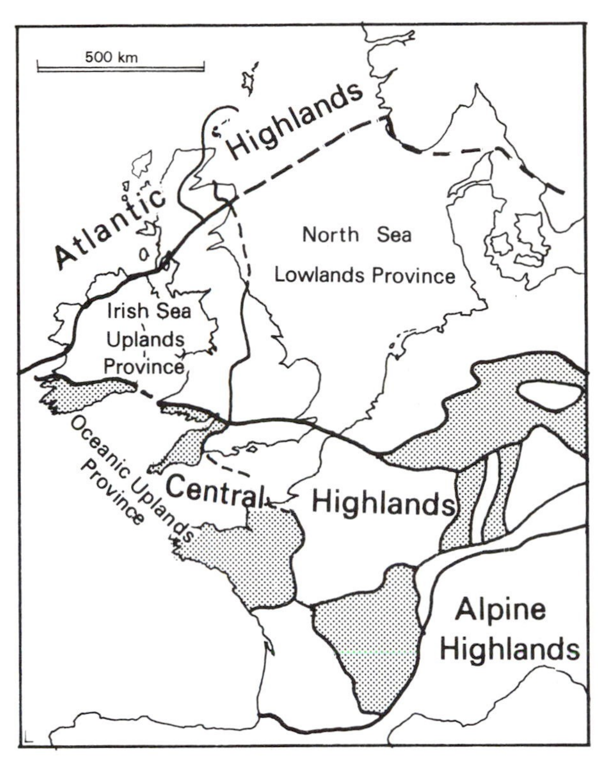


*Figure 1.4.* The major physiographic divisions of western Europe and their component physiographic provinces. (After Linton, in *London essays in geography*, 1950.)

February 1976). It is explained that the assessment of the complex natural regions of the USSR is 'essential for planning decisions and for constructive tasks on a regional or global scale' (Figure 1.3).

A second order relief feature would appear to be the primary continental sub-division which is referred to as a *major division* in the USA and as a *land* in the Soviet Union. These second order relief features are large and can only be applied to areas of sub-continental size. Examples are the central lowlands of America or the lowland of the European plain, areas of 500 000–800 000 km². The third order of relief recognised is the *province*, and the same term is used in both the USA and the Soviet Union. In the USA there are only three provinces which are smaller than the whole of England, so the dimensions of these third order features are still large, ranging up to 500 000 km².

If the concept of the highest morphological common factor is taken a stage further, a fourth order of relief can be determined: the *section*. A well-known example is the Black hills of Dakota, which form a structural dome of 8000 km² which, if further sub-divided, would lose the obvious unity of the area. The major physiographic divisions of western Europe as envisaged by Linton are shown in Figure 1.4.

It is at this point that there is a break in the classification and uncertainty exists, with various authors taking slightly different approaches. However, if a fresh start is made at the other end of the relief spectrum, with the indivisible flat or slope, it is possible to work upwards in size to meet the sequence established so far. A *site* is the indivisible flat or slope which Linton referred to as 'the electrons and protons' from which landscapes are built. The sites can be assembled into *facets*, units of the land surface which have similarity of surface form and uniformity of age and origin. In turn, facets can be assembled into characteristic associations referred to as *recurrent landscape patterns*. These are of such a size that they can be identified fairly readily upon air photographs and are much used in resource surveys of undeveloped regions. Several years ago, Linton suggested the terms stow and tract to

Table 1.2. *Orders of relief*

| Order of relief | Term | Approximate size |
|---|---|---|
| First | Continent or ocean basin | – |
| Second | Major division | 500 000 – 800 000 km$^2$ |
| Third | Province | 50 000 – 500 000 km$^2$ |
| Fourth | Section | 500 – 50 000 km$^2$ |
| Fifth | Recurrent landscape pattern | 50 – 500 km$^2$ |
| Sixth | Facet | 5 – 50 km$^2$ |
| Seventh | Site | 5 km$^2$ |

equate with facet and recurrent landscape pattern but his suggestions have not been widely accepted. The orders of relief are listed in Table 1.2.

In *Geography: a modern synthesis*, Haggett (1973) observes that geographers have to deal with objects which vary greatly in size. So, the examples used in his book have been allocated to one of five orders of magnitude: first order – the Earth itself down to features of 12 500 km$^2$; second order – 12 500 down to 1250 km$^2$, e.g. USA or Australia; third order – 1250 down to 125 km$^2$, e.g. a state such as New York; fourth order – 125 down to 12.5 km$^2$, e.g. a city; and fifth order features – less than 12.5 km$^2$. In the present account the emphasis is inevitably on large-scale features; consequently much of the analysis will be of second, third and fourth order features, which comprise the basic elements of world geomorphology.

## The Earth as a planet

The Earth is a sphere, slightly flattened at the poles (an oblate spheroid) which has a mean radius of 6380 km, a volume of $1083 \times 10^9$ km$^3$ and a surface area of $510 \times 10^6$ km$^2$. The area of the landmasses is $149 \times 10^6$ km$^2$ (29.2% of the Earth's surface) and the area of the oceans is $361 \times 10^6$ km$^2$ (70.8%).

It is clear from historical records and religious accounts that mankind has been interested in the shape, size and origin of the Earth throughout civilised time. Many ingenious theories have been propounded which attempt to show how the Earth as we know it today came into being, one of the more widely known being the Judeo-Christian account of the Creation. Man's ideas of the cosmos were initially Earth-centred with the Sun, planets and stars revolving around. As early as the third

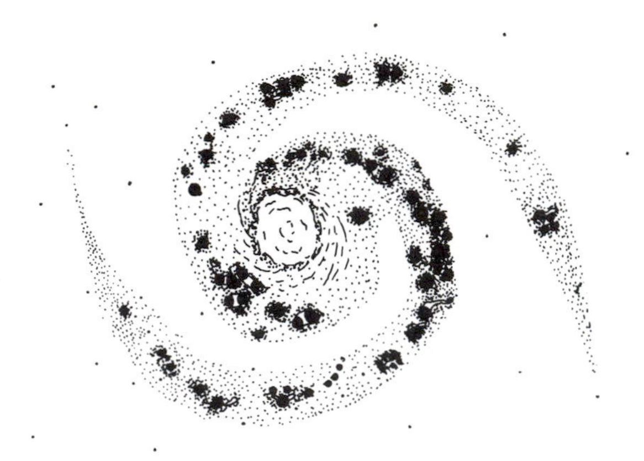

*Figure 1.5.* An impression of a spiral galaxy, one star of which is similar to the Sun.

century BC the Greeks were expressing doubts about this model and, although it was suggested at that time by Aristarchus that the Earth and planets revolved around the Sun, the idea only obtained grudging acceptance when it was again proposed by Copernicus in the sixteenth century. The Earth is a planet and part of the Solar System, which is where a discussion of the Earth's origin must begin.

## The origin of the Earth

The sun and its attendant planets, known as the Solar System, comprise a very small part of the galaxy in which they are situated. In astronomy, distances are measured in light-years, the distance light travels in one year at a speed of 300 000 km per second. One light-year is almost 10 million million ($10^{12}$) kilometres. The distance across the Solar System is a mere 11 light-hours whereas the distance across the galaxy is 80 000 light-years. The nearest star, Proxima Centauri, is only 4 light-years away from the Solar System and is one of the hundred thousand million ($10^{11}$) stars which make up the galaxy. The galaxy can be seen from Earth as the 'Milky Way' and is a flattened disc of gas, dust and stars arranged in two spiralling arms (Figure 1.5).

Many theories have been proposed to account for the origin of the Earth and the Solar System during the last 200 years. They fall into two groups: earlier theories advocated a 'hot' molten origin for the Earth but recent ideas have tended to favour a 'cold' origin. The first

*Figure 1.6.* Origin of the Solar System as a cloud of dust. Individual eddies coalesced to form the planets. (After C.F. Wiezächer.)

significant proposal is an hypothesis originally suggested independently by Kant (1755) and Laplace (1796) which stated that the Solar System condensed from a gaseous nebula. This nebula was very hot and rotating, with subsequent developments resulting from the effects of cooling. This approach has been conveniently called the 'hot' origin for the Solar System

Other information has gradually been acquired which supports the 'cold' or dust cloud theory for the origin of the Earth and which tends to refute the other theories (Figure 1.6) Urey, the American astronomer who proposed the dust cloud hypothesis, supports his ideas with evidence from the distribution of chemical elements which occur on Earth, compared with those known to occur on other planets and the moon. If a hot molten origin is proposed, then Urey argues that the denser iron mineral should be below a lighter siliceous crust which would form a 'scum' on the Earth's surface. Evidence from Mars indicates that it has a uniform composition which suggests that it was never in a molten state. In the case of the Earth, it has been argued that if it had been molten, a smaller amount of iron and a greater amount of silica should be present in the outer layers than is known to occur. From the evidence of impact craters on the moon, astronomers in America have suggested that fragments of nickel–iron have been falling on to the planets steadily since their origin. Nickel–iron fragments do float around in space, and when one comes into the Earth's atmosphere it glows brightly and is referred to as a meteorite. Occasionally one of these reaches the surface and forms a crater as has occurred at Meteor Crater, Arizona in North America and at Wolf Creek, Western Australia. On the Moon, which lacks an atmosphere, these fragments would crash directly upon the surface where they have caused grooves up to 80 km long and with final impact areas known as 'mare' or 'seas', where the moon's surface has been altered by the heat of impact.

If quantities of nickel–iron have arrived at the surface of the Earth in this manner, then there should be evidence of it in the outer mantle. It is thought that this iron would gradually migrate towards the centre of the Earth, and as it did so, the moment of inertia of the Earth would change, slowing the rotation of the planet. Astronomical measurements have demonstrated that the day length has lengthened by one or two thousandths of a second over the last 2500 years. Calculations indicate that there is agreement in the figures and that the change in moment of inertia as a result of the iron migration could be correct. Additionally, the amount of iron movement to form the present core of the Earth would have to be in the region of 50 000 tons per second from the mantle to the core, at which rate it would take between 500 and 2000 million years; a figure which begins to approach the age of the Earth. However, the proponents of these theories state that many more observations and measurements are necessary before the ideas are confirmed.

This is but a brief and elementary glimpse of some of the theories and hypotheses which have been put forward to account for the origin of the Solar System and our planet, the Earth. Most scientists seem to favour the condensation theory, in which a cold dust cloud came together and then proceeded to contract under its own gravity.

## The interior of the Earth

To understand the geomorphology of the continents and oceans it is necessary to begin with the Earth's structure and internal properties. The interior of the Earth can only be studied indirectly by remote means as, unlike Jules Verne's imaginative story, it is not possible to journey to its centre to see what it is like!

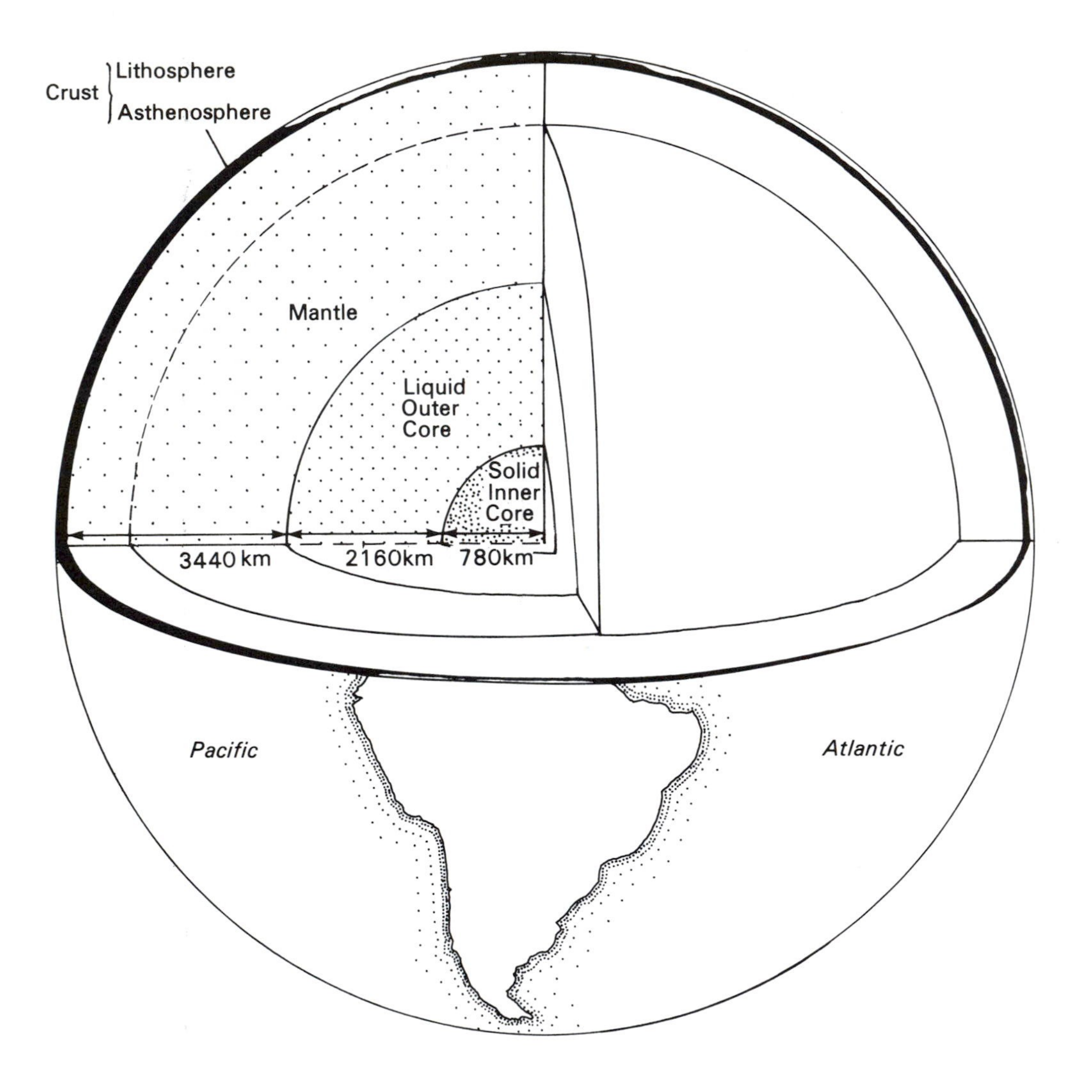


*Figure 1.7.* Section through the Earth showing its internal structure.

Information about the nature of the Earth's interior has come from the study of earthquake shocks. This study, the science of seismology, developed in the nineteenth-century after an Englishman living in Japan constructed the first seismograph and made it possible to record and measure earth tremors.

By the end of the nineteenth century three distinct types of shock waves, which emanate from the focus or epicentre of an earthquake, were recognised. Primary waves (P) take the form of expansion–compression waves, similar to sound waves in the atmosphere and although refracted at boundaries of different layers within the Earth, they can pass through molten material. Secondary waves (S), or shear waves, vibrate at right angles to the direction of travel of the shock and cannot pass through liquids because liquids have no shear strength. Surface waves (L) only occur in the surface layers of the Earth's crust not exceeding a depth of 32 km. These are known as Rayleigh waves when the ground has a motion like waves in water, and Lowe waves when the motion is horizontal and perpendicular to the direction of shock wave propagation. The speed of travel of each of these waves is different, the secondary waves travelling at about two-thirds of the speed of the primary waves. Their speed also varies with depth. Primary waves travel at about 5 km per second in the surface rocks and reach a maximum of 13.5 km per second at a depth of 2880 km.

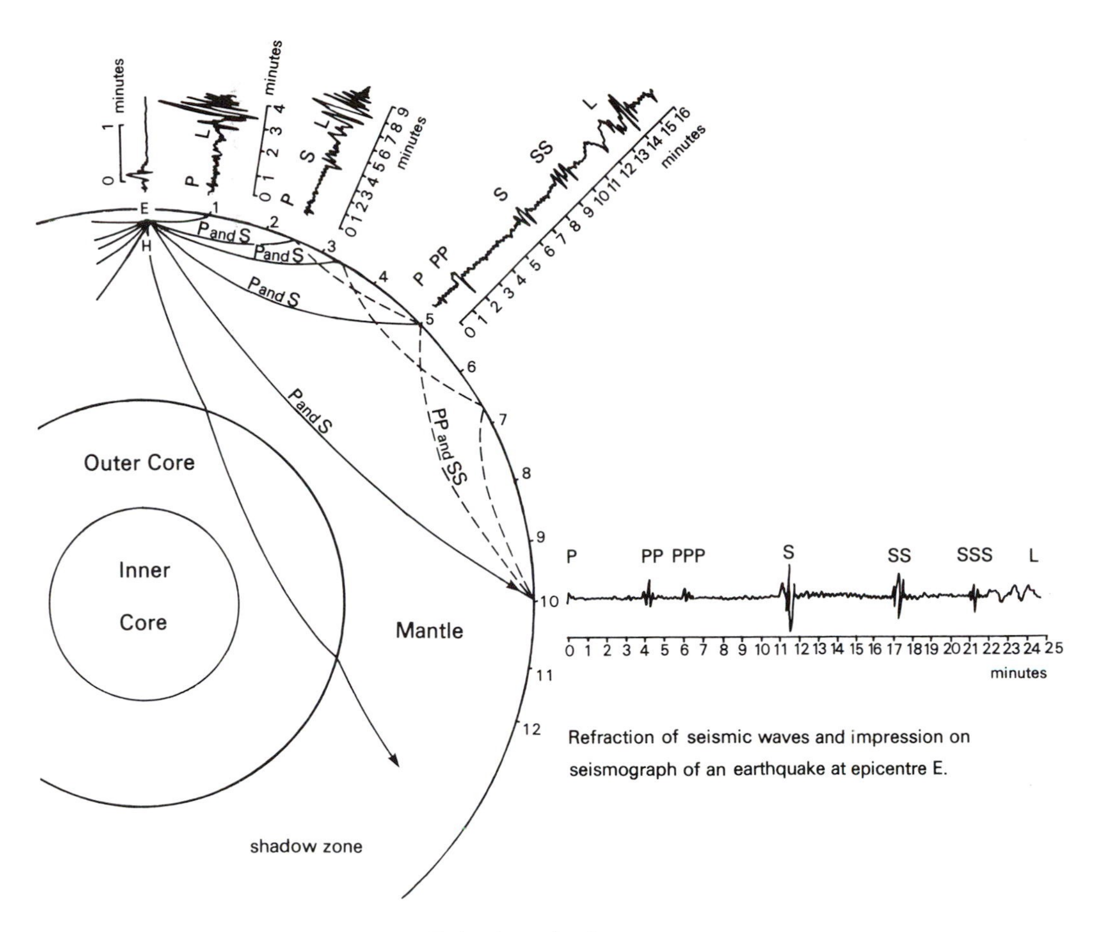


*Figure 1.8.* Seismic effects of an earthquake at E showing paths of P- and S-waves, the reflection of which leaves a shadow zone on the opposite side of the Earth.

These shock waves are reflected or refracted at the boundaries of layers with differing composition and physical properties, so it is possible to determine their path through the Earth's interior. In this way, it was discovered early in the present century that the earth possesses a central core with a radius of 2940 km. The radius of the Earth is 6380 km so the outer zone or mantel is 3440 km thick. As a result of subsequent investigations the core has been sub-divided into an inner core, radius 780 km, of extremely dense material reacting as a rigid body, and therefore probably solid, and an outer core (2160 km) which lacks rigidity and which acts more as a liquid (Figure 1.7). The composition of the inner core has been thought to be nickel–iron, but two German scientists have suggested recently that it may be highly compressed hydrogen (see note on 'expanding Earth', Chapter 2).

The principle whereby the presence of the core was determined is that there exist shadow zones where no direct seismic shock waves are experienced. Between 103° and 142° from a shock neither P- or S-waves occur; beyond 142° the P-waves reappear, but considerably retarded suggesting transmission through a liquid. The only S-waves which arise beyond 142° are those which have been reflected around the surface of the earth (Figure 1.8).

Outside the core, and extending almost to the surface, is the mantle with a thickness of approximately 3440 km. The mantle is thought to be composed of ultrabasic rocks, rich in magnesium–iron silicates similar to the common mineral olivine. At one time an ambitious project was put forward to drill down to this mantle material but it has since been realised that small fragments of these deep-seated rocks occasionally become thrust upwards and appear amongst the surface rocks. This has happened in Newfoundland, south-west Scotland and a particularly good exposure appears in the Troodos mountains of Cyprus.

The crust itself is a comparatively thin skin on the surface of the Earth, compared with the core and mantle, and on the scale of the diagrams employed in this book it would appear as less than the thickness of the line depicting the Earth's surface. However, it is the crust with which the geomorphologist is concerned, together with the layers immediately below, rather than with the deeper layers of the mantle and core. The division between the mantle and the crust is generally taken to be the Mohorovičić discontinuity, named after an eminent Yugoslav seismologist. Examining the records of an earthquake in Yugoslavia which occurred in 1909 Mohorovičić found the P- and S-waves showed two distinct bursts of activity from which he inferred that there was a shallow discontinuity which was later found to be about 32 km below the surface. This discontinuity (the Moho) is thought to represent changes in chemical composition or crystal structure, rather than the important change from upper, rigid crust to the mantle material beneath. Although the discontinuity lies at an average depth of 32–35 km below continents, it is only about 5 km below the ocean floors. Beneath high mountains, the Moho is depressed to over 65 km below the surface. A further, but less distinct, discontinuity lies between 10 and 25 km beneath the continental areas. This division of the crust separates the ocean floors and deeper parts of the continental crust from the upper continental crustal material. In effect, this discontinuity distinguishes between the basaltic sea floor (sima) and continental roots and the granitic continental areas (sial). At first it was thought that the Moho discontinuity demarked the surface material from the mantle, but recent discoveries have indicated that there is a plastic layer, called the asthenosphere, between 100 and 200 km below the surface, which is the significant boundary (Figure 1.9).

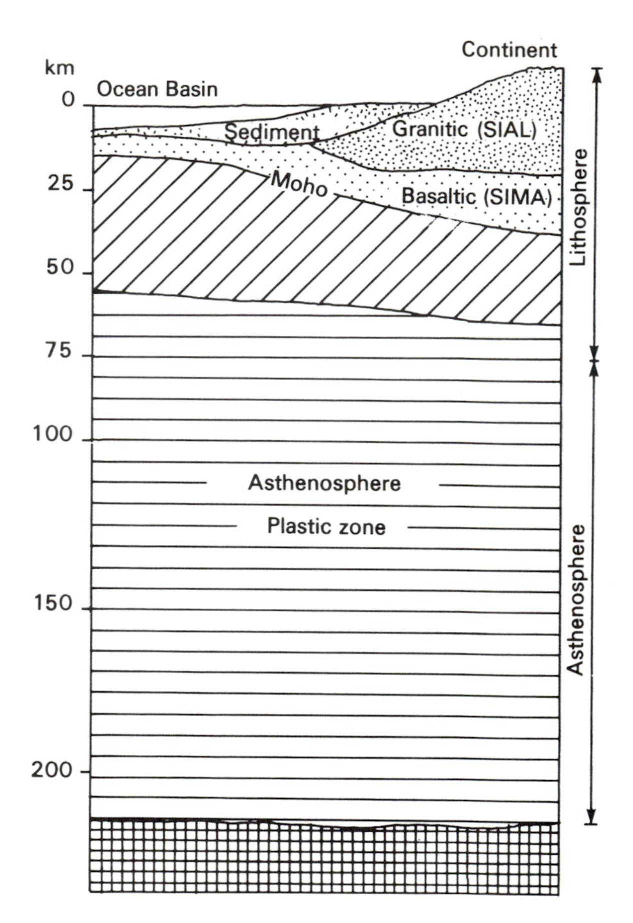


*Figure 1.9.* Section through continental (SIAL) and oceanic (SIMA) crust indicating the Mohorovičić discontinuity and, below the peridotite-rich base of the lithospheric plate, the asthenosphere.

The idea that rocks could become so hot that the centre of the Earth is molten was recognised many centuries ago as volcanic outpourings gave visible proof of molten rock from great depth. Descents into deep mines also show an increase of temperature with depth. The interplay of temperature and pressure with depth inside the Earth means that this idea is only partly true, for at about 200 to 250 km below the surface, the increase in pressure raises the melting point so that the material becomes rigid and is capable of transmitting seismological secondary waves. Between the rigid mantle and the rigid surface rocks the presence of a plastic zone at depths of 100 to 200 km is of utmost importance when the hypothesis of plate tectonics is considered.

The most convincing proof for the presence of the asthenosphere was provided by a Chilean earthquake in 1960. The shock was so violent that the whole earth vibrated and the frequency of vibration was such that the seismological records could only be interpreted by assuming the presence of a low velocity zone which transmitted horizontal and vertical shock waves at different speeds. Additional confirmatory evidence comes from the distribution of earthquake epicentres, most of which lie in the upper 60 km of the Earth's crust indicating the brittle state of the rocks. Below 60 km the number of shocks becomes less, suggesting that the rock is more plastic. However, in certain circumstances much deeper earthquakes are recorded in association with particular structures which will be described in the following chapter.

### Evolution of the atmosphere and hydrosphere

The atmosphere provides an important input to geomorphological systems as it supplies moisture and also controls the temperature régime of the Earth. Water takes part in weathering reactions which break down rocks the weathered products of which are carried away in the rivers and deposited as sediment or precipitated from water to form new geological deposits.

The atmosphere is approximately 300 km in depth and is composed of 78.1% nitrogen, 20.9% oxygen and small quantities of other gases. One of the most important minor constituents is carbon dioxide, at 0.03%, because in the presence of water it will dissolve to form carbonic acid, one of the aggressive factors in weathering.

It is doubtful whether it will ever be known what the original atmosphere or hydrosphere of the Earth were like but it can be said with certainty that both have changed. It has been suggested that the Earth may have been much larger in size with a molten core and a large gaseous envelope which it inherited from its origin in the nebula. At this stage a reducing atmosphere dominated by hydrogen, water and carbon monoxide surrounded the Earth. The original atmosphere was then stripped away and replaced from within the Earth by a process which is called 'out-gassing'. By the time the Earth's crust had solidified, the atmosphere consisted of water vapour, nitrogen, hydrogen, methane and carbon dioxide as well as small quantities of other rare gases. Hydrogen, being the lightest gas, gradually escaped from the Earth into space against the pull of gravity which held the denser gases more effectively. Losses of hydrogen, methane and ammonia seem to have taken place prior to the enrichment of the atmosphere with oxygen. The development of plant life, in which carbon dioxide is used in the photosynthesis of carbohydrates, had oxygen as a by-product, therefore making the role of plants in the development of the Earth's atmosphere very significant.

The atmosphere as we know it today is the result of a long period of evolution. But it is interesting to note that smaller planets in the Solar System have lost their gaseous envelopes whereas the larger planets have retained even the most volatile gases. The Earth occupies an intermediate position, retaining some gases, but losing others. If this is correct, and the composition of the atmosphere has changed through geological time, these changes must have influenced the geological processes of the past. The original atmosphere, lacking in free oxygen, would have resulted in very different forms of weathering. Evidence for the changing atmosphere early in geological time comes from the Pre-Cambrian banded ironstone formations; inter-layered iron oxide and silica deposits. The iron oxides were precipitated on the ocean floor as ferrous iron was oxidised to insoluble ferric iron as the early photosynthetic organisms such as stromatolites began producing oxygen.

A less reliable fragment of the same period of geological history is the presence of pyrite grains in Archean river deposits. The pyrite would be stable only in an oxygen-free atmosphere. The presence of uranite ($UO_2$) in Archean sedimentary rocks of southern Africa and Canada indicates an anaerobic atmosphere as it readily combines with oxygen and no longer accumulates today. The presence of carbon dioxide is indicated by the carbonates present in ancient sedimentary rocks.

It is probable that the atmosphere has remained relatively constant in composition through the last 650 million years. In geomorphological terms, the development of the atmosphere can be seen as an important feature of the Earth's development. Without an atmosphere, processes of weathering, transport and sedimentation would be very different from those which are active today.

These major changes in the atmosphere were closely linked to the evolution of the oceans. Before oxygen accumulated in the atmosphere through photosynthetic activity by plants, chloride and bicarbonate were dominant anions in solution in the oceans. Calcium, magnesium and sodium cations are derived from the products of weathering and brought to the sea by rivers. The chloride anion and the water itself probably originated as a primitive condensate or as a product of the outgassing process. The content of iron in sea water was probably up to 1000 times greater than at present. The pH of the early sea was probably lower, around pH 6.5, and the consequence of the increasing oxygen content of the atmosphere was that iron became insoluble and ceased to be a significant constituent; sulphate thus became the dominant anion in sea water. Such changes appear to have been completed by the beginning of Cambrian time and the composition of the oceans has remained much the same since then.

# 2

# Continental drift and plate tectonics

Science fiction is a literary form which enables authors to imagine amazing and sometimes frightening ideas of what life may be like on other planets or even on earth at some time in the past or future. Sometimes fact is stranger than fiction for one of the most intriguing stories of scientific discovery concerns our own planet, Earth, and the way the continental areas have evolved and assumed their present distribution.

For a long period geographers and geologists thought that there was a strong degree of permanence in the distribution and form of the continents and ocean basins. Although the highest mountain on land and the deepest parts of the sea were of great interest, it was significant that the average level of land occurs at 870 m above present sea level and that the average depth of the sea is 3800 m below sea level. This average picture, shown on the hypsometric diagram (Figure 2.1), also gives a good idea of the general shape of the ocean basins and their resemblance to a soup dish. The ocean waters slightly overfill the dish and extend on to the gentle slope of the rim. This gentle slope, known as the continental shelf, is bounded on the ocean side by a relatively sharp descent to the ocean depths of the abyssal plain. The North Sea and the shallow seas around Great Britain are a good example of the continental shelf at the present time. However, epicontinental seas such as this have varied considerably throughout geological time and many of the familiar sedimentary strata were laid down in them.

Besides the clear difference in relief, it has been appreciated for the last 50 years that the major landmasses were predominantly composed of lighter mater-

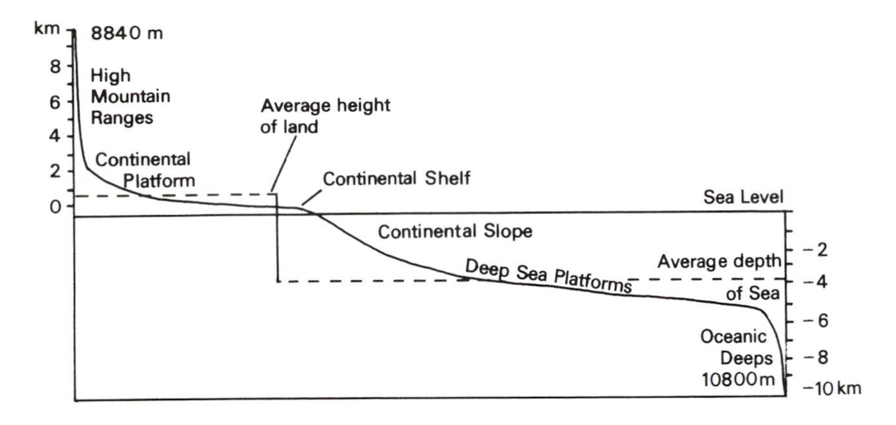


*Figure 2.1.* The extent of global relief lying at different elevations is shown by the hypsographic curve. The mean continental level is 370 m and the mean oceanic level is − 3730 m.

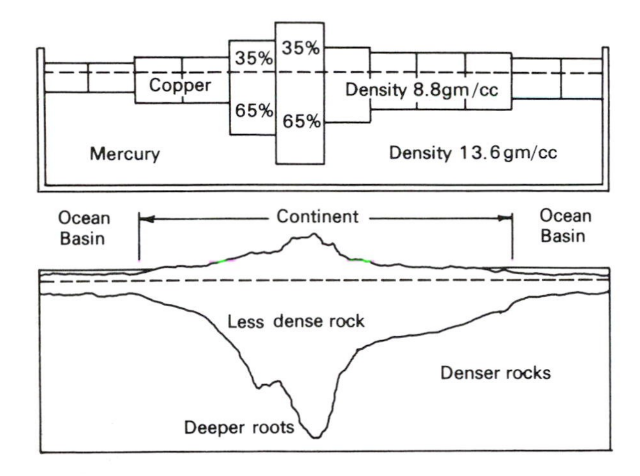


*Figure 2.2.* The principle of isostasy states that the Earth's crust is in a state of buoyant equilibrium. Blocks of the same density but different lengths will float at different levels; in the case of the Earth's crust the lighter sial 'floats' above the denser sima, but high mountains must be compensated for by deep roots.

ials such as granites, rich in silica and aluminium. Continental material, therefore, received the acronym 'sial', and as it had an average specific gravity of only 2.7 it was thought to 'float' on the denser material which made up the sea floor and which underlies the continental material at depth. The ocean floor material, rich in silica and magnesium became known as 'sima' and had a specific gravity of between 2.9 and 3.0. The discontinuity between granitic and basaltic material has already been mentioned in the previous chapter as lying between 20 and 25 km beneath the surface and is also known as the Conrad discontinuity.

In order to meet the physical requirements imposed by gravity, and for high mountains and deep sea basins to exist, the concept of 'isostasy' was introduced by the American geologist Dutton in 1889. If blocks of wood of different length are floated upright in water, or if blocks of copper are similarly floated in mercury, they assume different elevations in proportion to their length. They also protrude by different amounts into the supporting liquid (Figure 2.2). If this concept is applied to sial and sima, it becomes immediately apparent that the higher the mountain mass, the deeper roots it must have penetrating down into the earth.

Evidence for deep roots beneath large mountain masses was first noted by the French in a scientific expedition to the Andes in the first half of the eighteenth century. A smaller than expected deflection of the plumb-line was observed in the Indo-Gangetic plain in response to the considerable mountain mass of the Himalayas. Once adjustments have been made for elevation and for the effects of the pull of the Sun and Moon, the values of gravity should approximate to calculated values for the idealised ellipsoid Earth. That they do not is further indication that material of lower specific gravity penetrates deeply below major mountain ranges. Evidence from earthquakes suggests that the roots of mountains extend to depths of 40 km or more.

Before the ideas of continental drift were put forward by Wegener in his paper of 1912 and his book of 1915 *The origin of continents and oceans*, geologists were in a great dilemma. It was apparent that there had been considerable crustal shortening as could be seen where the rocks were folded and overthrust. This led them to the conclusion that the Earth had contracted and that the mountains represented the wrinkled crust of the shrinking sphere. A further difficulty was that the climatic conditions in which rock strata had been laid down had varied greatly with the passage of time; either the climate had changed greatly or the continent might have had a different disposition from that of the present day, but how could a continent move?

One ingenious, but unsupportable, scheme was that of Lothian Green. Shrinkage of the globe, he suggested, would lead to a tetrahedral shape with the oceans occupying the faces of the tetrahedron and the continents the edges. Although imaginative, this idea cannot stand up to several arguments; it would be gravitationally unstable and isostatic readjustments would take place causing the corners to sink back into the Earth, and, as can be seen from space, the major relief features of the Earth at true scale scarcely diverge from the circumference of the globe.

The absence of continental material, sial, from large areas of the Earth's surface gave rise to some other interesting theories. One of the most intriguing was that proposed by George Darwin, son of the famous natural scientist Charles Darwin. He suggested that the sialic rocks, missing from the whole of the Pacific basin, had been removed from the Earth and now formed the moon. The Pacific basin was the scar left behind when the Moon was separated from the Earth. Darwin calculated that the combined effects of the rotation of the Earth and the Sun's attraction would produce a tidal resonance which would increase until a mass separated from the Earth. The attraction of this idea was that the Moon has been found to be moving further away from the Earth with the passage of time. However, the tidal resonance theory of the origin of the moon and the resultant lack of sial on the areas of the Pacific Ocean floor was disproved because internal friction would prevent the tidal resonance proposed by Darwin.

Alternative theories now suggest that the Moon is most probably a captured body and that this capture must have taken place simultaneously with the formation of the Earth. Darwin's idea that the removal of material from the Pacific resulted in the movement of the other continental masses to give the ring of mountains and volcanoes which surround the Pacific must also be abandoned, and some other explanation sought for the so-called ring of fire.

Evidence for continental movement has been accumulating since the beginning of the seventeenth century when Sir Francis Bacon noted the complementary nature of the coasts of Africa and South America. A Frenchman in the seventeenth century concluded that America was separated from the Old World at the time of the biblical Flood; other scientists denied the whole affair and suggested that the Atlantic resulted from the foundering of the mythological Continent of Atlantis. Many geological events were seen to have taken place in a catastrophic manner in the early period of geological studies. The correspondence of geological features on both sides of the Atlantic, including the distribution of similar fossil plants in the coal measures of Europe and North America, was first described in detail in 1858.

## Continental drift

By the first decade of the twentieth century sufficient information had been accumulated for more definite theories on the movement of continents to be put forward. Two Americans were first in the field: F.B. Taylor (1908) and J.B. Baker (1911) used the idea of continental drift to explain the origin of mountains and the correspondence of relief on either side of the Atlantic. However, the most famous protagonist during this period was undoubtedly Alfred Wegener, who was not a geologist but a meteorologist. Wegener had adopted the idea of continental drift to account for the different climates which had affected continents in earlier geological times. He had amassed a wealth of biological, palaeontological, palaeoclimatic, tectonic and geophysical evidence for the existence of continental drift. Unfortunately, the concept was too much for his contemporaries, who picked holes in his arguments and rejected the hypothesis. In spite of some excellent positive evidence, his proposals really foundered on disbelief, as nobody at that time could envisage a force or mechanism which could move a continent across the face of the Earth.

The evidence collected by Wegener suggested to him that the present distribution of the continents is derived from the disruption of a super continent which he called 'Pangea' (all lands) which had existed in the late Carboniferous (Figure 2.3). The break-up of this large landmass took place in the Mesozoic, with the core areas of the future continents moving towards their present positions. Although Wegener and some of the early workers were trying to show how different climate might have affected the five continents, their maps did

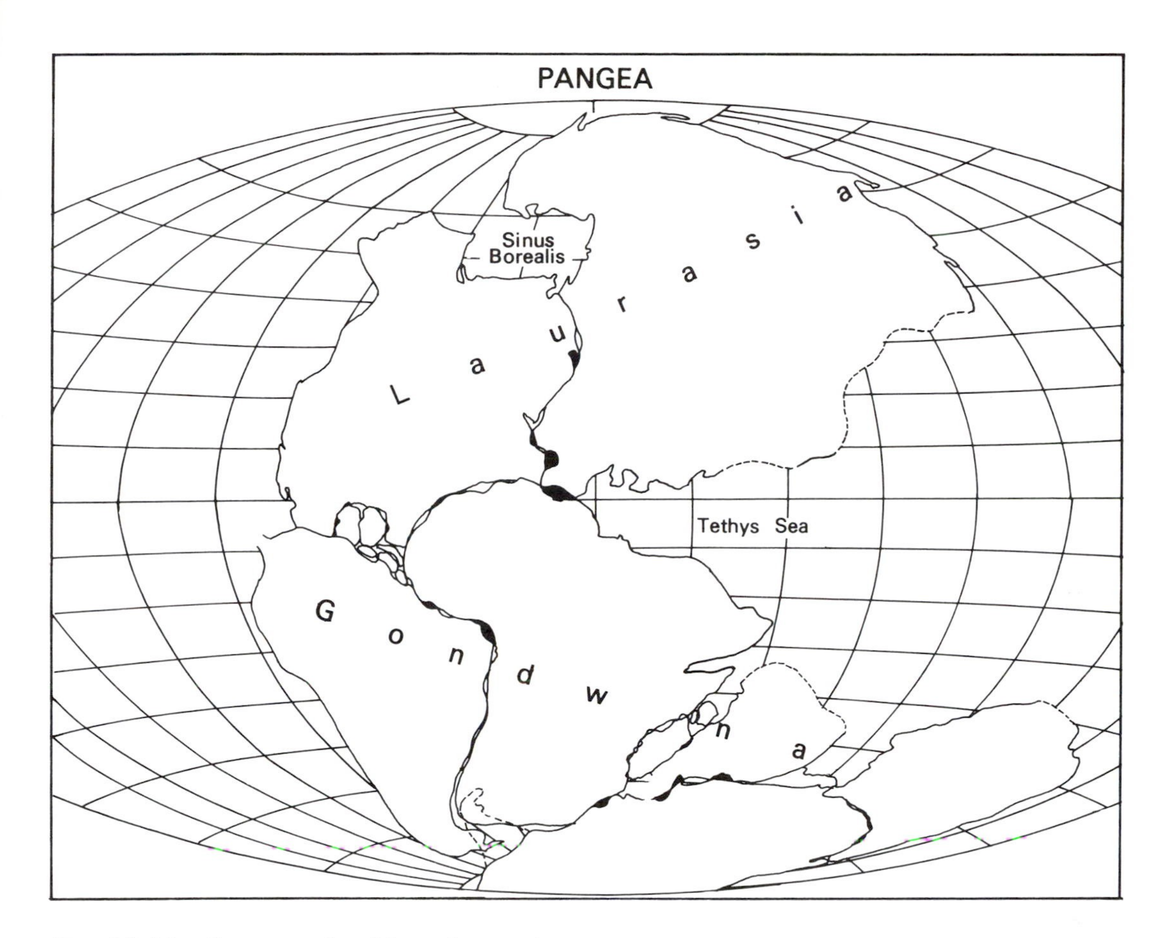


*Figure 2.3.* A best-fit reconstruction of the continents to form 'Pangea' (all-lands) as it may have been in Permian times. Laurasia and Gondwana were separated subsequently by rifting and movement of the tectonic plates.

not show Europe and North America sufficiently near the equator to have tropical forests. Wegener was well aware of the northward movement of Africa and India as he invoked a force called 'polflucht' to account for their motion. Such a force does in fact exist, but it would have to be many million times more powerful before it could move a continent. Additionally, such a force would not be adequate to account for the generally westward movement of North and South America, and certainly could not explain any reversals of movement which are thought to have occurred.

Other proponents of the continental drift theory, such as du Toit, distinguish between a northern group of continental nuclei, which they called 'Laurasia' and a southern group called 'Gondwana' (Figure 2.4). The name Laurasia is formed from elements of words describing part of North America and Asia. The name Gondwana is taken from a place in India, Gondwana, where fossil plants of the *Glossopteris* flora were first described, only to be found later in South America, Africa, Australia and Antartica. Between Laurasia and Gondwana a geosynclinal area developed beneath a sea known as Tethys. Into this subsiding zone the deposits were laid which eventually became the fold mountains of the Alps and the Himalayas.

Although opinion generally swung away from the idea of continental drift as a practical explanation of continental distribution, the idea was kept alive by one or two active proponents. In South Africa, Alexander du Toit remained convinced of the possibility of continental drift and wrote articles and a book supporting the theory. In Britain, Anthony Holmes can be credited with being the first to suggest a possible mechanism for moving continents. After working on methods of dating

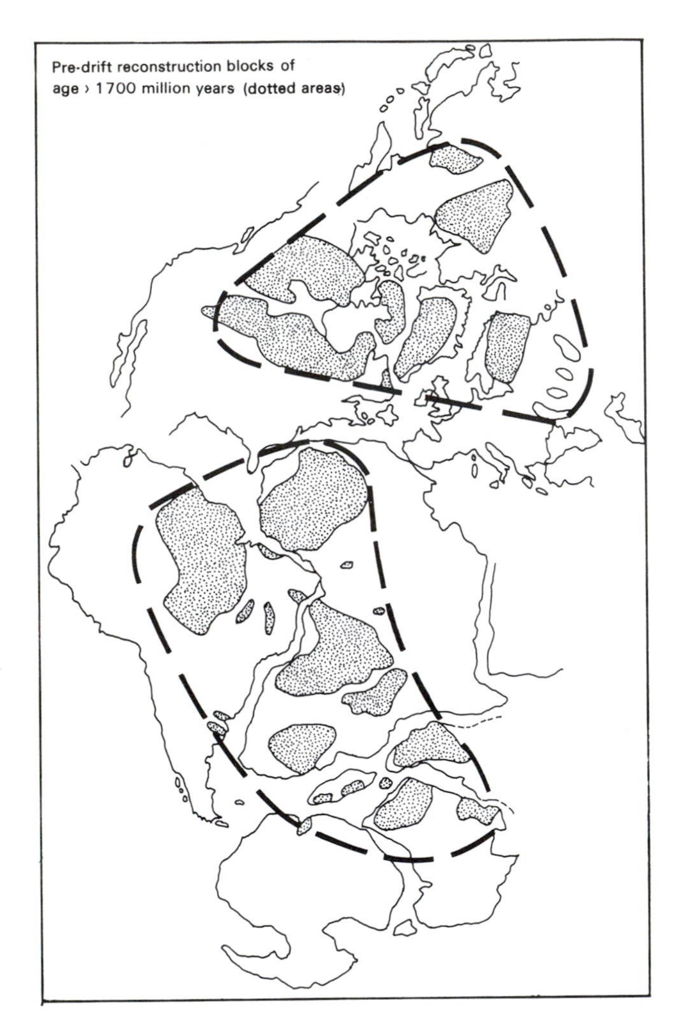


*Figure 2.4.* Continental nucleii – cratons of 1700 million years ago reassembled into their positions before disruption of Pangea. (After Hurley and Rand, 1969, *Science* 164, 1229.)

*Figure 2.5.* Simplified plots of continental movement shown by changes in the direction of magnetic north.

rocks by measuring radioactive decay, he proposed that the heat liberated by radioactive disintegration of mineral deep in the Earth's crust would cause convection currents which might raft the continents along.

In spite of the convincing nature of the evidence collected by those scientists who favoured continental drift, the geological establishment of the time was very reluctant to accept the concept of moving continents. Mysterious land bridges were conjured out of nowhere to account for the movement of plants and animals from one continent to another. In the majority of cases, no trace has ever been found of sunken land which could in any way have fulfilled such a purpose. Thus, despite some very good circumstantial evidence, the concept of continental drift went out of favour and remained so until the late 1960s when it was once again revived. A decade later there was virtually complete agreement that continents had moved and that the findings of Wegener concerning the climatic, biological, geological and physiographical characteristics were correct. The convictions and faith of a few scientists, mainly in the southern hemisphere, were finally vindicated in the new approach which came to be called 'plate tectonics'. The evidence which made this new approach possible is considered in the next section.

## Plate tectonics

The possibility of continental mobility received fresh impetus from a number of important discoveries made during the two decades, 1956 to 1976. Significantly, these discoveries were made by the very group of scientists (geophysicists) who formerly had doubted the theory of continental drift. Realisation of the significance of residual rock magnetism and the implications of mid-oceanic ridges did not appear separately to have immediate importance for continental mobility. When taken together, however, these concepts provided the opportunity for a complete reappraisal of the ideas of continental drift and for the first time gave a credible means of demonstrating continental motion.

Firstly, a paper published by a group of British scientists in 1956 revealed that the direction of magnetisation of rocks could be related to the position of the

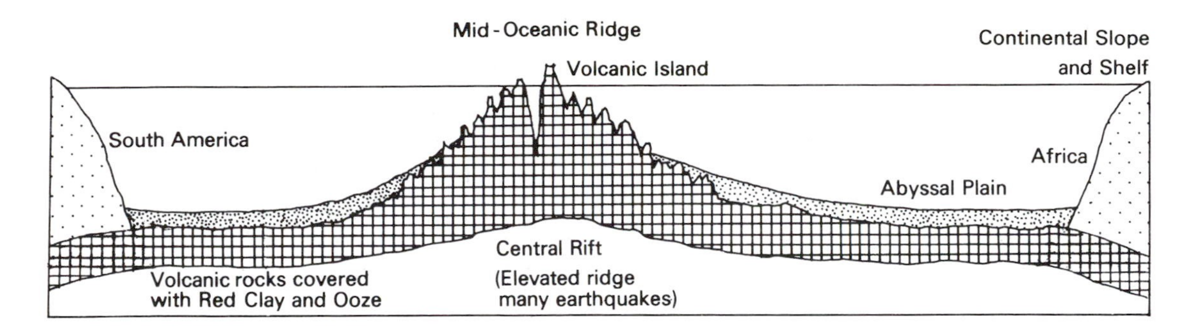


*Figure 2.6.* Diagrammatic cross-section of the Atlantic Ocean indicating the mid-oceanic ridge.

magnetic pole of the Earth at the time the rocks were laid down. When rocks cool from a magma, and even when sedimentary rocks are deposited, the iron minerals within them become slightly magnetised and their polarity is in sympathy with the Earth's magnetic field at the time and place of formation. By careful measurement it is possible to reconstruct the position of the magnetic pole at different stages of the past history of the Earth. When this was done, it was found that the results were consistent for any one continent, indicating that either the pole or the continent, or both, could have moved. However, when the previous positions of the magnetic pole were plotted for other continents, differences became apparent which can only be explained by a differential movement of the landmasses (Figure 2.5).

A second important hypothesis was proposed by an American, H.H. Hess in 1960, following the realisation that the mid-oceanic ridges were a characteristic feature of not only the Atlantic, but the other oceans as well. It was appreciated for the first time, that the largest and longest mountain range of the Earth's surface lay at the bottom of the sea (Figure 2.6). It possesses peaks which rise from the ocean depths to more than 4000 m above sea level, making a total height greater than Everest, and its length is more than 64 000 km. As this enormous feature of the oceanic floor is composed entirely of volcanic outpourings, Hess suggested that the sea floor cracked open along the line of the mid-oceanic ridge and that new volcanic material came to the surface and gradually spread on either side of the spreading centre. This proposal was attractive as it offered a means of solving the apparent youthfulness of the ocean floors, virtually all of which are post-Cretaceous in age, and explained at the same time why there was a lack of sediment covering much of the sea floor.

Thirdly, during the period 1965 to 1975 there was the realisation that the magnetism in the rocks on the sea floor could be related to the age of the lava and that together these three discoveries could be used to give the most convincing proof yet available that the continental plates had moved and are still moving. As Figure 2.7 shows, the reversals in the Earth's magnetic field are faithfully recorded in the bands of lava extruded from the spreading centre. Each side is pushed further and further apart by the continuing process of lava extrusion at a rate of between 1 and 9 mm per year. The original continental drift theory assumed that the lighter granitic or sialic rocks were floating upon the sima or denser basaltic rocks of the sea floor. The modern interpretation is that the continents are firmly embedded in the sea-floor material which is as rigid and brittle as the continental material. Both continental and sea-floor materials are being moved along together on a large fragment of the Earth's crust which is referred to as a lithospheric plate. Each of these plates is between 150 and 200 km thick and is bounded by a margin which is marked by localised seismic, volcanic and tectonic activity (Figures 2.8 and 2.9). If new material is constantly being introduced along the line of the mid-oceanic ridges and spreading out from them, it follows that it will meet material from another spreading centre. It will either come into direct collison with it, it will override it, or it will itself be overridden (Figure 2.10).

Three types of plate boundary are recognised:

1. At the mid-ocean ridge margin of a lithospheric plate new material is being injected into the centre of the spreading zone as the plates on either side move apart – this is known as a *constructive* margin.

2. Where island arcs, trenches and fold mountain belts occur there is a convergent movement of the plates

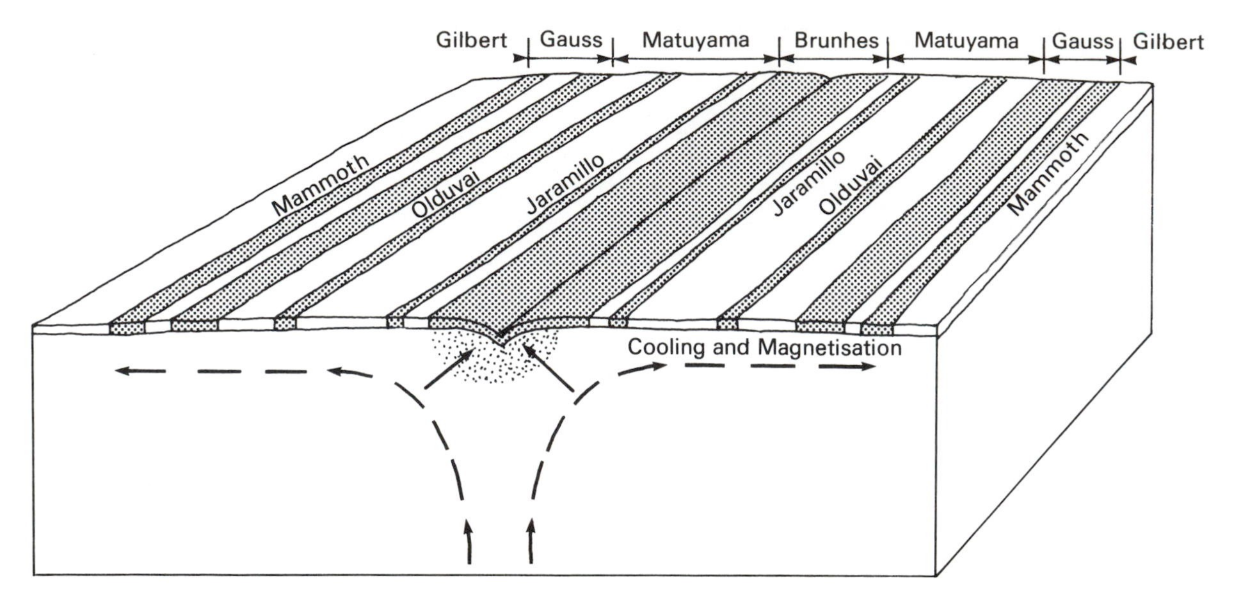


*Figure 2.7.* As new sea-floor material is produced at the spreading centre, it takes the polarity prevailing when it cools. The names Brunhes (normal), Matuyama (reversed), Gauss (normal) and Gilbert (reversed) refer to periods when polarity alternated during the past 4 million years.

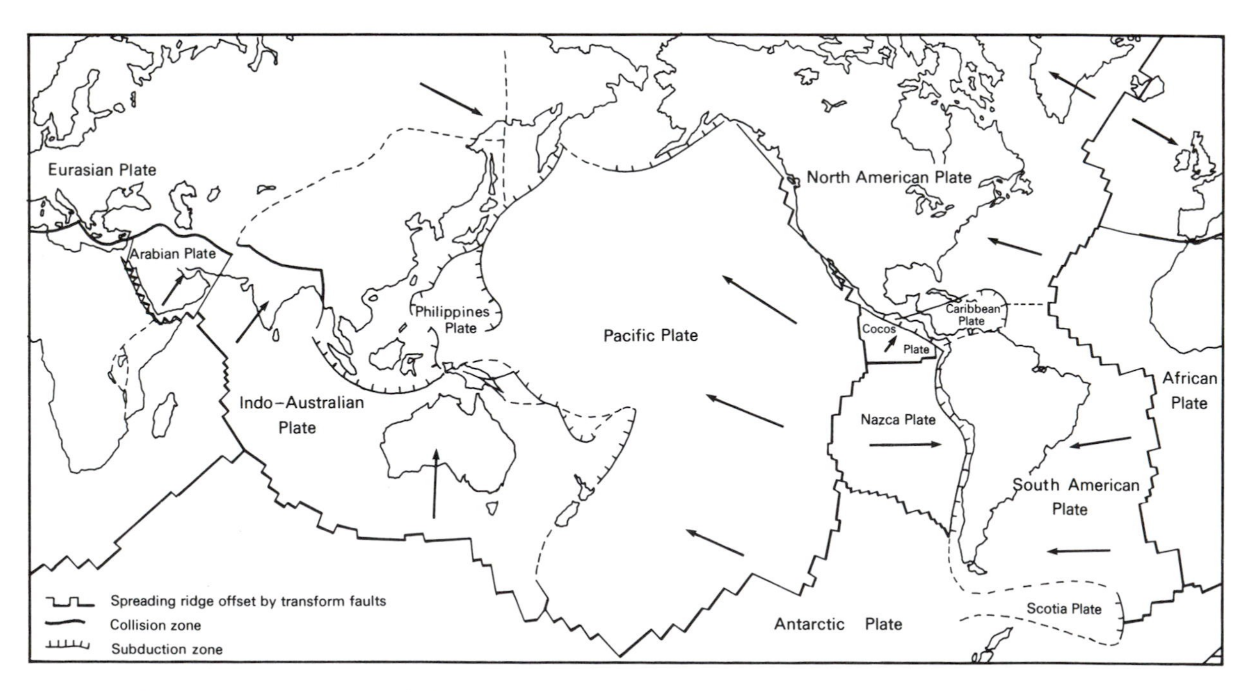


*Figure 2.8.* The major lithospheric plates of the Earth.

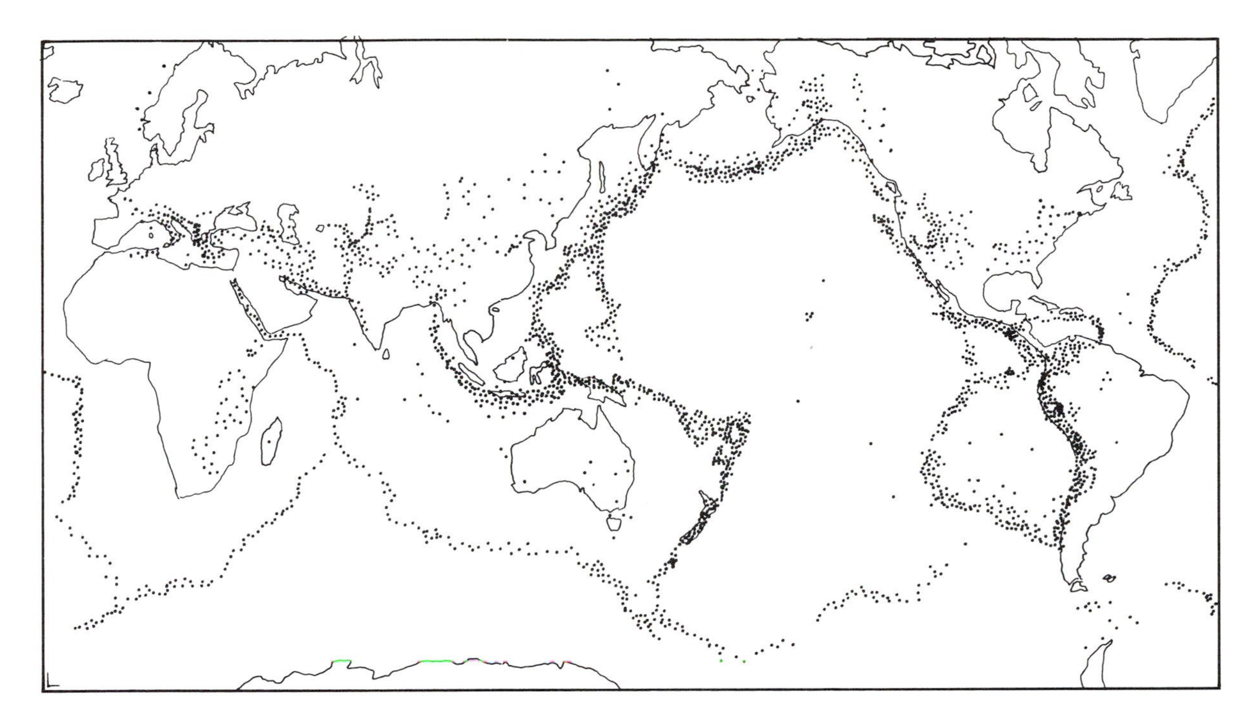

*Figure 2.9.* A map showing the location of major earthquakes outlines the major lithospheric plates.

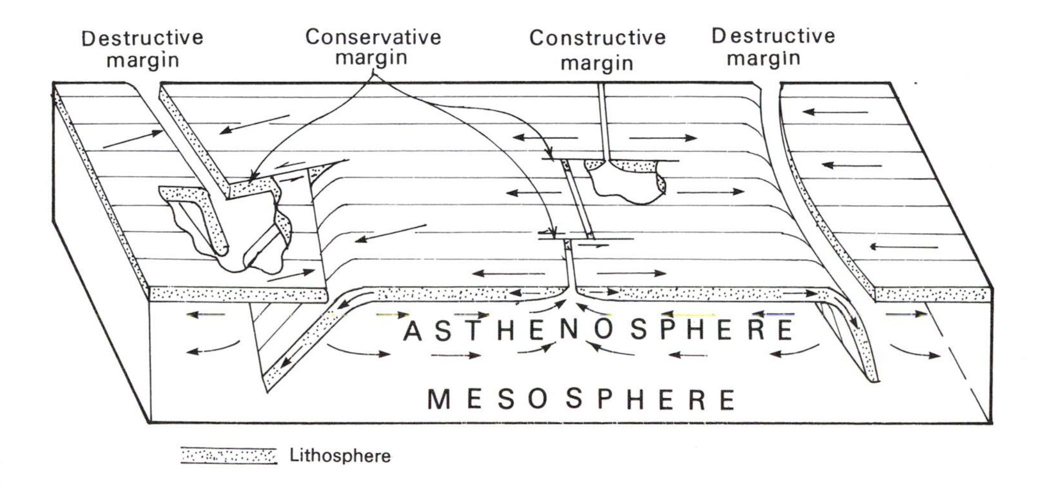


*Figure 2.10.* Types of plate boundary.

and the margin of one plate is being pushed down (subducted) into the asthenosphere – this is known as a *destructive* plate margin.

3. Where the plates are moving past each other, as along the famous San Andreas fault, and other transform faults which cross the spreading zones of the mid-oceanic ridges, lithospheric material is neither being created nor destroyed and so these margins have been called *conservative* margins.

The margins of the Atlantic Ocean were formed as the North American and African/European plates moved apart. Hence the name *Atlantic* or *trailing* margins are used. Around the Pacific the margins are mainly destructive with the plates moving together. These are referred to as *Pacific* or *collision* margins.

These names simply describe a number of different situations where the lithospheric plates interact. As the ocean-floor material is different from the material of the continents, it must be expected that the interaction of the plates would be different:

*Simple subduction*: the simplest case occurs where oceanic material meets continental material, as is taking place in the Andes. As continental material is so light, the denser, oceanic Pacific plate is being subducted beneath the South American plate where it is melted as it descends into the asthenosphere, so providing magma for volcanic activity.

*Island arcs*: these are a characteristic of the western Pacific Ocean and reflect a complex situation which is not altogether understood. A deep sea trench lies alongside a single or double arc of islands, the inner arc usually being volcanic. Island arcs differ in complexity for they are not all backed by continental materials as occurs in the Indonesian arc. In some cases oceanic material is in collision with oceanic material, as occurs in the South Sandwich arc of the south Atlantic. In the Philippines it is thought two arcs have collided to give the particular structures of that group of islands.

*Continent to continent*: where the Indian plate meets the Asian plate the Indian plate has been thrust beneath the Asian plate to give a double thickness of continental crust. A similar, but not so spectacular situation occurs in the mountains of Armenia where the Arabian plate is thrust beneath the Eurasian plate.

Within the lithospheric plates lines of weakness are indicated by a pattern of rifting seen in all continents, but particularly in the rift valleys of east Africa. In the past, fragmentation of Gondwana also seemed to have taken place, by rifting along similar lines to the east African rift valley with typical Y- or triple-junctions. This rifting was preceded by intrusion of alkaline volcanics which caused the doming of the area, and the extrusion of basaltic lava flows. It is uncertain whether the African rift valleys are a failed spreading centre. In many cases two arms of the triple junction are successful in developing a spreading centre and one arm fails; this failed arm is known as an aulacogen.

In the process of continental growth, the results of successive mountain-building episodes (orogens) have been attached to the older cratons of basement complex. In some cases materials of different ages are very firmly attached, but in other cases the line of weakness between them continues to be the seat of earthquakes.

Although the theory of plate tectonics has been widely accepted in the earth sciences and it provides answers to many of the problems of global tectonics, it still does not answer all of the questions. One such problem is the disparity between the length of the spreading centres compared with the length of the trenches where material is consumed. Similarly the pattern of earthquakes is not always consistent with the ideas of subduction, and it is apparent from the North American continent that it is riding over the margins of the Pacific plate regardless of the underlying structure. The link between mountain building and plate tectonic theory is not strong and the idea of compressive folding is losing ground to ideas of gravity sliding in the production of nappes and other folded structures seen in Alpine-like mountains. The simple model of a limited number of large lithospheric plates is breaking down as it is found that there are many micro-plates which are equally important in the determination of the large-scale morphology of the Earth.

Up to this point in the discussion of plate tectonics the size of the Earth has been assumed to remain constant. However, this may not be a correct assumption. It has been suggested that many of the problems of interpretation could be solved if the continents could be arranged on a smaller globe. The pattern of spreading outwards from the Atlantic, Indian and Southern oceans on an Earth of constant size suggests that much more compression should exist around the Pacific Ocean. In fact behind some of the island arcs, back arc basins have subsidiary spreading centres showing that the Pacific, too, has expanded. Although virtually impossible to calculate, it appears that more new sea floor is being created at spreading centres than is being consumed in the subduction zones. The roughly poly-

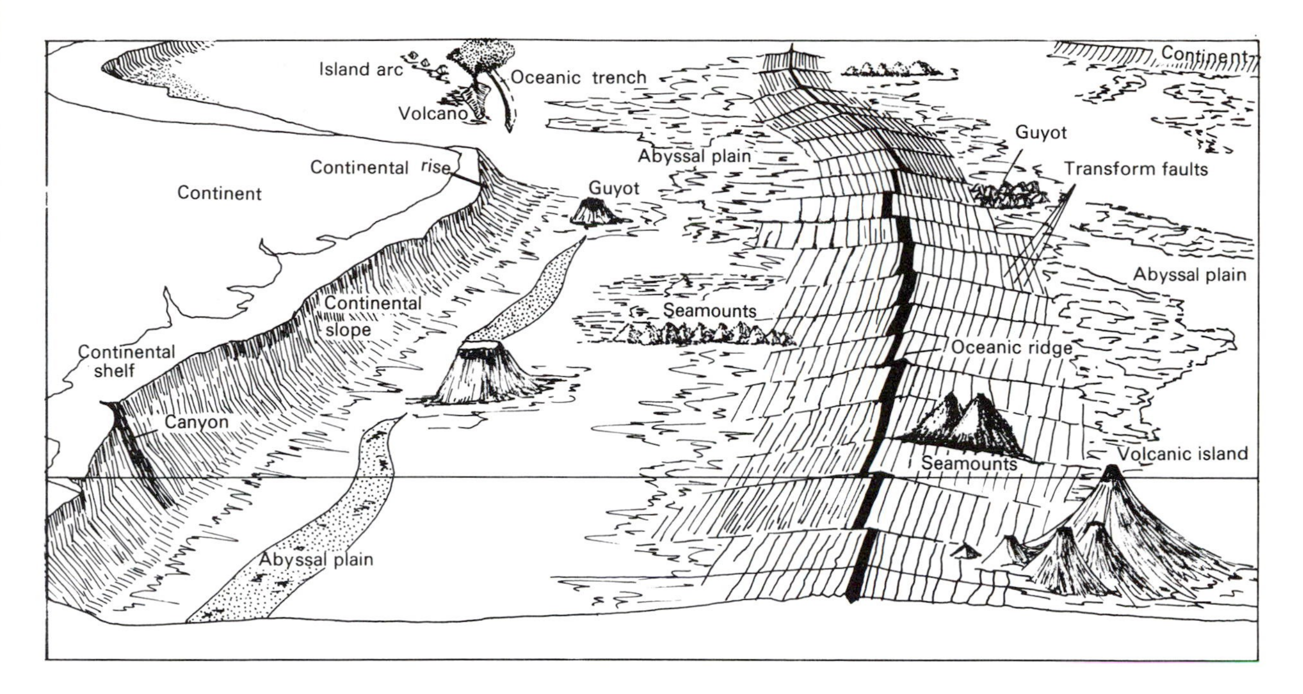


*Figure 2.11.* Features of the ocean floor.

gonal shape of the lithospheric plates is consistent with an expansion of the earth and, except for Antarctica, all the other plates have migrated northwards from the Southern Ocean spreading centre, again giving support to the idea of an expanding Earth. With these cautionary words, a consideration of the origin and development of the ocean floors and the continental areas follows in the next section.

## Origin of the sea floor

It has been demonstrated that ocean-floor rocks are different from the continental rocks. In general, this difference can be expressed by describing the sea-floor rocks as basaltic and the continental rocks as granitic. The geological origin and development of sea floor and continents is markedly different, so it is convenient to discuss them separately.

Recent discoveries have shown that the morphology of the ocean floor reflects a dynamic state and not a static situation. As the role of the sea floor in the broad pattern of the Earth's morphology is seen now to be of basic importance, it is clearly the most suitable place to begin a study of world geomorphology. There are three major elements in sea-floor morphology: the spreading centres; the ocean basins; and the trenches. Each of these features plays an important part in the understanding of the global geomorphological development.

### *The spreading centres*

The Atlantic, Indian and Southern oceans are characterised by mid-oceanic ridges which mark the location of spreading centres (Figure 2.11). The Pacific Ocean is slightly different as the line of the spreading centre swings from the south-eastern Pacific and eventually passes on to land in California as the San Andreas fault, eventually passing back to the ocean floor again off the coast of north-western USA. The only other major land area where the spreading centre emerges is in Iceland which lies astride the mid-Atlantic ridge.

As revealed by surveys, the form of the mid-oceanic ridge is of parallel strips of material extruded on either side of the spreading centre. The overall height of the mid-oceanic ridge depends partly upon the rate at which material is being extruded and partly upon the rate of spreading because away from the spreading centre the newly extruded sea-floor material settles back into the Earth's crust. If spreading is rapid, the mid-oceanic ridge will be broad with gently sloping sides, but if spreading is slow, the slopes of the ridge are much steeper, as in the South Atlantic. In some cases it has

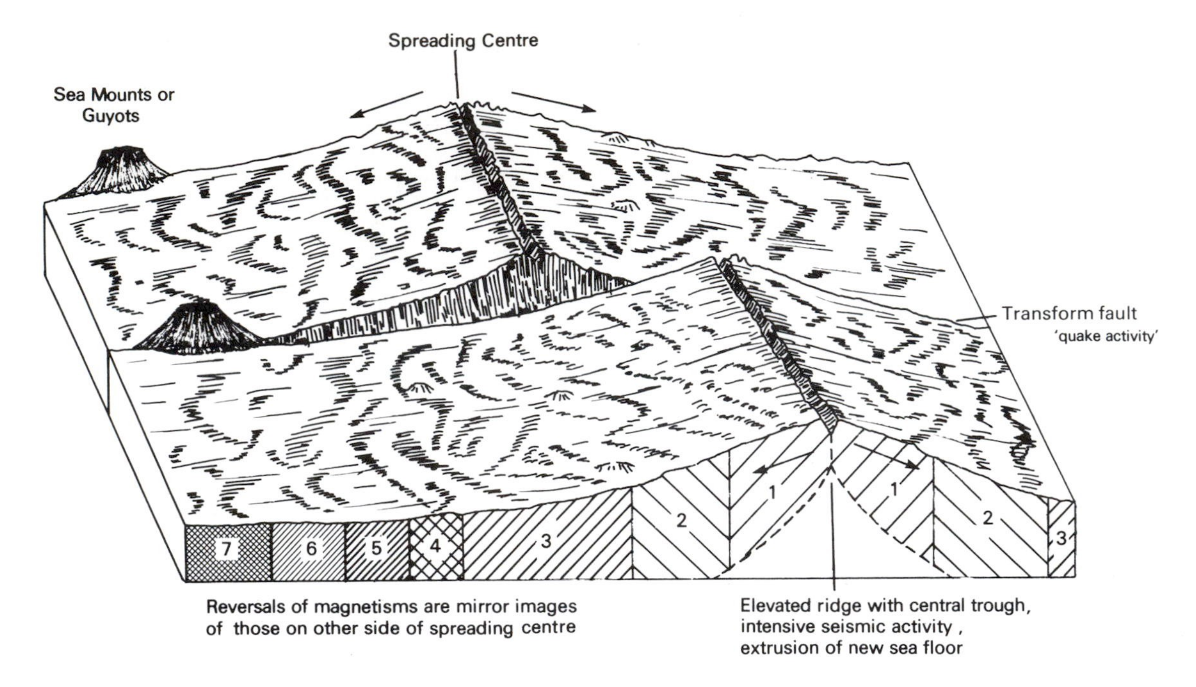


*Figure 2.12.* A spreading centre with a transform fault.

been possible to distinguish a rift-like valley between the two sides of the spreading centre; in Iceland it is 45 km wide, and in the Indian Ocean the central rift valley is 300 m deeper than the ocean floor on either side.

The mid-oceanic ridge is not continuous but comprises a succession of segments where it has been shifted out of line by transform faults oriented at right angles to the ridge. As Figure 2.12 shows, this can lead to situations where adjacent segments of the Earth's crust are moving in opposite directions. Where this occurs earthquakes are particularly common and they are often accompanied by volcanic activity. Occasionally, the amount of volcanic outpourings is sufficient to raise the level of the lava above sea level to form one or more volcanic islands. It is noticeable that islands further from the spreading centre are larger than those close by. This fits the general picture because the oldest volcanoes have obviously continued to grow as they have produced much greater quantities of lava over the longer period of their existence. Most volcanoes cease to be active after a period of 20 to 30 million years and only a few, such as the Canary Islands, remain active for up to 100 million years.

Extinct volcanoes on the sea floor are referred to as guyots or seamounts, many of these do not grow sufficiently to appear above the sea surface as islands and even some of those which did become islands later sank below the sea again as the crust settled away from the spreading centre. Often guyots are flat-topped as wave action has trimmed their summits. As they gradually sink out of sight these guyots sometimes form the foundations of atolls. The guyots are carried along on the tectonic plate and they are ultimately consumed when the edge of the plate passes into a subduction zone. One such guyot has been discovered on the side of the Tonga trench, appropriately tipped over as it is being carried downwards before finally disappearing for ever; another sits on the floor of the Aleutian trench, about to be consumed.

### *The ocean basins*

Away from the spreading centres, the ocean floor gradually descends to the level of the ocean basin floor, alternatively called the abyssal plain. This descent from the crest of the mid-oceanic ridge, at about 2000 m, to the abyssal plain at 5000 m takes place in a series of steps associated with successive outpourings of new material from the spreading centre. As new material is intruded along the line of the spreading centre, the previous intrusions are pushed further and further apart. As each intrusion became magnetised in sympathy with the

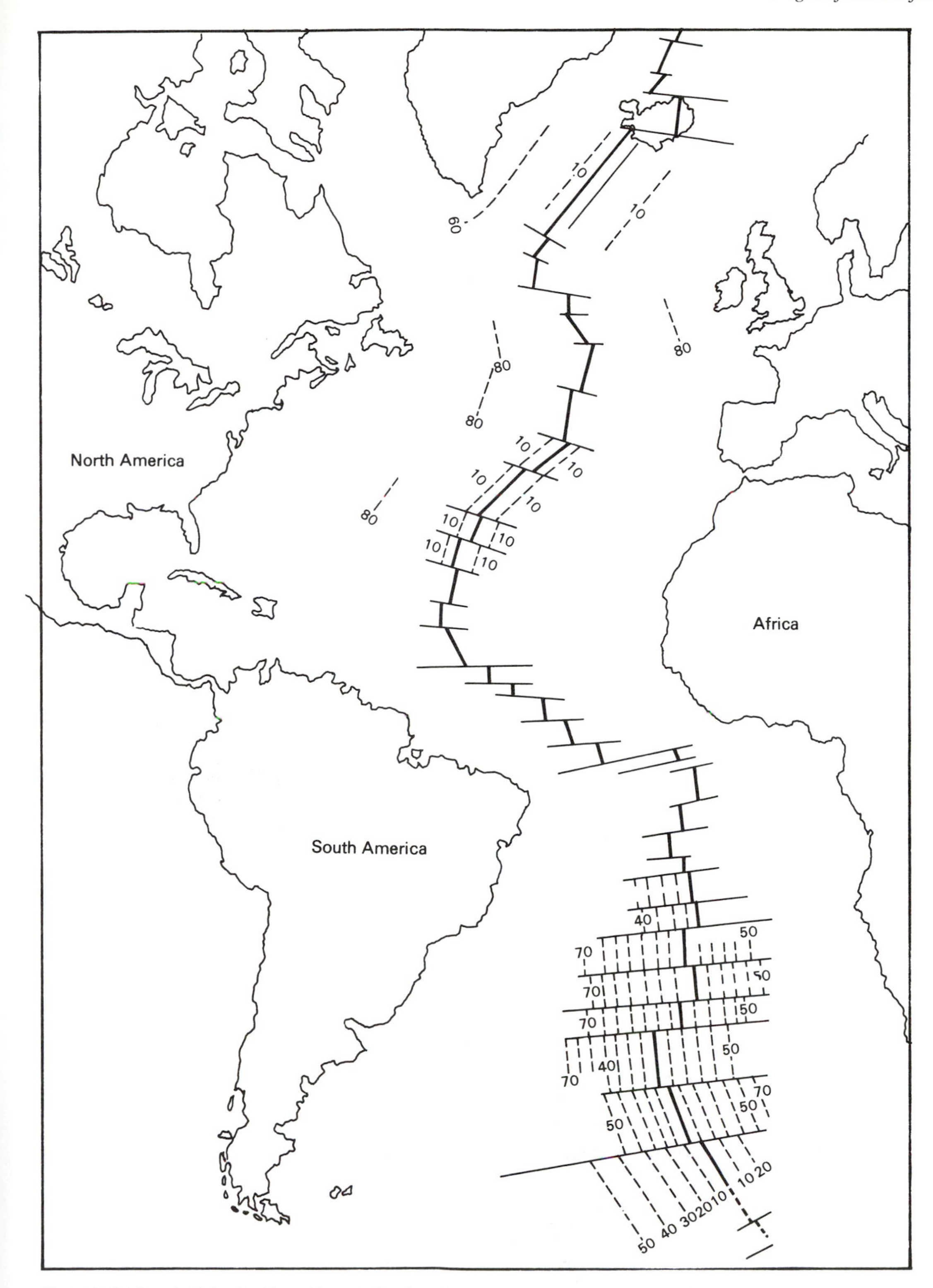


*Figure 2.13.* The mid-Atlantic ridge with parallel strips of new sea floor. The age of these strips on either side of the spreading centre is given in millions of years.

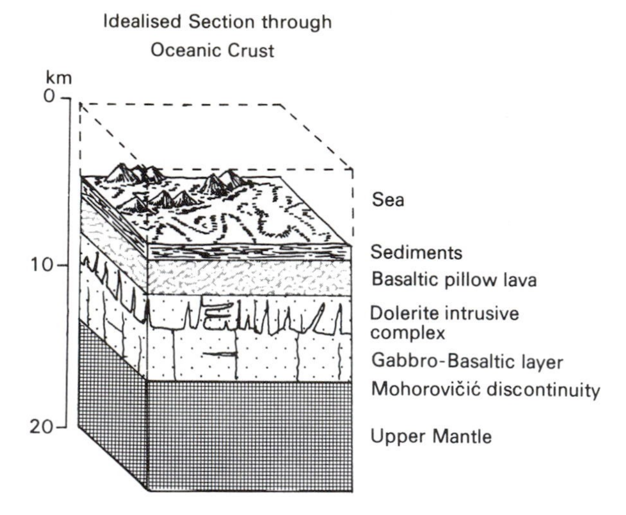


*Figure 2.14.* Idealised section through the oceanic crust.

Earth's contemporary magnetic field, it follows that the changes in polarity observed in the rocks are related to changes in the Earth's magnetic characteristics. Surveys have shown that this theoretical picture is correct, and that the sea floor is made up of a sequence of parallel strips of rock some of which have reversed magnetism. As polarity has changed many times in the past and at varying intervals, it is possible to compare the pattern of alternately magnetised strips from different oceans. On all ocean floors the evidence is the same, areas of new sea floor are symmetrical and the alternately polarised bands are found to agree in age and direction of magnetisation (Figure 2.13).

The floor of the ocean basins appears to have three major layers. Overlying the mantle and the asthenosphere (or weak plastic layer) is the 'oceanic layer' composed of a basic igneous rock complex with many doleritic dykes passing downwards into layered gabbros, making a layer 5 km thick. Upon this is a second layer, 1.5 to 2.0 km in thickness, composed of basaltic pillow lavas which have been extruded under water and rapidly cooled. Adjacent to continental areas, this, volcanic layer may become interstratified with consolidated sedimentary rocks such as limestone and shales (Figure 2.14).

## The trenches

If new sea-floor material is constantly being intruded along the lines of the spreading centres and the size of the Earth remains the same, it follows that an approximately equal amount of material must be destroyed or absorbed back into the deeper layers of the Earth's crust. This takes place in the deep oceanic trenches.

In the nineteenth century a British oceanographic survey vessel found depths greater than 8000 m (24 000 ft) near Tonga in the Pacific Ocean. Subsequently, it was found that the deepest parts of this abyss was 12 000 m (35 000 ft) deep and that it occurred in a long narrow V-shaped chasm where the sea was approximately seven times as deep as the Grand Canyon. Other trenches were subsequently found and in most cases they were found to be slightly over 12 000 m deep. Characteristically, these deepest parts of the oceans were not in the middle but around the periphery, near the Aleutian Islands, Japan, the Marianas Islands, the Philippine Islands, Java, Peru, Chile and Guatemala (Figure 2.15). Significantly, there is no trench off the western coast of the United States, but the structures associated with the San Andreas fault indicate a completely different geological situation. In the Atlantic Ocean, trenches are not so common, only being found in association with the island arcs of the West Indies and the South Sandwich Islands. The Indian Ocean has a trench south-west of Sumatra and extending towards the Andaman and Nicobar Islands.

The trenches are associated with negative anomalies of gravity and with earthquakes and volcanic activity on their landward side. The amount of heat flowing from the interior of the Earth in these regions is less than normal, suggesting that there might be a downward movement of cool rock which would reduce the outward flow of heat. As these trenches are the deepest part of the ocean, it might be thought that they would have an accumulation of sediment in them, but this is not so; only when trenches cease to be active do they accumulate sediments to any extent. It is not thought, therefore, that trenches develop into geosynclines, as the latter are characterised by large sediment accumulations.

The active role of the deep sea trenches concerns the leading edges of the tectonic plates. Previously it has been described how new sea floor is created at the spreading centres; it is also obvious that an equal amount must be removed in some way. As a crustal plate grows, its leading edge is destroyed at the same rate by reabsorbtion into the deeper layers of the crust. Where tectonic plates are moving together at speeds of more than 5 cm per year, one plate normally slides beneath the other and is consumed in a subduction zone

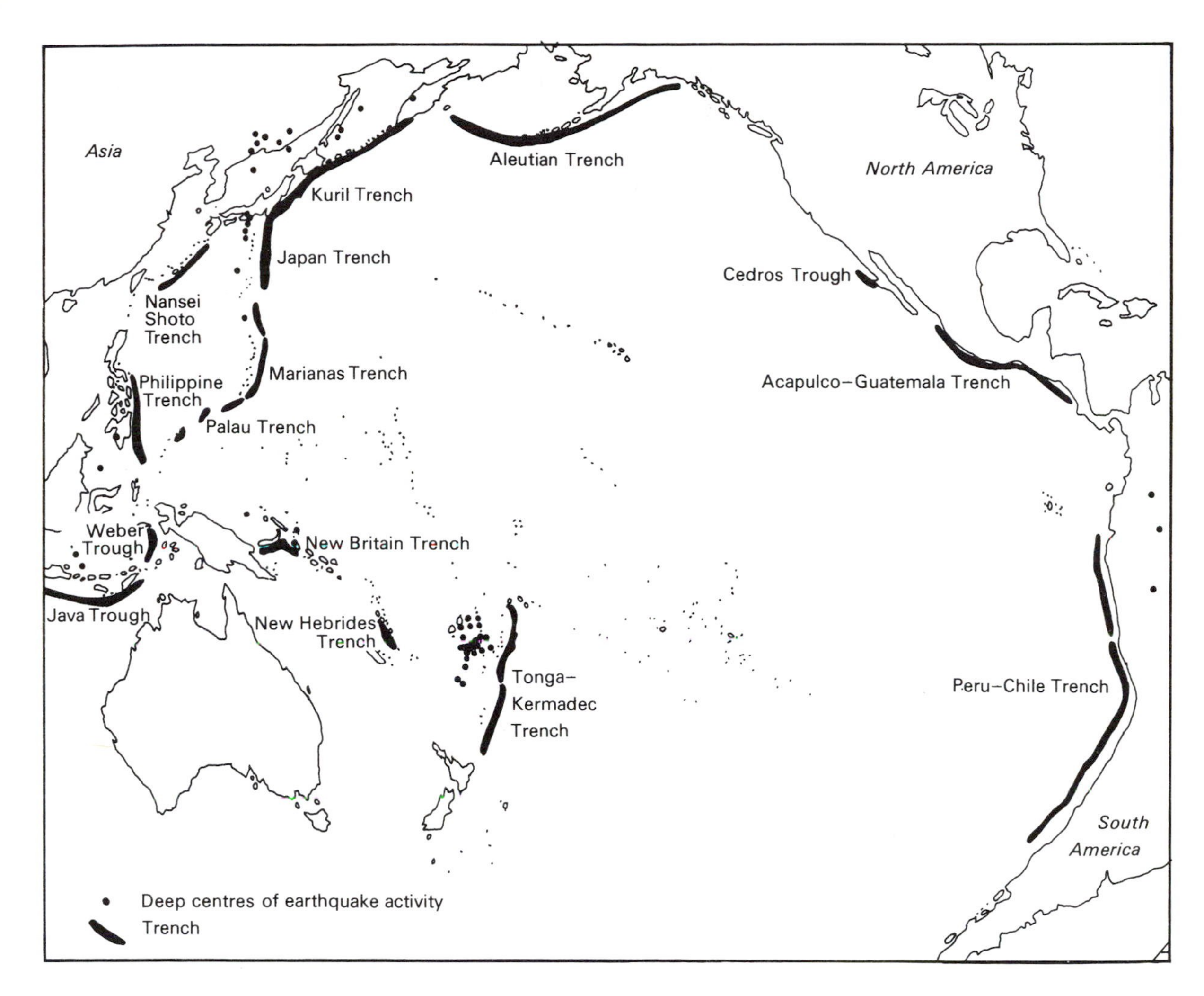


*Figure 2.15.* Trenches of the Pacific and location of deep earthquake foci. (After Fisher and Reville, 1955.)

(also known as a Benioff zone). The leading edge of the plate passes downwards into the asthenosphere and this zone is usually marked by many earthquakes and by intense volcanic activity (Figure 2.16). At lower speeds of approach it is possible for two colliding plates to buckle at the edges and produce fold mountain ranges between the two advancing plates.

## Origin of the continental areas

Uniformitarianism is a principle of geology which, put simply, states that the present is the key to the past. Observation of present-day processes leads to an understanding of what has happened in geological history throughout which a similar set of processes has operated. It is quite logical that if erosion takes place in one locality, there must be a corresponding deposition of material elsewhere.

For many years now, it has been understood that eventually all eroded material becomes deposited in downward-sagging belts of the Earth's crust known as *geosynclines*. This name was first given to these features by Dana in 1873. At first it was thought that the weight of sediment would initiate a downwarping, but this is now thought not to be the case and an independent downwarping occurs first, only to be reinforced later by the weight of accumulating sediment. The downward movement of the crust to accommodate the additional

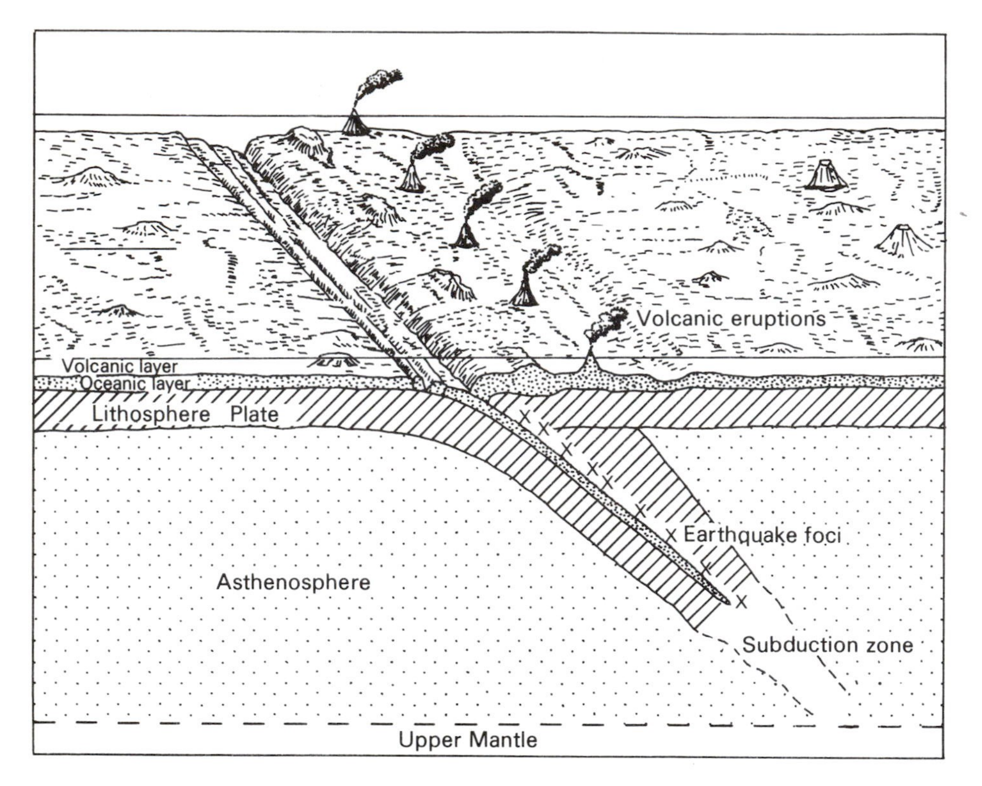


*Figure 2.16.* Section of deep sea trench with subduction zone and location of earthquake foci.

load of sediment is relatively slow. It has been calculated that 12 000 m of sediment accumulated between the Cambrian and the Permian in the Appalachian region and that this represented accumulation at the rate of 10 cm in 2500 years on average. It was assumed that as the downward-sagging sediments reached deeper and deeper into the Earth's crust they would be subject to pressure and heat which would produce metamorphic rocks such as schists and gneisses. Finally, movement of the forelands on either side of the geosyncline brought compression of the geosyncline resulting in uplift, folding and faulting. Igneous activity accompanied this activity with volcanoes and the instrusion of sills and dykes. In the deeper parts of the geosyncline plutonic intrusion invaded the sedimentary rocks to give granitic cores beneath the mountain ranges. These were only revealed subsequently when the mountains were eroded down to their roots.

Thrust-faulting or nappes (and less complicated folding) are greatest on either side of the geosyncline whereas in the centre considerable median masses may be uplifted, or in some cases downfaulted (Figure 2.17). The Tibetan plateau and the Hungarian basin are examples of uplifted/downfaulted but relatively unaffected land masses which were affected by the Alpine orogeny. The Tyrhennian Sea is an example of a large downfaulted basin. The ideal picture of a geosyncline given in many textbooks suggests a symmetrical upthrusting resulting from an inward movement of the forelands on each side. However, these movement are not always evenly matched. In the case of the Tethys geosyncline it appears that the African plate moved against the European foreland over which the northern folds have been thrust.

Although the concept of plate tectonics has revolutionised our understanding of the ocean floor, it is perhaps not quite so obvious how the theory can be related to the origin and growth of the continental areas. Fortunately, many aspects of the geosynclinal theory previously described can be linked to the ideas of plate tectonics to give a satisfactory explanation of the growth of continental masses.

Radioactive dating of the basement rocks indicates that there are nuclei or core areas in the continents

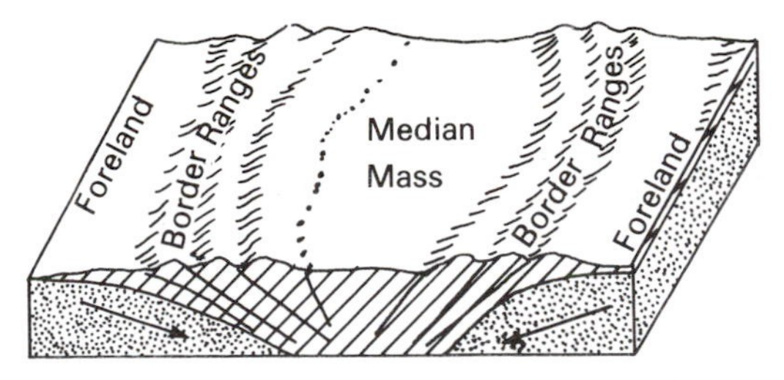


*Figure 2.17.* Diagram of a mediean mass (zweisengebirge.)

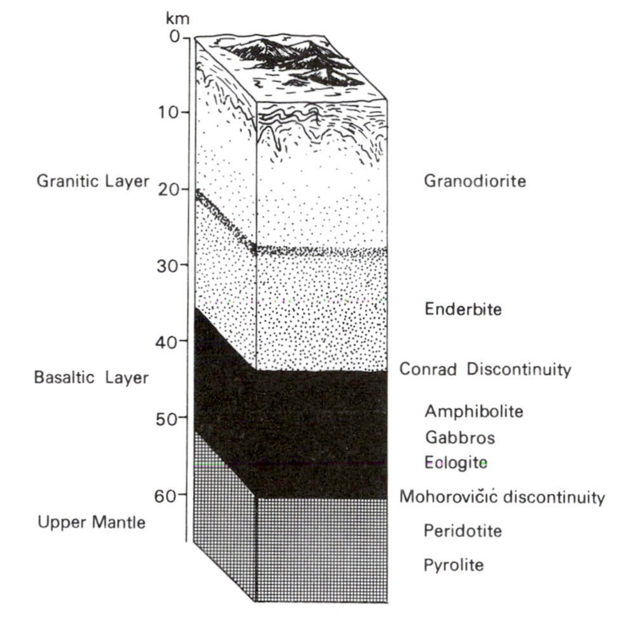


*Figure 2.18.* Idealised section through continental crust.

which are more than 2000 million years old, around and upon which successively younger deposits have accumulated. The landmasses have a different structure and composition from the ocean floor, already described. Overlying the mantle and the Mohorovičić discontinuity is a layer of basaltic composition similar to that found beneath the oceans (Figure 2.18). It has been described in Chapter 1 how this layer crops out only at a small number of locations and it appears to be formed of amphibolite, a metamorphosed form of basalt. Normally, it lies between 35 and 50 km below the surface except for the few fragments which have been brought to the surface in the process of mountain building.

The upper part of the continental crust lies above the Conrad discontinuity; it is composed of granitic rocks which extend to a depth of 35 km. These rocks are

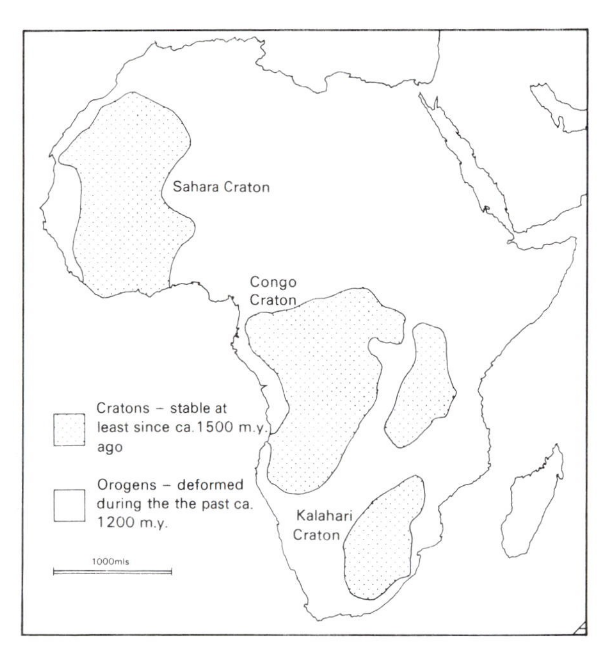


*Figure 2.19.* Cratons of Africa.

comprised of 92% igneous and metamorphic rocks with only 8% sedimentary material such as limestones, sandstones and shales, which lie near to the surface. The continental material lying above the Conrad discontinuity is formed from a combination of structurally complex, older shield areas, bordered by younger, intensively-folded sedimentary rocks.

Approximately 80% of the continental areas are underlain by the older, Pre-Cambrian rocks, which are called the basement complex. These oldest parts of the continents have grown by accretion as tectonic activity metamorphosed former sediments into crystalline rocks. Growth did not take place uniformly, but in phases associated with mountain building and igneous activity, which reached peaks around 2700, 1800 and 1000 million years ago. Most of the rocks of these older shield areas have experienced pressures and temperatures unknown in the younger fold mountains. The approximate ages of different parts (cratons) of the basement complex of Africa are shown in Figure 2.19.

Some parts of the basement complex are covered extensively with undeformed or slightly deformed sedimentary rocks. Others have accumulated in deep persistent downwarps which extend to 15 km depth below the surface. Where the Pre-Cambrian basement complex is exposed at the surface, it can be seen to be both structurally and chemically complex. Intense fold-

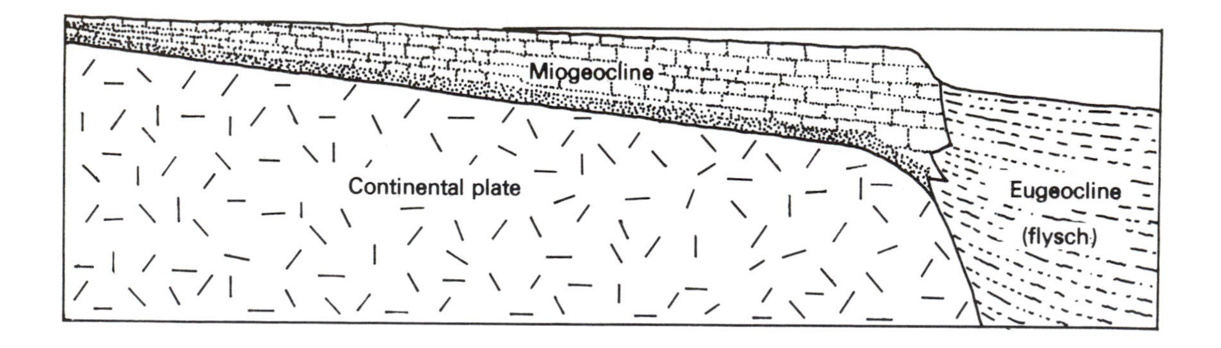


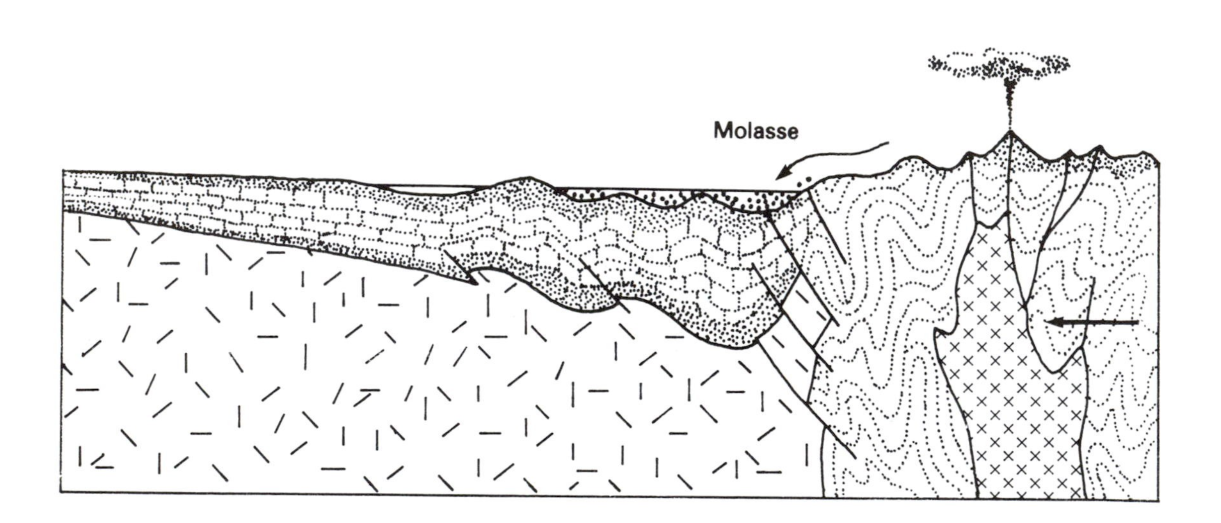


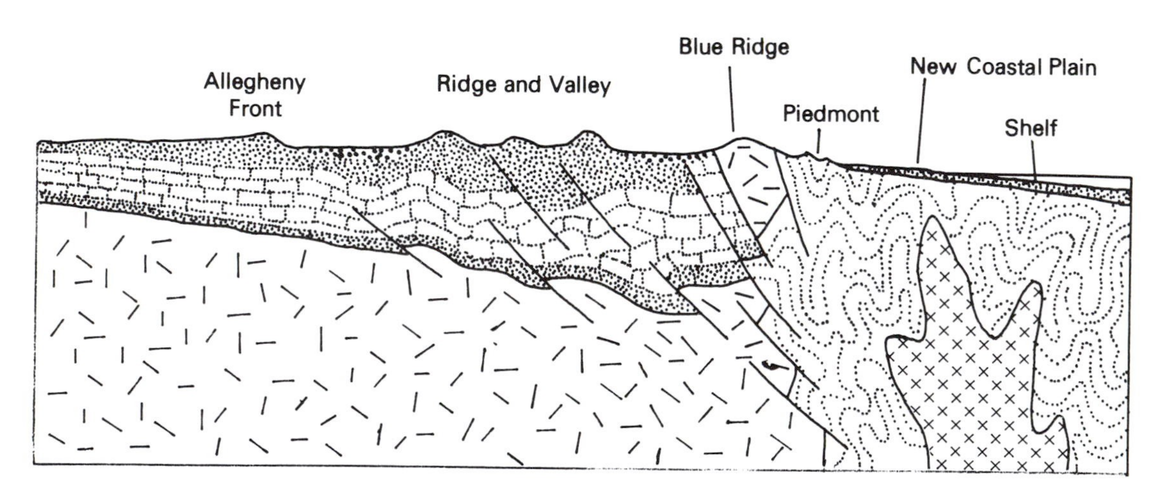


*Figure 2.20.* Eugeosyncline and miogeosyncline couplet.

ing is characteristic and may be associated with acid or basic igneous rocks, the latter possibly being former pieces of the oceanic floor caught up in orogenesis. Rocks from the basement complex, too, can be caught up in later orogenic activity and may form significant parts of the roots of mountain ranges.

An earlier view of orogenesis was given at the beginning of this section in the description of a geosyncline. The theory left many points unanswered which the recent discoveries associated with plate tectonics helped to answer. Recent observations have led to the suggestion that there are two parts of a geosyncline (a geosynclinal couplet), lying side by side and each playing a distinct role (Figure 2.20). According to Deitz (1972) deposition of shallow-water beds occurs in a miogeosyncline (lesser geosyncline) where crustal downwarping occurs at the edge of a continental mass. These sediments increase in thickness seawards and end at the edge of the continental shelf. Alongside the continental margin is the eugeosyncline (true geosyncline), comprising a wedge of sediments accumulated on the ocean floor and embanked against the continental rise. The sediments of the eugeosyncline are derived from muddy suspensions, known as trubidites, which have the appearance of thinly-bedded, poorly-sorted sands and silts interbedded with clays. Such sedimentary rocks have also been referred to as graywacke or flysch.

One of the most likely places for a geosynclinal couplet would be on the trailing edge of a lithospheric plate. It is suggested that as new sea floor was intruded an initial sagging took place into which the deposit of the eugeosyncline began to accumulate. Once started, subsidence continued under the weight of the sediments as isostatic compensation took place. On the landward side, this downward flexing provided the continuing shallow water conditions observed in the miogeosyncline. Altogether it is thought that a depth of up to 15 km of sediment could accumulate.

As the ocean basins are not fixed, but continually altering according to the lithospheric plate, it is possible that a subduction zone could be initiated by the deeply sagging eugeosyncline. Then as the ocean floor advanced towards the landmass, the sedimentary content of the eugeosyncline would be crumpled against it as well as being thrust deeply into the Earth's crust. Isostatic readjustment would eventually follow to raise the folded material as a mountain mass.

Alternatively, the eugeosyncline could be squeezed between two approaching continental masses. Once again crumpling would occur and the sediments would be strongly compressed, leading to the production of metamorphic rocks and to a thickening of the upper layer of the crust. Isostatic readjustment would then come into operation as in the previous case, associated with volcanic intrusion and emplacement of batholiths in the core of the folded sediments.

The miogeosyncline with its sedimentary deposits would be less strongly compressed and only affected by volcanic intrusions. The degree of folding would be greatest at the seaward edge of the miogeosyncline and would decrease inland until unfolded beds mantled the basement rocks formed earlier. As compression and mountain building took place, contemporaneous continental deposits of conglomerates, sandstones and shales, known collectively as molasse, accumulated to cover partially the slightly folded miogeosyncline beds as erosion attacked the rising mountain chain.

The theory of plate tectonics proposes that the surface of the Earth is composed of eight major lithospheric plates and a number of smaller areas, known as micro-plates. The major plates are the Eurasian, African, Indo-Australian, North American, South American, Pacific, Nazca and Antarctic plates. The micro-plates include smaller sub-continental areas such as occur in the Caribbean, south Pacific and Anatolian areas.

This account of world geomorphology attempts to discuss the geomorphology of the lithospheric plates which form the primary sub-divisions of the Earth's morphology. Each major plate has a submarine and a terrestrial component. Knowledge of the terrestrial geomorphology of each major plate varies greatly in detail; of the submarine geomorphology, which is less easily observed, only the briefest outlines are known.

# 3

# Africa

The area of Africa is approximately 29.78 million km$^2$. It lies astride the equator, extending for 8000 km in a north-south direction from latitude 37°N to 35°S. In the northern hemisphere it is 8000 km from east to west between longitude 18°W and 52°E but in the southern hemisphere it narrows to less than 4000 km. The highest point in Africa is Mount Kilimanjaro (5895 m) and the lowest point is Lake Assal (−150 m), near Djibouti. Africa has a mean elevation of 580 m above sea level, largely because of the extensive areas of crystalline shield which occupy 37% of the land surface area. Erosional plains cover 30% and depositional plains 26% of the continent's area. Volcanic plateaux amount to 4%, but young and old mountain belts together only amount to 3%.

In the past, geomorphologists have tended to assume the structure of Africa was simple, comprising only one ancient, rigid Pre-Cambrian shield with a coastline lacking in major indentations. In reality, the structure of the continent is complex with younger deposits laid over a composite Pre-Cambrian shield. Various ancient earth movements have affected the shield and on its northern and southern extremities fold mountains trend in a roughly east-west direction.

Africa is now thought of as the terrestrial part of a larger fragment of the Earth's crust which includes the surrounding areas of oceanic crust beneath the Atlantic, Southern and Indian oceans (Figure 3.1.). The western boundary of the African plate is the mid-Atlantic ridge from the Azores southwards to St Paul's Rocks and Ascension Island to Bouvet Island. The southern and eastern boundary, with the Antarctic plate, can be

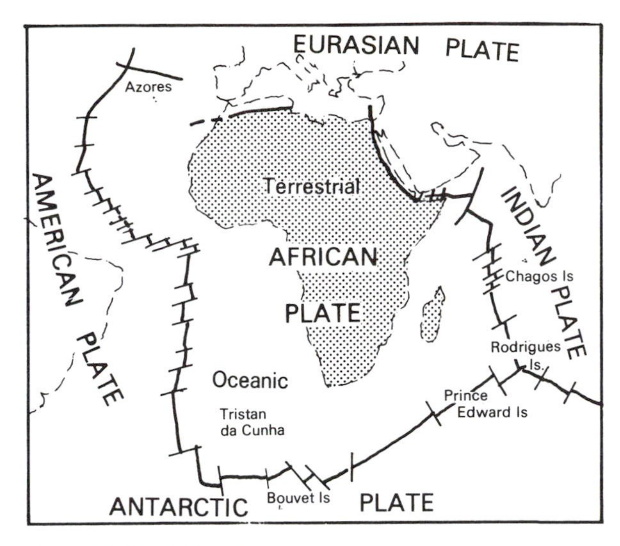


*Figure 3.1.* The African plate.

traced from Bouvet Island to Prince Edward Island and it joins the mid-Indian Ocean ridge south of Rodrigues Island. The eastern boundary, with the Indian plate can be traced from Rodrigues Island to the Chagos Islands, 1600 km south of India, and then along the Carlsberg ridge to the Gulf of Aden. By a series of transform faults and short segments of spreading centre the boundary of the African plate passes into the Red Sea and the Gulf of Aqaba.

The northern boundary of the African plate is not so easily traced as the interaction of the African and European plates has been complicated by several independent micro-plates, including Turkey, the Cyclades, Italo-Dinaric, Corsica–Sardinia and Iberia. A spreading centre is postulated in the eastern Mediterranean south of Crete which would form a natural boundary. Structurally the boundary of the African plate coincides with the southern boundary of the Atlas mountains (part of the European Alpine mountain system) in the Mahgreb. Problems of delimitation again exist with the boundary between north-east Africa and the Azores. If the line of the southern Atlas is taken it passes westwards through the Canary Islands, but if the line of greatest seismic activity is taken, this passes from Cape San Vincent directly to the Azores.

The oceanic floor of the African plate is thus extensive on the eastern, southern and western sides of the continent. Although below the ocean, some information is known about its broad geomorphological features. It has long been common knowledge that the coasts of Africa have virtually no continental shelf, the sea floor rapidly descends to great depth close inshore. The deep sea floor has considerable relief which may conveniently be described as ridges and basins.The basins, which descend to the depths of the abyssal plain, include the Somali, Agulhas–Natal, Cape, south-east Atlantic, Guinea, Sierra Leone, and Cape Verde basins. The depth of the sea in these basins is between 5000 and 6000 m and the age of the sea-floor rocks ranges back as far as early Cretaceous. They are not older, as the north Atlantic only began rifting between 180 and 280 million years ago and South America only began to separate from Africa 120 to 130 million years ago. The south-east coast of Africa was formed when it broke away from the Antarctic plate 120–130 million years ago. The most recent fracture occurs along the Red Sea where separation of the Arabian and African blocks began in the Pliocene.

Further evidence of the age of the sea floor is given by the oceanic islands. These either lie on the spreading centres or on ridges between the abyssal basins and are related to volcanic eruptions along these lines of crustal weakness. Ascension and Bouvet Islands lie on the mid-Atlantic ridge and the age of their oldest rocks is 1 million years. The Azores, St Helena and Gough Island lie slightly off the mid-Atlantic ridge and have rocks which have been dated at 20 million years old. Madeira and the Cape Verde Islands furthest from the mid-Atlantic ridge have rocks which are 90 and 120 million years old. Extrusion of volcanic material from certain 'hot spots' has given an indication of the relative movement of lithospheric plates. The Walvis hot spot has poured out magma for 140 million years. It remained stationary but the continental masses of Africa and South America have moved apart and also northwards as is indicated by the line of the Walvis ridge between the Cape and south-east Atlantic basins.

## Geological evolution

The geological evolution of terrestrial Africa is dominated by it having been part of Gondwana. The continent which we know today was situated towards the centre of this former landmass. The rocks of the Pre-Cambrian shield which crop out over at least one-third of Africa are referred to as the basement complex. This term includes igneous, metamorphic and sedimentary rocks of Archean and Proterozoic age, which form the cratons shown in Figure 2.19. The latter are of great economic wealth as structures associated with former active tectonic zones contain minerals such as gold, copper, nickel, lead and diamonds.

Little evidence remains of any succeeding Lower Palaeozoic formations because erosion rather than deposition dominated at this time. In northern Africa during the Lower Palaeozoic, a shallow sea transgressed southwards over what is now the Sahara and parts of west Africa from which limestones, sandstones and shales were deposited. Such a transgression was made possible by the proximity of these areas to the Tethys Sea. Elsewhere, with South America, India and Australia attached to Africa, the possibility of marine transgression was limited until the break-up of Gondwana occurred.

During the Upper Palaeozoic and Lower Mesozoic, the surface of the Pre-Cambrian basement rocks was subject to prolonged erosion. After this period of erosion, accumulation of rocks of the Karroo series began. These sediments were deposited in a continental environment, mainly during Carboniferous times. Although subsequently eroded, the Karroo series still covers extensive areas of southern Africa where they attain a maximum thickness of 8000 m.

In northern Africa following the retreat of the Lower Palaeozoic sea, continental deposits of the Nubian series accumulated. The Nubian series of the north-east are mainly sandstone, but further west, conglomerates and clays of the 'continental intercalaire' accumulated and now form the major aquifers beneath the surface.

Marine deposits were laid down during the Jurassic in peripheral basins after fragmentation; these are fairly extensive in the north but become restricted to narrow coastal areas in the south. It is from these Jurassic strata that oil and natural gas are now obtained in Libya and Algeria. Southern and eastern parts of Africa were subject to a prolonged episode of erosion during the Jurassic when the pre-existing relief was much reduced. Thus, towards the end of the Jurassic the appearance was plateau-like, and this surface extended over parts of South America which, until the end of the Jurassic, was lying adjacent to Africa. Fragmentation of Gondwana took place during the Cretaceous leading to the formation of the great escarpment around the periphery of southern Africa. The surviving remnants of the land surface at this stage of development of the African continent have been called the *Gondwana surface* (King, 1962).

Drainage of the Gondwana surface was disrupted as fragmentation took place and many rivers found new, shorter and precipitate courses to the sea. Erosion resulted in a lowering of the landscape close to the rivers and the production of a new land surface. King refers to this surface as the *African surface*, with remnants of the Gondwana surface remaining on the interfluves. Formation of the African surface continued well into the Tertiary. It is well developed in southern Africa, but in north Africa differences in height between the two surfaces are less significant. By the Quaternary, most of the Gondwana surface had been consumed leaving the African surface as a widespread summit level between 300 and 500 m above sea level. A lower, late-Tertiary surface was eroded also and rivers now are incised in gorges mainly cut during the Pleistocene.

Tectonic activity affected Africa throughout its history with uplift and depression of its surface, but the break-up of Gondwana and the development of the rift valley system represent more dramatic events which are apparent from the large-scale geomorphology of the continent. Hercynian folding in the Cape region along an east–west axis and Alpine folding in north Africa complete this outline of the tectonic history which has had the following geomorphic effects:

1. Africa was elevated in the south and east to between 1300 and 1700 m but much less in the north, with some tilting towards the oceans around the continental edges. The uplift of the interior resulted in rejuvenation of the rivers and development of erosion surfaces at lower levels separated by steeper escarpments, each surface corresponding to a cycle of erosion. The plantation of each surface was attributed to a lengthy period when folding, warping and faulting were quiescent. However, it must be admitted that the processes which result in extensive planation of the landscape are not fully understood and therefore still controversial.

2. The tilted surface referred to above has been subjected to downwarping in certain areas to form major basins with floors at about 300 m above sea level separated by upwarped rims (Figure 3.2).

3. A series of rift valleys has been created, beginning in the early Tertiary. The great rift valley of east Africa extends from Lake Malawi to the Ethiopian highlands where it divides to form the Red Sea and the Gulf of Aden. Other major faulting (rifting) has occurred in the Beneu trough of Nigeria and in the separation of Madagascar from the rest of Africa.

4. Volcanic landscapes with both basaltic lava flows

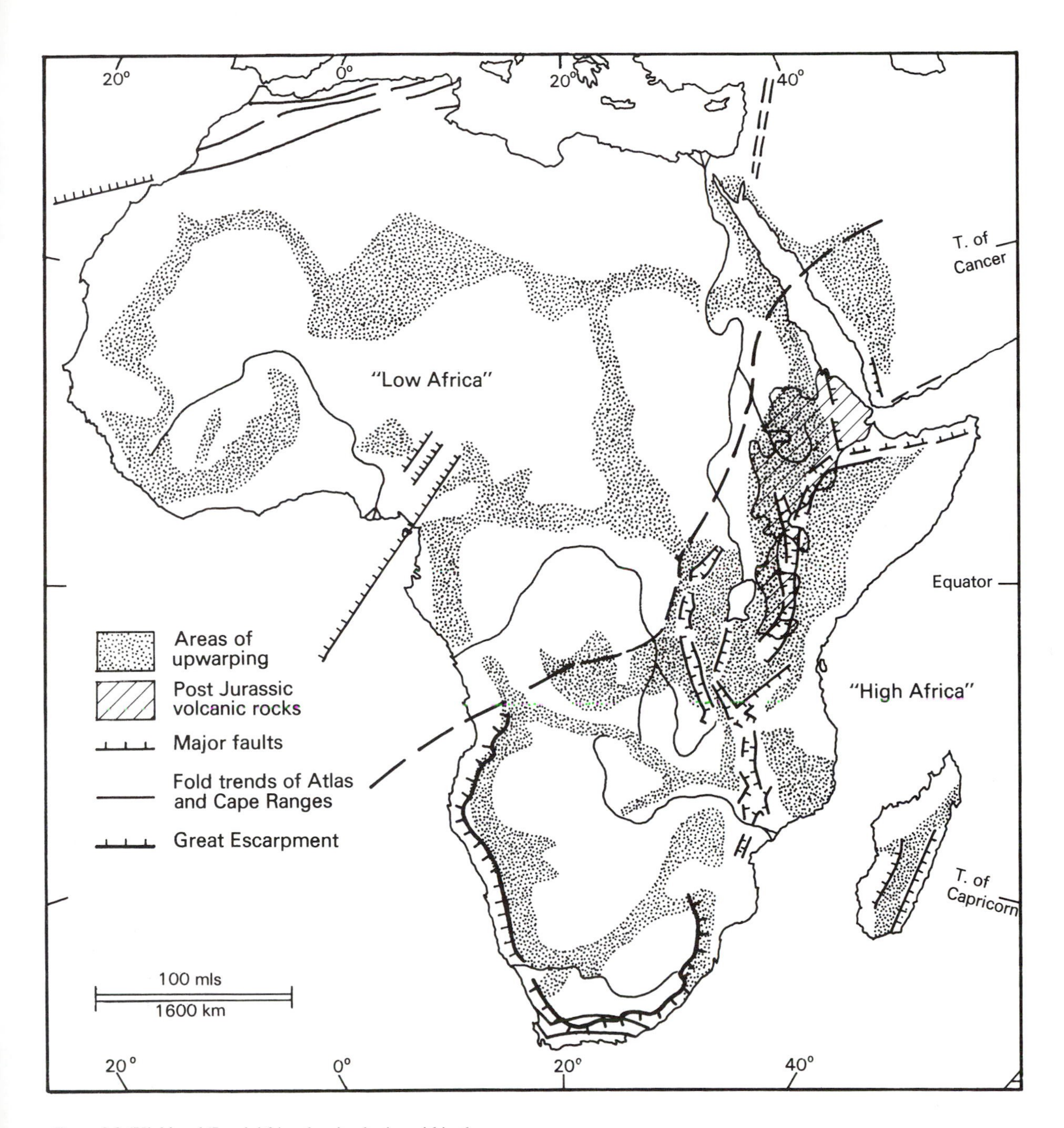


*Figure 3.2.* 'High' and 'Low' Africa showing basins within the basement complex and the line of the rift valley faulting.

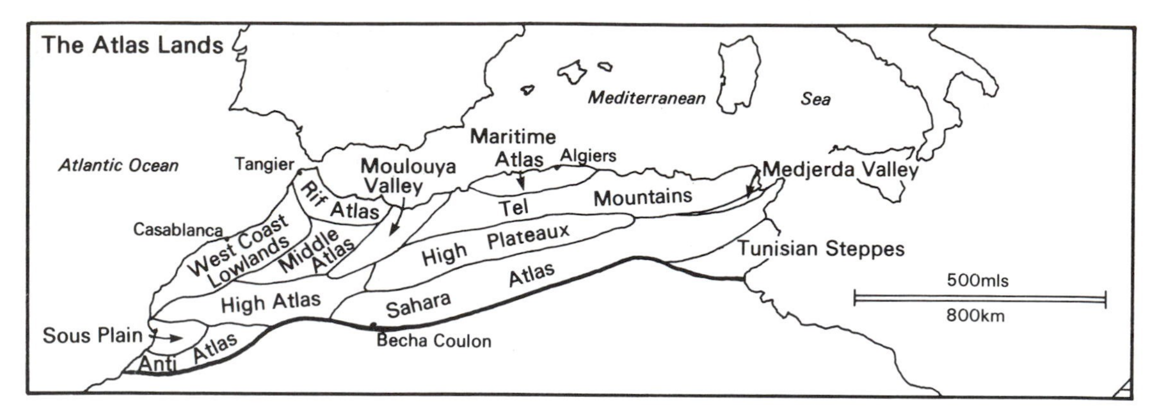


*Figure 3.3.* The provinces of the Atlas lands.

and volcanoes are associated with the great rift valley, particularly in Kenya and Ethiopia. Vulcanism is also present along the line of islands which include Sao Tomé, Fernando Poo and the Cameroon mountains as well as in the central Saharan Tibesti and Air mountains.

These geological events and the range of climate experienced gives the African continent great variation in its landforms. An approach commonly adopted is to describe Africa in terms of:

*High Africa*: those areas corresponding to the remnants of King's Gondwana and African surfaces, mainly to the south and east. Other elevated areas comprise volcanic summits.

*Low Africa*: the lower plateau surfaces cut on the basement complex and its superficial deposits, resulting from the incision and planation of the higher plateaux.

*Basins and ridges*: superimposed on the landscapes of high and low Africa are the basins and intervening ridges which characterise so much of the continent and which largely control the size and direction of flow of the larger rivers.

*Fold mountains*: at both northern and southern extremities of the continent the basement complex is caused by later folded strata which form the Atlas mountains and the Cape ranges respectively.

The sub-division of the African lithosperhic plate into geomorphological regions will follow the lines described in the introductory chapter. Seven major divisions are identified:

The Atlas lands
The Saharan lands
Plateaux and mountains of west Africa
The central African basin
East African plateaux
South African plateaux
Oceanic basins

**The Atlas lands**

The mountainous region north of the Sahara desert forms the part of Africa known as the Maghreb. It consists of Mesozoic and Tertiary sedimentary strata which have been folded during the Alpine orogeny into ranges of mountains with an east-north-east to west-south-west alignment. Plateaux of various elevations occur between the mountain ranges.

Some folding occurred during the Hercynian orogeny in Morocco and western Algeria, resulting in some Pre-Cambrian basement rocks and Palaeozoic strata being incorporated into the Anti-Atlas mountains. Alpine earth movements are the most important episode of folding as far as the present landscape is concerned. They began during the late Cretaceous when the Rif Atlas and the Tel mountains of the north coast were raised. Renewed folding and uplift took place in mid-Tertiary when the Cretaceous and Triassic rocks of central Algeria and Tunisia were folded and the high plateaux of central Algeria elevated. The area is discussed in the following eight provinces which are shown in Figure 3.3:

Rif Atlas
Middle Atlas
High Atlas
Saharan Atlas

Tel mountains
High plateau
West coast lowlands
East coast lowlands

*The Rif Atlas*

Extending eastwards from Tangier, the rugged Rif Atlas mountains run almost parallel to the coast of Spanish Morocco. These mountains are formed of Jurassic and Cretaceous strata folded at the end of the Cretaceous in one of the earliest episodes of mountain building in the Alpine orogeny. The Rif Atlas is a mirror image of the Baetic Cordillera of southern Spain. Thrusting occurs southwards over Lower Palaeozoic rocks and flysch deposits. Glacial features occur on the higher peaks.

*The Middle Atlas*

The Middle Atlas comprises a branch of the High Atlas extending from the main range in a north-easterly direction and attaining elevations of 1200 m. These mountains are formed of Lower to Middle Jurassic limestones but beneath them metamorphosed Carboniferous marine deposits occur into which plutonic intrusion has occurred. In the Pleistocene, volcanic cones were produced and basalt flows occurred near Ifrane.

In the upper Ouiounane valley, the Ksouatene and Afriroua poljes have developed with other karstic features, such as sinks and blind valleys. The sides of the poljes are steep and in the Afriroua polje there are two small residual limestone masses (hums). As occurred in the Rif Atlas, the higher peaks of the Middle Atlas have corries and below them valley outwash terraces. Periglacial features have also been observed.

*The High Atlas*

Extending inland from Agadir on the west coast of Morocco for 400 miles, the High Atlas has several snow-capped peaks including Mount Toubkal, the highest, which reaches 4165 m. The rocks comprising these mountains are Jurassic, Cretaceous and Eocene. They rest unconformably on the Pre-Cambrian basement complex which is exposed by erosion in several places. As the strong earthquake at Agadir in 1960 indicated, this area is under tectonic stress and the Canary Islands, with active volcanicity, lie only 160 km offshore. The higher peaks of the High Atlas also were glaciated during the Pleistocene. South of Marrakech, a southern branch of mountains extend south-westwards to Ifni, enclosing the Sous lowlands.

*The Saharan Atlas*

The Saharan Atlas range can be traced from near the town of Bechar Coulon on the south-west border of Morocco, north-eastwards across Algeria, eventually to become the Dorsale or central mountains of Tunisia. The Saharan Atlas is formed mainly from limestones and marls of Cretaceous age. In Tunisia the folding is mainly of an open Jura type and phosphates are found in association with Eocene beds on the southern flank of the range. At its eastern extremity this tectonic unit is squeezed between the semi-stable Kerkennah block and the Tel mountains resulting in large-scale faulting which bring up the underlying Cretaceous limestone to form Djebel Zaghouhan (1417 m).

*Tel mountains*

These mountains form a double line of peaks separated by a series of basins, the formation of which dates from the Miocene. The littoral Tel is composed of Jurassic, Cretaceous, Eocene and Oligocene strata which in Tunisia have been quite strongly thrust-faulted. The inland Tel is also formed of Jurassic and Cretaceous limestones, in which lead, zink and mercury are mined and a small amount of oil occurs. A series of faulted basins, drained by the Medjerda river, lies between the littoral and inland Tel.

The name Maritime Atlas is given to the mountains on the Algerian coast east of the capital city. The coast is rocky and plunges steeply into the Mediterranean, where cliffs are formed from Tertiary sedimentary rocks which overlie metamorphosed Lower Palaeozoic rocks.

*High plateaux*

Between the Tel and Saharan Atlas in Algeria, elevated plateaux occur at 750–1000 m above sea level. These elevated plains are often without external drainage, having ephemeral salt lakes (chotts). The plateau surfaces are mantled with Neogene marls and unconsolidated sands. The western end of the high plateaux country is drained by the Moulouya river which flows to the Mediterranean between the Rif Atlas and the Tel mountains in eastern Morocco.

### *West coast lowlands*

The low-lying land between the Rif Atlas and the Middle Atlas forming the hinterland of Rabat and Casablanca is referred to as the central plateau and rises to an elevation of between 1100 and 1300 m. In the south-west part of the area, phosphates are mined.

A smaller area of lowland lies between the ranges of the High Atlas and the Anti-Atlas. This is drained by the Sous river, occupying a triangular-shaped, alluvial-filled structural depression.

### *East coast lowlands*

The east coast of Tunisia is composed of a low plateau of less than 300 m formed upon the Kerkennah block, a semi-rigid fragment of the African shield, covered with Mesozoic and Tertiary sediments. The inland margin of the low plateau is marked by a line of subsidence occupied by salt lakes, of which the Chott Sidi el Hani is the largest. To the west, Jurassic and Cretaceous rocks crop out extensively to form the High Plains of Tunisia, 300–500 m above sea level.

The Tunisian east coast is low with a ridge of sand dunes backed by salt marshes interrupted by some areas of cliffed coast where the Tertiary rocks crop out, as at Monastir. Faulting has disturbed the Monastirian raised beach level. North of Tunis, the Medjerda river has built its delta into the sheltered waters of the Gulf of Tunis, but a recently constructed flood relief channel now takes all the flow of the river direct to the coast where a new delta is forming.

## The Saharan lands

The most extensive desert area in the world extends from the Atlantic coast 5000 km to the Red Sea and continues eastwards through Saudi Arabia. From north to south the Sahara is 1600 km across. As north Africa is situated beneath the sub-tropical high pressure cell, some 80% of the Sahara receives less than 20 mm rainfall per year. The landscape today is dominated by aridity but considerable evidence exists to show that it has not always been as dry as it is at present. Equally, the extension of dune landscapes, now no longer active, suggest a greater extent of desert conditions at other times in the past.

Small areas of the Sahara have been eroded below sea-level, but over extensive plateaux the average elevation is 500 m. These low-lying plateau surfaces are interrupted by the central Sahara mountains of the Ahaggar and Tibesti which reach elevations between 3000 and 4000 m. The greater part of the desert surface is formed of either bare rock surfaces (hamada) or surfaces covered with gravel (reg). Sandy desert (erg) only covers about 12% of the surface of the Sahara.

Wind-sculpted rocks, yardangs, have long been part of the folklore of geographical knowledge about desert landforms. They are best-developed in silty deposits of former lake floors, but evidence from satellite photography has revealed that low-lying wind-sculpted rocks are far more extensive than realised previously. Such areas are described as a *crest-couloir* system and they are best developed around the Tibesti mountains, the Adrar, the Tademait plateau and around Jebel Uweinat in west Sudan.

Movement of sand by winds has concentrated the sand desert in certain places mainly around the periphery of the Sahara, leaving the central mountain areas free of sand accumulations of any size. As has been suggested in the past, the sands are not always in former alluvial basins to which they have been brought by fluvial action and then re-sorted by the wind. The ergs, or sand deserts, occur where winds have brought the sands together.

Another aspect of wind action in the Sahara is the export of dust from the desert. Occasionally dust from the Sahara is precipitated in northern Europe and Britain, but satellite photography has traced dust clouds across the Atlantic reaching the West Indies in summer and Brazil in the winter. It is estimated that between 25–35 million tons of dust per year are blown westwards (Figure 3.4). In one period of 6 hours, an estimated 400 000 tons of dust passed over one 100 km stretch of the Atlantic coast of the Sahara. A further 25 millions tons of dust is annually exported eastwards.

Despite its many provinces, briefly outlined below, the Saharan region of Africa has a unity and has existed for much of geological time as a major geomorphological division. Sedimentation in Cambrian and Ordovician times took place on a uniform erosion surface inclined gently to the north. This surface developed in arid conditions where surface rill-wash led to the formation of gravel-covered plains with faceted peb-

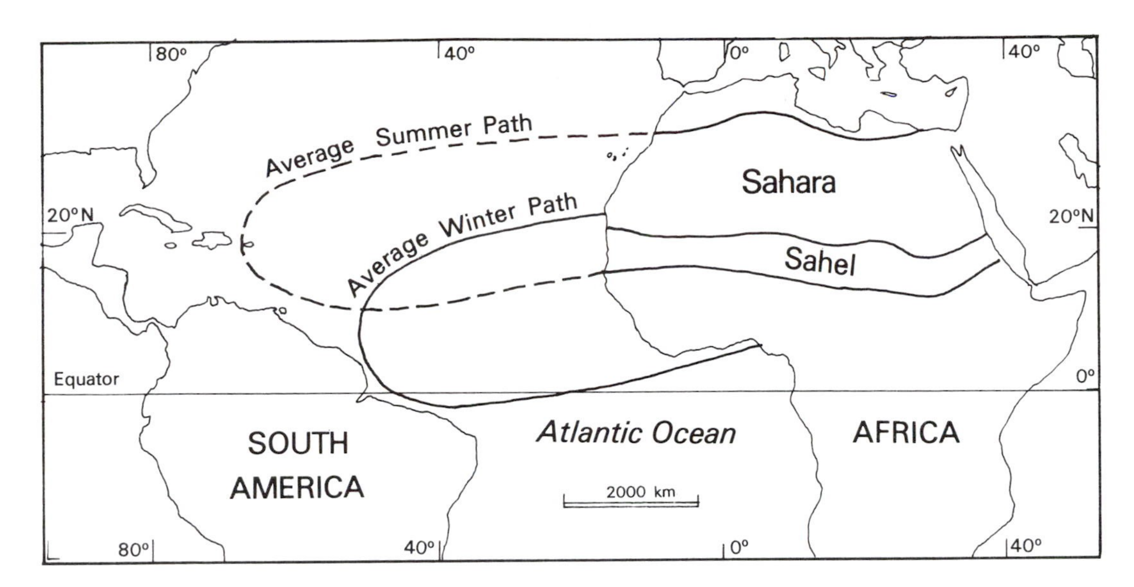


*Figure 3.4.* Dust plumes from the Sahara.

bles. Between the Ordovician and Lower Silurian, glaciation affected part of the Saharan area (when Africa was near to the South Pole). It left tillites as convincing evidence on the margins of the Tibesti, the Ahaggar massif and the Tassili-n-Ajjer.

Throughout a long period of geological history, from the Carboniferous to the Cretaceous, continental deposits accumulated on the surface of the Pre-Cambrian basement. These deposits are known as the *continental intercalaire* and include the Nubian sandstones which today are important aquifers. The northernmost part of the African shield underwent a marine transgression in the Cretaceous and early Tertiary which left behind calcareous sediments including the Nummulitic limestones of Egypt. Throughout the rest of the Tertiary, the sea withdrew and deposits known as the *continental terminal* accumulated. Volcanic activity also affected parts of the Saharan region during the period from the Tertiary to the Quaternary, especially Tibesti, Jewel Marra and Jebel el Soda.

Although the Sahara has been drier and more extensive than at the present day, it has also been wetter. The present trend is one of increasing aridity. Evidence from desiccated river systems, soil profiles and biogeographical distributions all indicate an earlier wetter phase. Taking the structural and surface features into consideration it is possible to describe the geomorphology of the Saharan major geomorphological division in the following provinces (Figure 3.5):

The Atlantic coastal plain
Requeibat ridge
Tindouf basin and Iguidi erg
Taoudenni basin and Chech erg
Eglab plateau
Tanazrouft plain
Ahaggar massif and Tassili
Air and Adrar plateaux
Tibesti mountains
Ennedi plateau
Iullemeden basin
Chad basin
Fezzan and Kufra basins
Hamadas of Tademait, Tinrhert and El Hamra
North Saharan basin
The western desert
Plateaux and hills of Cyreniaca
North and south Sudanese plains
Jebel Marra and the Nuba mountains
The Nile valley and delta
The Nubian desert
The basins of the White and Blue Nile
The Red Sea hills and the Arabian desert

### *The Atlantic coastal plain*

Along the Atlantic coastal lowlands the Pre-Cambrian basement rocks form the Rio del Oro basin, which has been infilled mainly with Cretaceous sediments overly-

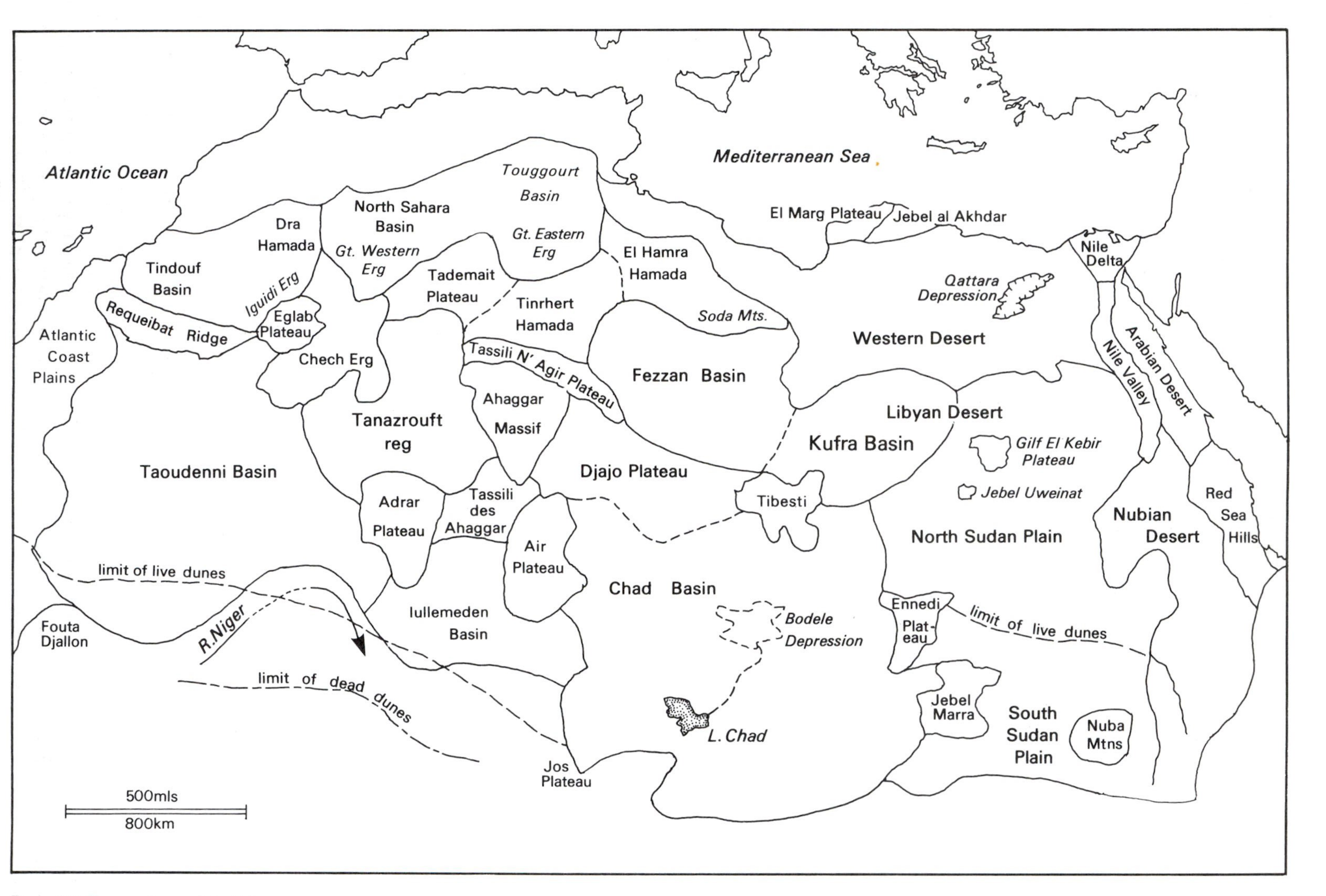


*Figure 3.5.* The provinces of the Sahara.

ing Lower Palaeozoic rocks. Adjacent to the coast Eocene and continental intercalaire deposits occur; the Eocene rocks containing phosphatic limestones. Calcareous sandstones and limestones are characteristic of the Cretaceous rocks. With the low rainfall and high rock permeability, few wadis of any size reach the coast in the northern part of this coastal plain. A pedimented surface occurs on the outcrop of the Palaeozoic rocks of the inland part of this region.

Towards the south, the Atlantic coastal plain becomes the Senegal plain which has a surface formed of the continental terminal deposits comprising sandstones, shales and unconsolidated sands and clays. Although wadis are shown on maps to reach the coast north of Nouakchott with dunes on the interfluvial areas, the first permanent stream of this coast is the Senegal river which has its headwaters in the Fouta Djallon mountains of Guinea. South of Cape Verde, the appearance of the coast changes from desert to mangrove swamps and inlets.

*Requeibat ridge*

In northern Mauritania the granitised Pre-Cambrian rocks come to the surface to form the Requeibat ridge. At the western end non-metamorphic rocks of upper Pre-Cambrian age occur, but further east the basement is covered by the sands of the Iguidi erg.

*Tindouf basin and Iguidi erg*

In this region the basement rocks of the Pre-Cambrian have been downwarped in the Tindouf basin, infilled with Palaeozoic strata and covered by the continental terminal deposits. The northern part of this region is covered by the Hamada of Dra but towards the south the landforms are dominated by the dunes of the Iguidi erg. The eastern margin of this region is the Ougata range formed from folded Palaeozoic strata. These slightly higher ridge and valley landforms extend towards the Tademait plateau. Satellite photographs indicate strong lineation of the dunes of the Iguidi erg in a north-east–south-west direction.

*Taoudenni basin and Chech erg*

A very large proportion of west Africa is made up of this basin which extends 1500 km from east to west. A broad sag in the Pre-Cambrian basement begins in the Requeibat ridge in the north and disappears beneath slightly metamorphosed Cambrian and Ordovician rocks, eventually to re-emerge on the southern side at the Sahara in the Fouta Djallon hills of Guinea. The northern part of this huge basin is occupied by the dunes of the Chech erg which, like the Iguidi erg further north, have a north-east–south-west lineation.

*Eglab plateau*

The crystalline rocks of the basement complex emerge from beneath the sands of the Iguidi and Chech ergs to form a plateau 500 m above the sand deserts on either side. The Eglab plateau forms a feature which is clearly identifiable from satellite photography and forms part of the unwarped zone between the Taoudenni and Tindouf basins.

*Tanazrouft plain*

The Tanazrouft plain lies west of the Ahaggar massif and forms an extensive surface covered with a regolith of pebbles and coarse sands. All descriptions of this area stress its lack of relief. The northern part of it includes the Sabkhas of Mekerrhane and As el Matti at the geographical centre of Algeria.

*Ahaggar massif and Tassili*

The central Sahara is dominated by the presence of the Ahaggar mountains, which rise to 3280 m. This is an area of crystalline rocks of Pre-Cambrian age surrounded by Palaeozoic strata which form the Tassili plateaux. The rocks of the Ahaggar include gneisses and mica schists which are overlain by middle Pre-Cambrian sedimentary strata including conglomerates, sandstones and some limestone. The Palaeozoic strata of the Tassili close to the Ahaggar include Devonian marine sandstones and marls with Carboniferous shales, sandstones and limestones further away. This area was glaciated in Silurian times leaving tillites and other glacial features such as striations and eskers which were buried by later sediments, lithified and subsequently revealed by erosion. Evidence of Pleistocene periglacial activity has been claimed by some authors. Drainage of this rugged area is centrifugal by a series of wadis which are normally dry but flood once or twice each year. Oases occur around the periphery of the mountain massif where water is perennially available.

The Tassili plateaux are the dip slopes of the Lower Palaeozoic rocks which slope gently away from the core of crystalline rocks. Their surfaces are variably covered with desert detritus and alluvial spreads occur along the wadis.

### *Air and Adrar plateaux*

Pre-Cambrian outcrops occur also in the Adrar and Air plateaux which extend south and south-westward from the main massif of the Ahaggar. The Adrar des Iforas is a very flat plateau surface with bare rock or erg surfaces. Relief on the Air plateau is more varied as it was affected by volcanicity and possesses some volcanic cones. The elevation of these plateaux is about 1000 m above sea level with the highest peaks attaining 2000 m.

### *Tibesti mountains*

Some of the most spectacular scenery of the Sahara occurs in the Tibesti mountains of the central Sahara where peaks rise to 3300–3500 m above sea level. As elsewhere in Africa, the basement rocks of the Pre-Cambrian are the foundation of these mountains but in the Tibesti, they are overlain by volcanic rocks of Tertiary and Quaternary age. The volcanic rocks include basalt, dolerite and labradorite, and are to be found capping the tops of hills. The Trou au Natron is a caldera 600 m deep and 8 km in diameter. Emi Koussi, 3735 m, and Tousside are rhyolite cones. Periglacial features have been observed in the Tibesti mountains.

### *Ennedi plateau*

The south-eastern area of high land in the central Sahara is the Ennedi plateau. It is formed of Lower Palaeozoic strata (Cambrian–Ordovician and Devonian) of a similar nature to those of the Tibesti massif. Sandstones predominate and any precipitation is soon absorbed into the ground or evaporated. Strong north-east winds are funnelled between the Tibesti and the Ennedi plateau to give striated rock landforms and longitudinal dunes.

### *Iullemeden basin*

This structural downwarp between the central Saharan massifs and the west African plateau lies in eastern Mali and south-western Niger centring on Niamey. The continental intercalaire outcrops in the northern part of the basin and in the south-east, younger deposits of the continental terminal occur upon Cretaceous calcareous

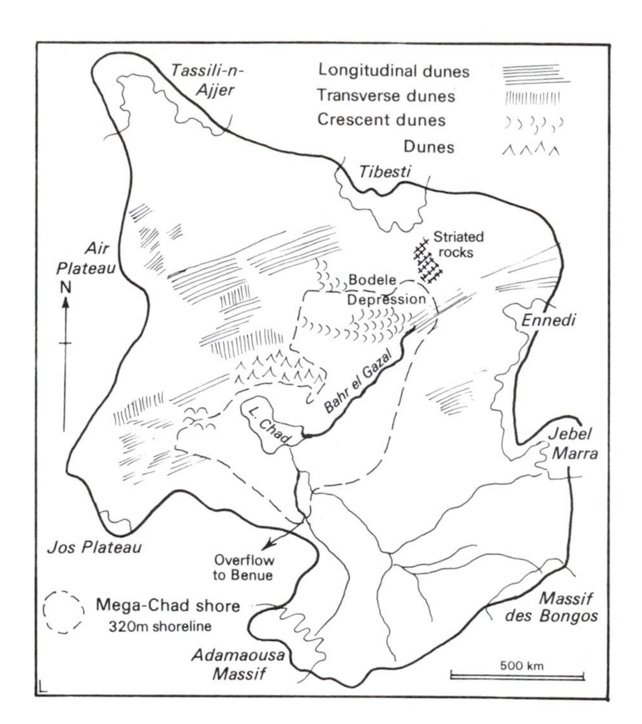


*Figure 3.6.* The Chad basin.

strata. The River Niger, having passed through the inland delta region, resumes its course in a south-easterly direction.

### *Chad basin*

Downwarping of the Chad basin took place mainly in the Quaternary, consequently it does not contain deposits of the earlier continental terminal. Quaternary materials overlie the basement to a depth of 100 m. Much of the floor of the basin is at about 300 m above sea level, but the present Lake Chad lies at 200 m and the even lower Bodele depression lies at 170 m (Figure 3.6). The present Lake Chad does not occupy the lowest part of the basin, but a channel, the Bahr el Gazal, connects the two lower parts. Between 5000 and 10 000 years ago a much larger lake, Mega-Chad, occupied the basin up to the 330 m level. The floor of this former lake now forms a monotonous desert plain with dunes, shallow wadis and saltflats. Water for the present Lake Chad originates from the savanna lands of the Ubangui–Chari plateau in the Central African Republic on the watershed between the interior basin of Lake Chad and the Congo. The level of Lake Chad varies annually according to the amount of rain received during the wet season.

*Figure 3.7.* The Great Eastern Erg, southern Tunisia.

### *Fezzan and Kufra basins*

These two tectonic basins are situated on the northern flanks of the central Saharan mountains, in southern Libya. Downwarping in the Fezzan basin can be dated to the Palaeozoic after which the continental intercalaire and continental terminal rocks have been deposited. Subterranean supplies of water are available and oases occur in the sand sea which forms the surface of these regions. Between the Fezzan and Kufra basins an extensive outcrop of volcanic rocks (basalts) occurs. From beneath the dunes rocky ridges emerge enclosing depressions, but toward the Mediterranean Sea are the slightly more elevated areas of the Soda Mountains and El Hamra plateau. The Kufra basin is covered by the sand sea of Calanscio.

### *Hamadas of Tademait, Tinrhert and El Hamra*

These features are a series of plateaux situated north of the central Saharan mountains and inland from the lower coastal plain in Libya and south of the great ergs in the central part of Algeria. As their name suggests, these are rocky surfaces, the aridity of which is increased by the underlying permeable calcareous rocks of Cretaceous age, upon which Tertiary limestones have also been laid down. The general elevation of these hamadas is less than 1000 m above sea level.

### *North Saharan basin*

South of the Saharan Atlas in Algeria and Tunisia, the Pre-Cambrian basement is downwarped to between 3000 and 5000 m and the resulting basin is infilled with continental intercalaire sandstone and evaporites upon which Cretaceous clayey sandstones, dolomites and anhydrite occur, followed by the Nummulitic limestone of the Tertiary. The sands of the Great Western Erg and the great Eastern erg (Figure 3.7) cover the surface. The area in the east, known as Touggourt is of lower elevation with, on its northern margin, the great Chott lakes at or below sea level (Chott Djerid −17 m; Chott Melrir −30 m). Terraces around the Chott Djerid indicate former higher water levels, but at present the Chott surface only floods during the winter when evaporation is decreased. Pipelines bring oil from Hassi Messaoud and gas from Hassi Rmel to the Tunisian coast at Es Skhira.

### *The western desert*

The coastal region of north Africa from Tripoli to Libya to Alexandria in Egypt has a similar geological history to that of the north Saharan basin. The Pre-Cambrian basement is depressed and covered by continental intercalaire, the most significant member of which is the Nubian sandstone. These deposits were inundated by the Cenomanian transgression which deposited the Cretaceous dolomites and marls. In turn these have been covered by the Tertiary Nummulitic limestones. The desert surface is plateau-like and partly covered with shifting sands. The southern margin of this region has been taken as the furthest penetration of the Cretaceous or later marine transgressions. This region widens eastwards to form the western desert of Egypt where the Qattara depression has been eroded to − 134 m below sea level. It is generally accepted that the Qattara depression has resulted from deflation of salt-weathered silts and sands derived from the limestone-capped Miocene Moghera formation. Material has been eroded down to the water table and sands from this depression form seif dunes extending south-eastwards from the indistinct southern margin of the depression. The northern margin is a pronounced scarp.

The elevation of the western desert rises from the Mediterranean to the Gilf el Kebir plateau at 1000 m and Jebel Uweinat (2000 m) which marks the point where Egypt, Sudan and Libya meet. Oil occurs inland from the Gulf of Sidra on the fringes of the Calanscio Sand Sea.

### *Plateaux and hills of Cyreniaca*

In the hinterland of Benghazi, the El Marg plateau is broad and rolling but becomes dissected further east and merges with an elevated region known as the Jebel al Akhdar, comprising a series of ridges and discontinuous coastal lowlands. The underlying rocks result from the marine transgressions which affected this area, specifically in the Tertiary, when marine sands, clays and limestones were laid down. The Mediterranean climate with greater rainfall introduces a greater variety of fluvial action and a more vigorous vegetation cover than found elsewhere in the Sahara.

### *North and south Sudanese plains*

The apparently level plains of the Libyan desert continue southwards without interruption into Sudan. Rocks of the continental interclaire extend slightly south of Wad Medani in the Nubian sandstone facies. However, the continental terminal overlying Pre-Cambrian basement complex rocks forms the surface of the southern part of the Sudanese plain where the Umm Ruwaba alluvial sands and clays are extensive. The basement complex protrudes through these later deposits to give upstanding areas such as at Sabaloka where the Nile's Sixth Cataract is located, and inselberg landforms can be observed (Figure 3.8).

West of Khartoum, in Kordofan, sands derived from the Nubian sandstones and the Umm Ruwaba deposits form large areas of fixed dunes known as Qoz which were emplaced during the Quaternary. The division between the northern and southern Sudan plains is drawn at the limit of contemporary live dunes. The southern plains increasingly have the appearance of savanna rather than desert, but desertification is currently extending southwards.

### *Jebel Marra and the Nuba mountains*

Two major upland areas interrupt the Sudanese Plains. Jebel Marra (3300 m), is a Tertiary volcanic area in the extreme west of Sudan and the Pre-Cambrian basement is exposed in the Nuba mountains (2000 m). Numerous smaller exposures of the basement complex interrupt the plains forming low flat-topped jebels, some of which have laterite cappings.

### *The Nile valley and delta*

The Nile receives 1900 km$^3$ of water annually from rainfall, 94% of which is lost by evaporation. Annual discharge at Aswan ranges from 42 km$^3$ (1913–1914) to 151 km$^3$ (1878–1879). Rainfall amounts reach a maximum in the Ethiopian highlands in the months June to September causing high water levels in the Blue Nile, the flood peak occurring in September (Figure 3.9). The White Nile has a dampened flood hydrograph curve as much water is lost by evaporation in the Sudd.

*Figure 3.8.* Tors, near Sabaloka, Sudan.

Downstream from the confluence of the Blue Nile with the White Nile at Khartoum, the river occupies a shallow valley as it crosses the desert plains. At Sabo-loka the Nile has been superimposed on to the basement complex and igneous rocks and flows in a gorge. Rapids in the gorge are known as the Sixth Cataract (Figure 3.10). The broad shallow valley with low terraces continues around the great bend with other cataracts at intervals until the Nile enters the body of water held back by the Aswan Dam. At Isna, 200 km north of Aswan, the valley broadens to about ten miles wide, and is bounded by pronounced bluffs. The delta begins north of Cairo just over 100 miles from the sea where the Nile divides into its main tributaries, the Rosetta and Damietta branches. Low mud banks separate the delta from the sea. Salt marshes, brackish lagoons and lakes occupy about half of the delta area. Since the control of flood waters by the Aswan High Dam, salinisation has occurred in the cultivated delta lands and the fisheries offshore have declined.

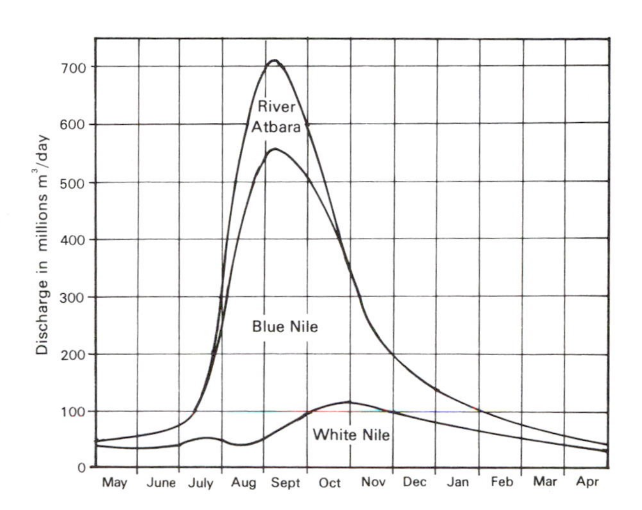


*Figure 3.9.* Hydrograph of the flow of the Nile.

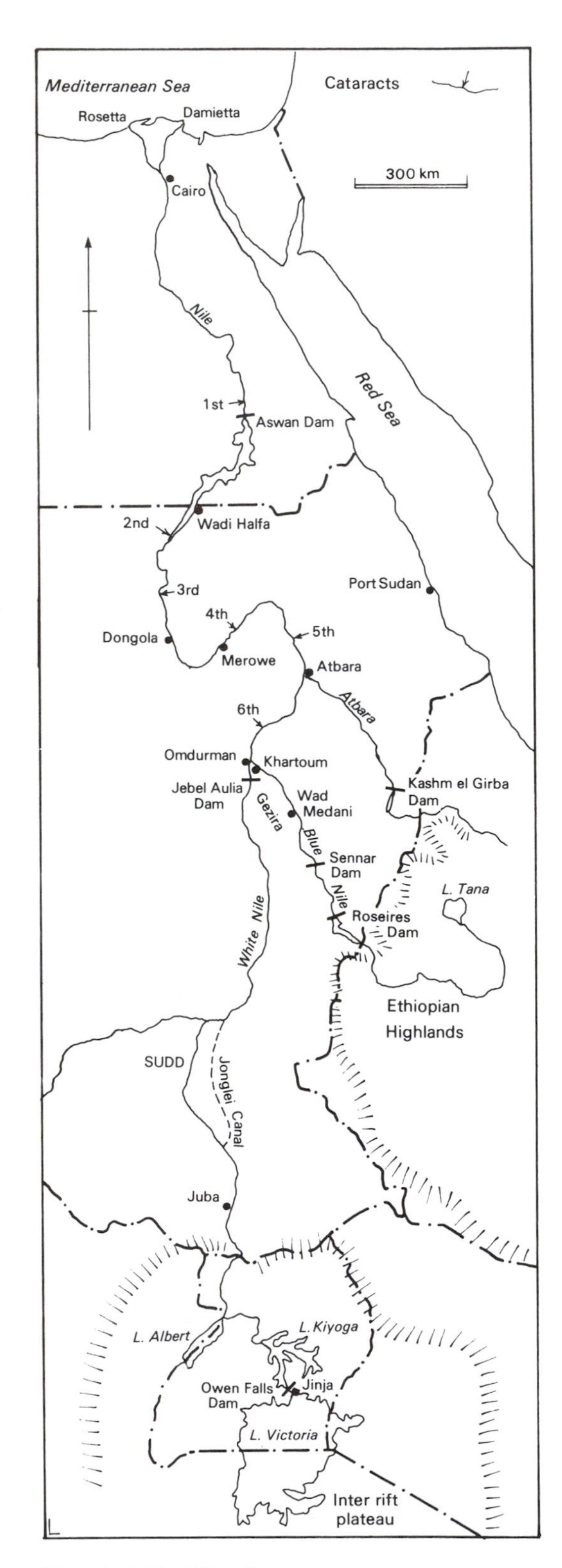


*Figure 3.10.* The Nile valley.

In the Nile valley between Khartoum and Atbara there are three terraces 3.5 m, 11 m and 17 m above low water level. Sedimentary material comprising these terraces is mainly from the Blue Nile. The older and higher terraces are more gravelly and eroded than the lower one. Three terraces have also been identified in Southern Egypt.

In the region of Shendi in Sudan, an early Tertiary surface occurs, developed in the Nubian formation and laterised. The oldest deposits of the Nile are different in mineralogical composition from later alluvial materials. This has led to suggestions that the proto-Nile may not have received the Ethiopian tributaries and that these only became part of the Nile during the Pleistocene. The Nile ancestor may or may not have been along the course of the present river. The Hudi chert deposits in northern Sudan are silcretes which were formed on a peneplain surface subsequently dissected by a predecessor of the present Nile. In the late Eocene, when the Mediterranean became dry land, the Nile cut a deep canyon (− 170 m) at Aswan which has since been buried by alluvial infilling.

### *The Nubian desert*

Between the Nile and the Red Sea hills, the area of the Sudan plain is known as the Nubian desert. The Pre-Cambrian basement is covered by marine Jurassic deposits and the Nubian sandstone further west. Southwards, the basement is covered by the volcanic rocks of the Ethiopian highlands.

### *The basins of the White and Blue Niles*

A thick series of continental terminal deposits in southern Sudan implies that this area has been a sedimentary basin throughout the Pleistocene. Unconsolidated clays and sands named the Umm Ruwaba formation are evidence of this period of the geological development. In the Sudd, sedimentation continues to the present day and toward the northern boundary of this province fixed dunes of the Qoz sands cover the older rocks. Extensive areas of alluvial clays in the Gezira and Blue Nile basin have developed black cracking soils known as vertisols. The landscape is very flat but occasional flat-topped hills protrude through the alluvial deposits.

In the Blue Nile valley a terrace has been cut 6–10 m above low water level. Downstream it appears to merge with the Er Roseires terrace. The Er Roseires terrace is

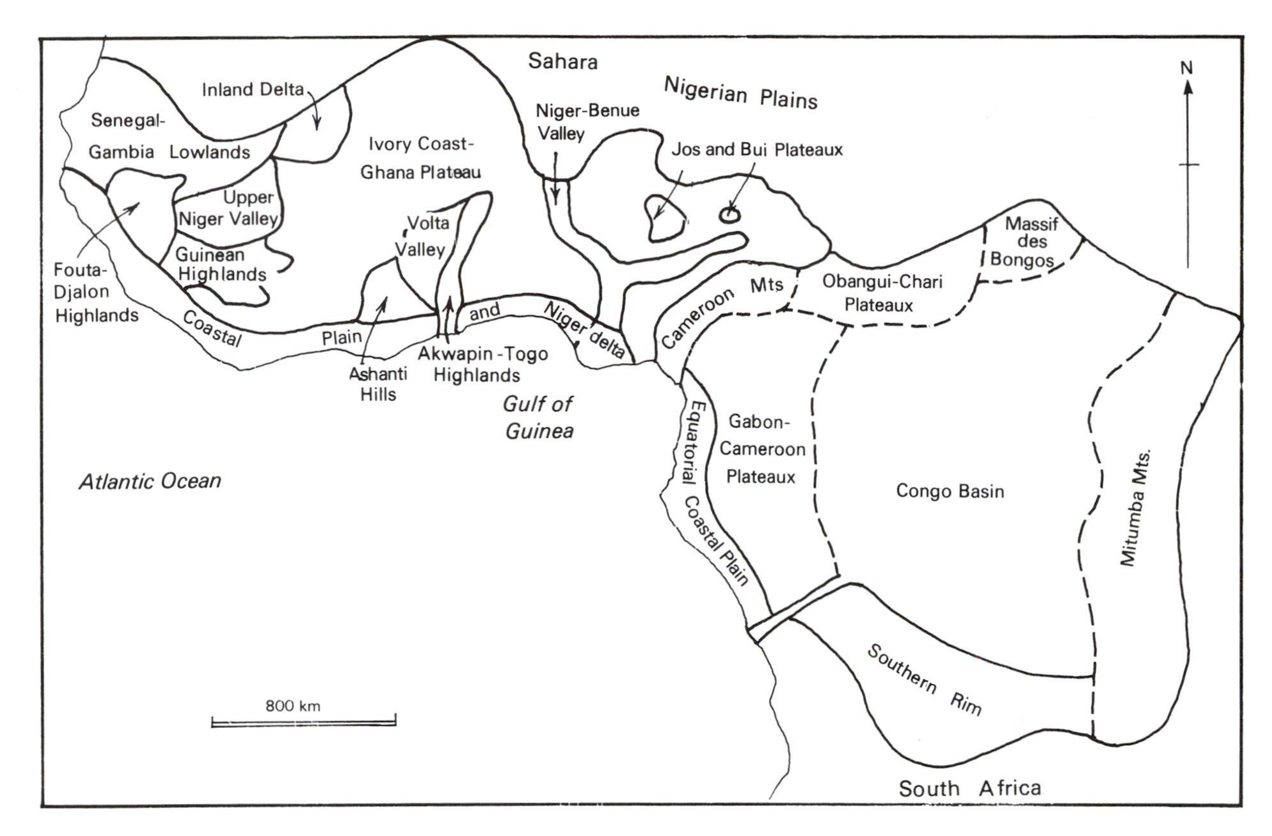


*Figure 3.11.* The provinces of west and central Africa.

thought to be a former floodplain into which the Blue Nile has incised. It was formed in the period 35 000–11 000 BP. The terrace deposits include increasing amounts of sand downstream.

### *The Red Sea hills and the Arabian desert*

The African plate is abruptly broken along the line of the Red Sea by rifting, a process which continues at the present time as the Red Sea widens by about 6 mm per year. The Red Sea hills of Sudan and the Arabian desert of Egypt as well as the rocks of the Sinai peninsula are made up of the basement complex. Both are dissected by numerous wadis.

## Plateaux and mountains of west Africa

The landforms of west Africa, south of the Sahara are dominated by erosion surfaces cut across rocks of the basement complex. In the Volta basin of Ghana, Lower Palaeozoic rocks are important, but Mesozoic and younger strata occur only in coastal areas and down-faulted troughs. In a number of places, large-scale upwarping brings the basement complex rocks to sufficient elevation to enable some of the older erosion surfaces of the continent to be preserved. Remnants of the Gondwana and African surfaces occur, separated by pronounced breaks of slope.

Unlike the arid Saharan major physiographic division to the north, west Africa has according to location, a tropical or equatorial climatic régime which provides sufficient moisture and warmth for rapid rock weathering and sufficient water for permanent or at least seasonal streams. Many of the older erosional and depositional surfaces of the west African landscape have developed ferruginous crusts which often result in low, mesa-like topography where they are being eroded.

The landforms of west and central Africa may be described in the following provinces which are shown in Figure 3.11.

Senegal–Gambia lowlands
Fouta Djalon highlands
Guinean highlands

Upper Niger valley
Inland Niger delta
Ivory Coast plateau
Volta basin
Ashanti hills
Akwapin–Togo–Atakora hills
The Jos plateau
Northern plains
The Bui plateau
Lower Niger valley
The coastal plains and Niger delta

*Senegal–Gambia lowlands*

The coastal lowlands from the Senegal river south to Guinea-Bissau are a continuation of the comparable lowlands of the Rio del Oro basin in the Saharan division. The downwarp in the basement complex has been infilled with sediments which begin with the Palaeozoic, include the continental intercalaire of Jurassic and Cretaceous age and the sandstones, shales, marls, sands and clays of the continental terminal. The river valleys are wide and shallow and the interfluves between are sandy plains less than 50 m above sea level.

*Fouta Djalon highlands*

The Fouta Djalon highlands are a deeply dissected upfaulted plateau with an average elevation of 1000 m, rising in its highest parts to 1800 m above sea level. This highland plateau has steep west and north slopes, but the eastern slopes are more gentle. The underlying sandstone rocks are of Ordovician age and have been intruded by dolerite and gabbro sills. The plateau surface is remarkably level, but the streams which dissect it have cut deep valleys with waterfalls, such as the Kinkon Falls near Pita in Guinea where the river plunges over an intruded volcanic rock.

*The Guinean highlands*

The Guinean highlands are developed on basement complex, reaching an elevation of 1947 m in the Loma mountains and 1752 m at Mount Nimba. This highland along the borders of Sierra Leone and Liberia is one of the northernmost occurrences of King's Gondwana erosion surface. In contrast with the Fouta Djalon highlands, these hills have rounded relief with few plateau surfaces. However, the presence of quartzite on Mount Nimba has allowed the preservation of erosion surfaces at 1600 m, 1300 m, 800–900 m, and 550–600 m. The southern part of these highlands is formed by the plateau of Man, a granite massif which gives rise to a gently undulatory plateau at 1300 m. In these highlands, precipitation is equatorial in régime with two maxima giving 2000 mm of rain per annum. A short dry season is also experienced.

*Upper Niger valley*

The Niger river is formed from the confluence of several tributaries which originate in the Guinean highlands. As the Niger crosses the plain towards Bamako, it is slightly incised into the plain, which has an elevation of 500 m above sea level. The plain, which has been equated to King's African surface, is interrupted by a few residual hills which rise 180 m above it. The Upper Niger has a complex régime which reflects the passage of the monsoon seasons. The floods resulting from the July to October rains in the Guinean highlands take until December to reach the delta in Nigeria.

Formerly the upper Niger turned westwards to flow into the Atlantic at the Gulf of Senegal, but during a dry phase of the Pleistocene its course was diverted by sand dunes into an enclosed basin to form Lake Araouane. This was one of the largest pluvial lakes in Africa, its extent is known from deposits of salt and diatomite which remain as evidence. When the last pluvial period arrived, 10 000–15 000 years ago, the Niger filled the lake basin to form Lake Araouane which overflowed at Bourem to establish the present course of the Niger along the line of the Tilemsi valley, a wadi which originates on the Adrar plateau in the Sahara. Although the climate has now become drier again, the Niger has been able to maintain its flow along its 'new' valley, but the frequent rapids between Gao and Niamey point to the relative youthfulness of this part of the river (Figure 3.12).

*Inland Niger delta*

Beyond the town of Segou the Niger enters an area of very low relief where it develops a braided system of distributaries. The river is no longer constrained by the incision present in the upper valley and so it frequently floods. This 'inland delta' which extends a further 320 km to Timbuktu has formed since the Araouane pluvial lake drained about 5000 years ago. North-west

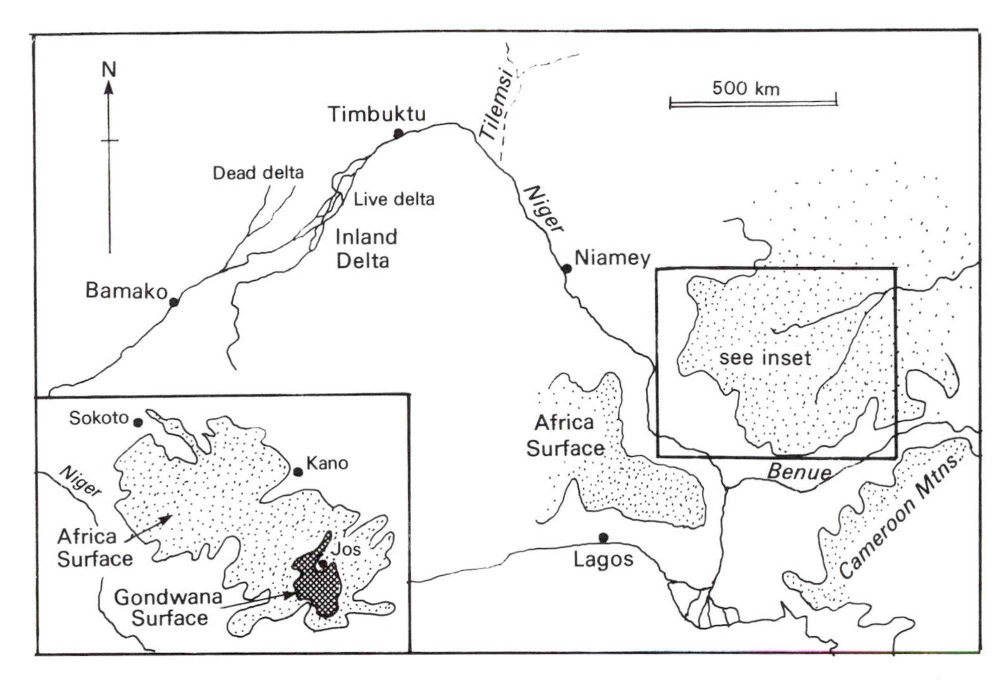


*Figure 3.12.* The inland delta of the Niger and the Jos plateau.

of the present delta, distributaries of an earlier delta can be seen which are no longer flooded. The reason for this is not known but isostatic readjustments may be responsible following the removal of the water from Lake Araouane (Figure 3.12).

### *The Ivory Coast plateau*

Between the Guinean highlands and the Volta river basin a large area of basement complex rocks is exposed in an extensive post-African surface throughout the Ivory Coast, Western Ghana and the southern parts of Burkino Fassu. This plateau surface is tilted southwards so the major rivers, the Sassandra, Bandara and Comou flow directly to the Gulf of Guinea. Numerous flat-topped residual hills, remnants of former surfaces, now covered with lateritic crusts occur. In the northern part of this region, the Bobo Dioulasso district, Cambrian rocks overly the basement complex giving further plateau-like landforms.

### *Volta basin*

The underlying structure of this province is synclinal and the downfolded basement complex has been covered with the Voltain sandstones, shales, conglomerates and dolomites which range from Pre-Cambrian to Ordovician in age. These rocks form outward-facing escarpments around the periphery of the basin and are most impressive in the south where they form the Kwahu plateau, almost 500 m above the surrounding plains.

The basin is drained by the various tributaries of the Volta river. The Black Volta receives the water from the eastern part of the Ivory Coast plateau, but the Red and White Volta rivers rise in Burkino Fassu, become confluent and flow southwards through Ghana. The waters of the combined Volta rivers are held back by the 113 m high Akosombo Dam to form the 8482 km$^2$ Lake Volta. The dam is constructed where the Volta, a superimposed river, crosses the line of the Akwapim hills, 130 km from the coast.

### *Ashanti hills*

South-west of the escarpments of the Voltain sandstone are the Ashanti hills, a region described as having ranges of hills with wide flat-bottomed valleys. An interesting feature in this region is the presence of the Bosumtwi crater lake, formed by a meteor impact. Diamonds are mined from alluvial sediments in the Birim valley and gold is obtained from mines near Tarkwa. Bauxite is present at many locations and manganese ore was mined until 1975 at Nsuta. Two terraces have been identified along the Birim river at 270–300 m and 300–315 m above sea level.

*Akwapin–Togo–Atakora hills*

The eastern outcrops of the Voltain sandstones form these hills which extend from south-east Ghana across the international boundary into Togo. The crests of the ridges form a plateau surface into which 300 m deep narrow valleys have been cut. Tertiary sandstones and clays have been deposited nearer the coast to form low undulating country less than 100 m above sea level. The major economic wealth of the district comes from the phosphate mines at Akoumapé.

*The Jos plateau*

The Jos plateau, 1200–1400 m above sea level, is the most elevated region of Nigeria. It is a remnant of the Gondwana surface, bounded by escarpments 600 m high in the west and south which clearly separate it from the surrounding African surface at 450–475 m (Figure 3.12). The plateau which slopes gently northwards is formed of a number of Jurassic ring dykes emplaced in the Pre-Cambrian basement complex. The plateau surface is interrupted by inselbergs, some in the form of domed bornhardts, others as tors. Remnants of ancient lava flows have become capped with ironstone duricrusts, but basaltic lavas have given rise to a reddish-chocolate coloured soil. Cinder cones up to 100 m high, some of which have not long been quiescent, occur in the Panyam district, 65 km south-east of Jos. Associated with the volcanic activity has been mineralisation with tin, lead, zinc, and columbite (niobium). Alluvial tin mining has led to considerable disturbance of the valleys, choking the streams with sediment.

*Northern plains*

North of the Jos plateau and extending from Sokoto in the west around Kano and along the Hadejia valley are the northern or Hausa plains. The landscape is an extensive erosion surface developed over the Pre-Cambrian basement rocks, forming part of King's African surface. Lateritic ironstone caps low hills and up to 50 m of regolith lies on the weathered basement rocks. The streams, which drain this area flow towards Lake Chad, are dammed to provide irrigation, especially in the Hadejia valley. Considerable areas of Kano, northern Bauchi and Borno states are covered with an ancient erg of sand dunes laid with an east-south-east to west-north-west alignment during a dry period in the Pleistocene. The sand of the dunes is coarser in the east but towards Kano and Zaria the grain size becomes finer. They are extensively cultivated, but are liable to erosion in the summer rains.

*The Bui plateau*

The Bui plateau was an area of volcanic activity during the Tertiary. Basaltic flows filled in valleys on the granite plateau but subsequent weathering and erosion have resulted in 'inverted' relief as the granite weathers more rapidly than the basalts, so that the basalts originally in the valleys now form the hills.

*Lower Niger valley*

As explained previously, the Niger downstream of the inland delta took advantage of the dry valley of the Tilemsi when it overflowed from Lake Araouane. Although the climate has since become drier again, the river has maintained its course across the sill at Bourem. Throughout this section the valley is broad and open, but it narrows at Jebba as it enters the downfaulted trough of its lower valley. The lower valley of the Niger–Benue occupies a tectonic trough which has been infilled with Cretaceous and later sediments, estimated to be between 4500 and 6000 m thick. This trough formed when South America split away from Africa and is an example of an aulacogen or failed arm of a triple junction. In northern and western parts of the trough, sandstones extend over the trough margins and onto the plateau formed by the basement complex.

*The coastal plains and Niger delta*

The Senegal coast is low-lying and bordered by sandy beaches and lagoons. A sandspit has diverted the Senegal river 40 km southwards. Records since 1658 indicate that the spit has grown at the rate of about 100 m a year. The most westerly point of Africa, the Cape Verde peninsula, is an example of a tombolo where two sandspits link a former island to the mainland. The southern part of the cape is formed from columnar-jointed basalt of Miocene age but the northern part is younger with Pleistocene flow basalts and dolerite lavas.

Banjul, the capital of Gambia is situated on a

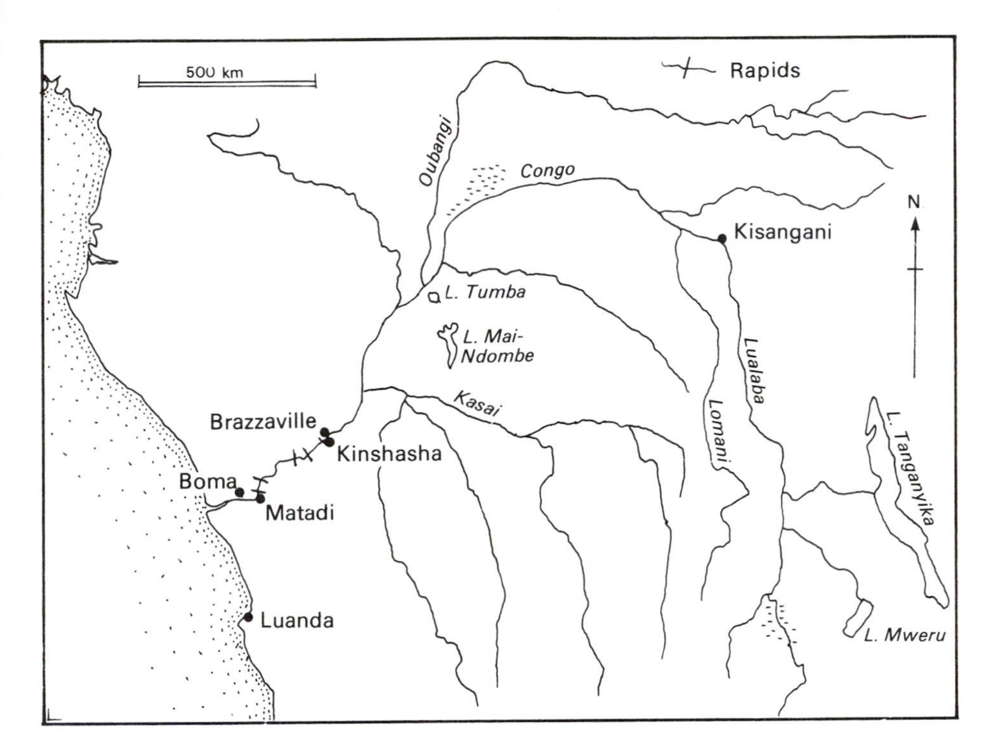


*Figure 3.13.* The Congo valley.

recurved sandspit backed by coastal marshes with mangroves at the mouth of the Gambia river. In addition to the presence of terraces alongside the rivers, evidence of tectonic uplift and depression can be seen along the west African coast. Raised beaches have been identified on the Sierra Leone peninsula at Freetown with elevations of 10–12 m and 40–50 m above sea level, cut in Tertiary sediments and uplifted during the Quaternary. The core of the Sierra Leone peninsula at Freetown is a gabbro lopolith.

The Ghanaian coast is low-lying with a sand dune and lagoon structure at Abijan and near the Volta delta. At Accra, there are low cliffs and a wave-cut shore platform cut in Devonian sandstones. The Volta delta is not large, only 115 km across. It is infilling a former drowned estuary, but slowly as it does not have a large sediment load. The Niger river, in contrast with the Volta, has a large delta, extending from 480 km westwards in an arcuate shape from the town of Opobo. It has numerous channels to the sea and has a thickness of 12 000 m of accumulated sediment.

Mineral deposits of value along the west African coastal plain include bauxite, mined in Liberia, bauxite and diamonds mined in Sierre Leone and phosphates which are obtained from Taiba in Senegal and at Akoumapé in Togo.

## The central African basin

This physiographic division comprises the great downwarp of the Congo basin. It is centred upon the Republic of Zaire, but the rim of the basin extends into Cameroon, the Central African Republic, Zambia and Angola. The Congo basin is 1800 km across in an east to west direction and the same distance from north to south. It is characterised by the presence of the one large trunk stream, the Congo or Zaire river, which curves in a great arc through northern Zaire before cutting through the western rim in the gorge between Kinshasha and Matadi. It rises in the southernmost part of the Katanga province in the Mitumba mountains and is known as the Lualaba in its upper reaches. Its major tributaries are the Lomani and Kasai on the left bank and the Oubangi on the right bank (Figure 3.13).

The geological history of the basin began when the basement complex rocks started to subside during the Palaeozoic, a process which continued throughout the Mesozoic era. Sediments from the Dwyka tillite of the

Carboniferous glaciation of Gondwana through to the Karroo continental deposits, including the continental intercalaire and the continental terminal deposits, are present in outcrops around the periphery. Recent research has shown that the sands of the Kalahari were formerly much more extensive than the present desert. Fossil dunes may be observed on air photographs to extend beyond the equator well into the tropical rain forest areas of the present day. These fossil dunes have been modified by fluvial processes and were silicified into silcretes. In the centre of the basin, the older deposits are covered by 150 000 km² of Holocene alluvial sediments.

Formerly the Congo is thought to have flowed over the rim of its basin in the south of Cameroon, but this outlet to the sea was closed by tectonic uplift of the edge of the continent. A large lake must have developed during the Tertiary but this was drained by a stream cutting back from the coast in the Crystal mountains. As the lake overflowed, about 15 000 years ago, it cut the gorge between Kinshasha and Matadi to form the course which has been maintained ever since. The lakes Tumba and Mai-Ndombe are the small remnants of this formerly large body of water.

The geomorphology of the central African division may be sub-divided into a series of plateaux which constitute the basin rim province, surrounding the basin floor province with the trunk stream of the Congo (see Figure 3.11). The plateaux of the rim, clockwise from the Congo gorge, are:

The Gabon–Cameroon plateau
The Cameroon mountains
The Oubangi–Chari plateau
The Mitumba mountains
The southern rim
The Congo basin

### *The Gabon–Cameroon plateau*

The waterside rim of the Congo basin in Gabon and southern Cameroon is formed by a plateau on the basement complex at an elevation of between 600 and 1000 m above sea level.

### *The Cameroon mountains*

The watershed between the Congo and the Benue in northern Cameroon is formed by the line of the Cameroon mountains along a major fault-line in the basement complex. The upward edge of the continent is capped by a range of volcanic peaks which extend inland from the islands of Annabon, Sao Tomé, Principé, Bioko (Fernando Poo). Mount Cameroon is an active composite volcano with eruptions in recent times in 1909, 1922 and 1954 and 1959. Lava flows from the 1922 eruption spread right down to the sea. The base of the volcano dates from the late Cretaceous and the southern Etinde peak is late Tertiary. Fako, the highest point (4070 m) is on the rim of an extinct crater. North-east of Mount Cameroon lies the Eboga caldera, one of several collapsed volcanoes in this range. Lakes occupy many of the craters and in 1986 an escape of dissolved gases from Lake Nyos, resulted in the deaths of 1700 villagers living nearby.

### *The Oubangi–Chari plateau*

North and north-east of the alluvial basin of the Congo the land rises to between 1000 and 1200 m on an ill-defined watershed between the drainage to the Congo and that to the Nile. The Massif des Bongos rises to 1378 m in the east of the Central African Republic but the Congo basin rim is only at 600 m in the central part of that country. It is placed by King on his late Tertiary erosion surface. Drainage of this section is to the Congo by the Oubangui river, but the northern facing slopes of the basin rim drain to Lake Chad through the tributaries of the Chari river.

### *The Mitumba mountains*

The eastern rim of the Congo basin is defined by the 3000 m highlands on the west side of the rift valley. The upwarping of the crust here in association with the process of rifting has produced the highest part of the basin rim. Most of this uplifted rim has been allocated to the African surface by King, but the summits are thought to be remnants of the Gondwana erosion surface.

### *The southern rim*

Confusingly, the high ground on the south-east of the Congo basin rim is known as the Chaine des Mitumba which lies in the south-east part of Zaire. The summits of these mountains are thought to be part of the Gondwana erosion surface and the lower parts are graded to the African erosion surface of King. West of these higher lands, the rim of the Congo basin is lower and the superficial sands of the Kalahari extend over the watershed and into the Congo basin (see Figure 3.11). The circuit around the Congo basin rim is completed by the Bangou–Mokaba plateau which lies south of the Congo gorge in northern Angola. Elevation is between 600 and 1000 m and has been placed in the late Tertiary erosion surface by King.

The plateaux surrounding the Congo basin have valuable mineral resources. In the Katanga region, sulphide ores of copper, lead and zinc, gold, manganese and tin are mined and diamonds are found in alluvial deposits of the upper reaches of the Kasi river and its tributaries. Uranium is found in the Central African Republic, north of the Oubangi river and also south-east of Gabon. Iron ores and bauxite occur frequently in Cameroon and Gabon, but probably the most significant resource is the oil discovered at the mouth of the Ogowe river in Gabon and at Point Noire in Congo. Phosphates and potash occur in the immediate hinterland of Pointe Noire.

### *The Congo basin*

The Congo river drains an area of 3.7 million km$^2$ and is 4350 km in length. Its major headwaters are found on the Lualaba which draws its waters from Lake Mweru on the Zaire–Zambia border and from other lakes and swamps north of Bukama in southern Zaire. It also takes water overflowing from Lake Tanganyika along the Lukuga river. Numerous right-bank tributaries join the Congo from the Mitumba mountains of the western rim of the rift valley. Except for the Lomami, the left-bank tributaries of which the Kasai is the largest, tend to take more westerly courses. Between Kisangani and Kinshasha the average gradient of the river is 7 cm per kilometre and its broad course is frequently braided in a channel which can be up to 12 km wide. As the Congo turns southwards north of Mbandaka (formerly Coquilhatville), extensive swamps occur on the right bank where the Congo is joined by the Oubangi. On the left bank, Lakes Tumba and Mai-Ndombe are surrounded by low-lying swampy ground between the Kasai and Congo. Upstream from Kinshasha to Kisangani, the bed of the Congo only descends 100 m over a distance of 2000 km. From Kinshasha to Matadi, less than 300 km, the Congo drops 275 m in a series of rapids as it flows through the gorge it has cut in the western rim of the basin. The river finally enters the estuary at Boma where it is still over 100 km from the open ocean.

## East African plateaux

The more elevated parts of Africa lie south and east of a line which runs from the Red Sea coast, around the Ethiopian highlands and the east African plateaux with their rift valleys and south around the Congo basin to reach the Atlantic coast in the northern part of Angola. The landscapes of 'High Africa' are dominated by extensive plateaux, which in east Africa have been split by rifting. The rifting was preceded and accompanied by volcanic activity which continues until the present day in several locations. Many of the rivers draining this part of Africa have interesting courses affected by tectonic activity. Some have reversed flows and others have impressive waterfalls where the rivers plunge over the edge of the plateaux into the younger landscape of a Quaternary gorge downstream.

The geomorphology of this part of Africa can be described in terms of the following provinces (Figure 3.14):

The Ethiopian highlands
The Ethiopian rift valley
Danakil lowlands (Afar triangle)
The plateaux of the Horn of Africa
The east African plateau
The African rift valleys
The inter-rift plateau and Lake Victoria
Madagascar

### *The Ethiopian highlands*

The Ethiopian highlands comprise two elevated areas divided by the Ethiopian section of the rift valley. Bounded by spectacular scarps, the highlands have been deeply dissected by rivers in gorge-like valleys leaving plateau areas, called ambas, along the interfluves (Figure 3.15). The average elevation in the north is between 2400 and 2700 m with peaks reaching 4600 m, but in the south the average plateau surface is at a height of less than 3700 m above sea level.

Mesozoic limestones of Jurassic age overlie the basement complex in the Ethiopian region but the major landforming rocks are the lavas which occurred in three episodes of vulcanicity. Older lavas, called the Ashangi

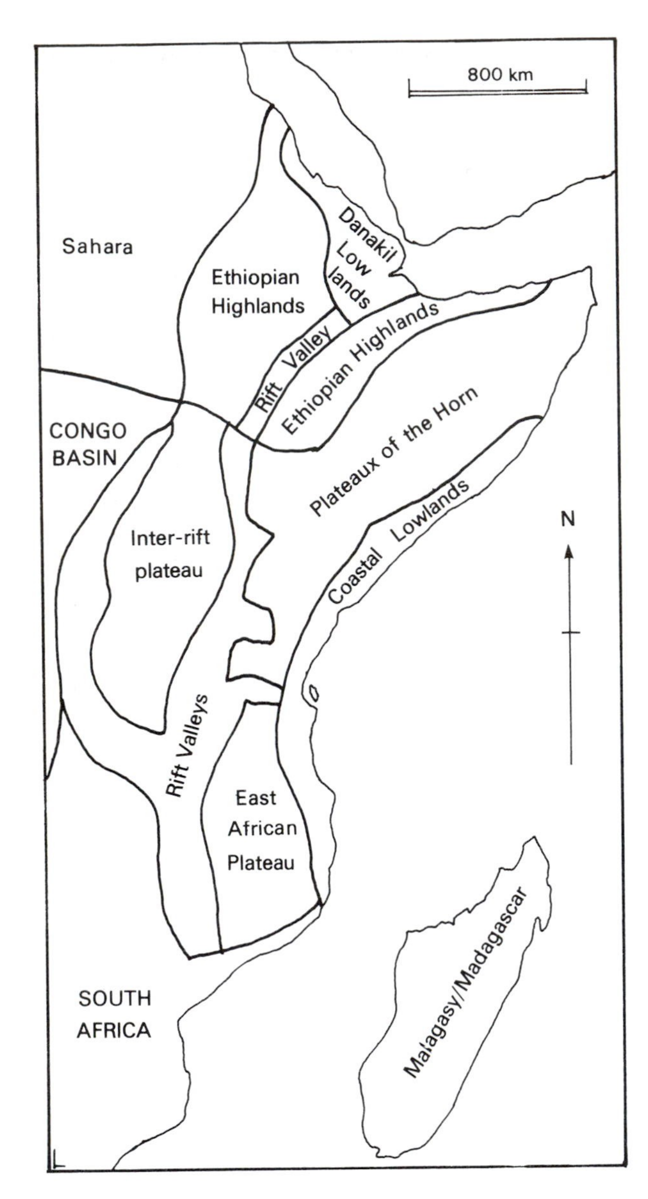


*Figure 3.14.* The provinces of east Africa.

*Figure 3.15.* The Ethiopian highlands. Level-bedded basaltic lava flows result in plateau landscapes with steep, stepped valley sides.

group, poured out onto the Mesozoic rocks. The area was uplifted during the late Eocene accompanied by further effusion of lavas of the Magdala group. Since the formation of the rift valley, vulcanicity has continued into the Pleistocene with the deposition of the Aden volcanic series, which only occurs within the rift valley.

Rainfall of over 1000 mm is received in the Ethiopian highlands and rivers such as the Blue Nile, Atbara, Sobat and Awash rivers rise on the plateau and descend to the plains over numerous waterfalls, e.g. Tesissat Falls, to reach the plains through valleys often 2000 m deep, such as the Dubai Gorge of the Blue Nile. Lake Tana on the Blue Nile, north-west of Addis Ababa results from damming of the valley by lava flows.

### *The Ethiopian rift valley*

The rift valley in Ethiopia is a trough approximately 50 km wide, overlooked by 1000 m scarps formed where the Ethiopian plateau has been split apart. It is drained by the Awash river which rises in the western highlands, descends into the rift and flows north-east to the Danakil lowlands where it ends in Lake Abbe. There are several small lakes on the floor of the rift including Lake Shala which occupies a 25 km diameter caldera feature surrounded by 150 m walls. The Awash river has been utilised for irrigation to grow sugar cane and cotton.

### *Danakil lowlands (Afar triangle)*

This arid area lies at the foot of the 4000 m scarp of the Ethiopian highlands. Although the rift valley leads into it, this lowland is not part of the rift as much as part of the floor of the Red Sea which happens to be above sea level. The alternating magnetic patterns seen on the oceanic floor have been observed here on dry land, the only place where this is known to occur. Block-faulting has thrown the area into a strongly accented relief with upthrust horsts, the Danakil mountains, and downthrust graben, the Kobar sink which is 120 m below sea level. As this area is one of exposed sea-floor material, there are no sedimentary rocks present other than evaporites which in places amount to 1000 m thickness. Many volcanic features occur, such as the Erta'Ale volcano and recent fissure eruptions have occurred. Mount Asmara is a former sea mount (guyot) resulting from an underwater eruption which is now elevated and on dry land.

### *The plateaux of the Horn of Africa*

The large escarpment which forms the northern edge of the eastern portion of the Ethiopian highlands continues eastwards towards Cape Guardafui, the 'Horn' of Africa. The southern margin of these highlands around Harar descends to the plateau of the Haud and Ogaden, semi-desert areas based on Mesozoic limestones and Tertiary sands. The major drainage channels cross these plateaux to reach the sea in the southern part of the Somali Republic; the Shebali flowing parallel to the coast from Mogadishu to Kismayu. There is a narrow coastal plain in the north-east, but south of Mogadishu it is over 96 km wide. The coast is emergent and a line of ancient sand dunes lies inland of the present barren coastal dunes. The south-western margin of this province lies along the highlands which form the scarp of the eastern rift, the Matthews range at 3100 m and the Kitui hills.

### *The east African plateau*

This landform extends from southern Tanzania into Mozambique, occupying the bulge of the coast south of Dar-es-Salaam. It is mostly less than 1000 m above sea level in the Nyasa plateau overlooking Lake Malawi. The rocks forming this province are metamorphosed upper Pre-Cambrian schists and gneisses into which granites have been intruded. Near the eastern rim of the rift valley, these rocks have been covered by basaltic lavas. Mesozoic and Tertiary rocks occur along the coast. The islands of Pemba and Zanzibar are small tilted blocks of coral rock with cliffs of 100 m on the seaward side. The Kenya coast has many inlets with mangrove swamps, sandy beaches and coral cliffs.

### *The African rift valleys*

These valleys are one of the best known features of the African continent and one of the Earth's larger geomorphological features. From the Jordan valley in the north to beyond the mouth of the Zambesi near Beira, the rift valley extends for 7200 km. In the central part of the system, the valley divides into two arms enclosing the inter-rift plateau and Lake Victoria.

Controversy exists over the cause and mechanics of rifting. In 1921 Gregory ascribed the rifts to tension, supporting Wegener's theory that the rifts were associated with the drifting continents theory, but Wayland proposed that the rifts were the result of compressional forces. At a later date Dixey suggested that the initial crustal weaknesses were caused by compression and that later tensions gave rise to rifting. The downfaulted rift valley has an average width of between 40 and 62 km. The sides are normally steep and in places precipitous rising to the adjacent peaks more than 2000 m above the valley floor. Evidence on the ground is often confusing but the concensus of opinion appears to be that the pattern of upwarping and downwarping of different parts of Africa hold the key to the problem of the origin of the rifts. Detailed studies have shown that the rifts lie along a line of faults in the Pre-Cambrian basement rocks and that an arching of the crust, followed by outpouring of basalts and rifting took place at the end of the Cretaceous and has continued through the Tertiary. Even at the present time some volcanic activity is experienced along the rift valley floors and earth tremors are a common feature.

The floor of the rift valleys has been covered with sediments and lavas to a considerable depth, often exceeding 1500 m. The original floor of the rift valley is below sea level in places; the bed of Lake Malawi is at −700 m. Elsewhere, lavas have raised the floor level to 1800 m near Lake Naivasha in the eastern rift and, in the western rift, block-faulting has elevated the Pre-Cambrian basement to over 5000 m above sea level in Mount Ruenzori (5709 m), the highest non-volcanic mountain in Africa. Ruenzori has glaciers above 4200 m and, below the present ice-covered areas, glacially-deepened valleys with lakes (Lac Vert and Lac Noire), morainic deposits and cirques.

The northern part of the western rift is drained by the

upper Nile through Lakes Amin (Edward) and Mobuto (Albert). The central part of the eastern rift drains to Lake Tanganyika and via the Lukuga to the Congo. In the south drainage finds its way to Lake Malawi and the Shire valley to join the lower Zambesi. In the western rift drainage is to internal lake basins such as Lake Turkana (Rudolph) (370 m) and the smaller lakes, Natron, Magadi (soda production is second largest after that of the Salton Sea, California) and Naivasha, which are saline.

Successive layers of basaltic lavas which are erupted from fissures cover the rift valley rim adding 100 m to the escarpment in the Aberdare range, the Mau range and at many other locations along the rift. The volcanoes of east Africa occur either on the plateau, e.g. Kilimanjaro, Meru and Mount Kenya (5199 m) or on the floor of the rift valley, e.g. Ngorongoru, Longonot, Ol Dinyo Lengai. Kilimanjaro is a dormant composite volcano which has been built up from three volcanic peaks, Shira, Kibo and Mawenzi, the highest point of which is Uhuru peak on Kibo (5895 m). Glacial troughs occur on Kibo's south and west slopes with accompanying features of glaciation. Mount Kenya, a frost-shattered volcanic plug, also has a number of small glaciers including the Lewis Glacier (0.31 km² at 4680–4960 m); but glaciers are retreating, having reached as low as 2000 m in the Pleistocene. There is less volcanic activity associated with the western rift; semi-active volcanoes occur between Lake Amin and Ruenzori and more than 200 shallow explosion craters of Pleistocene age are present.

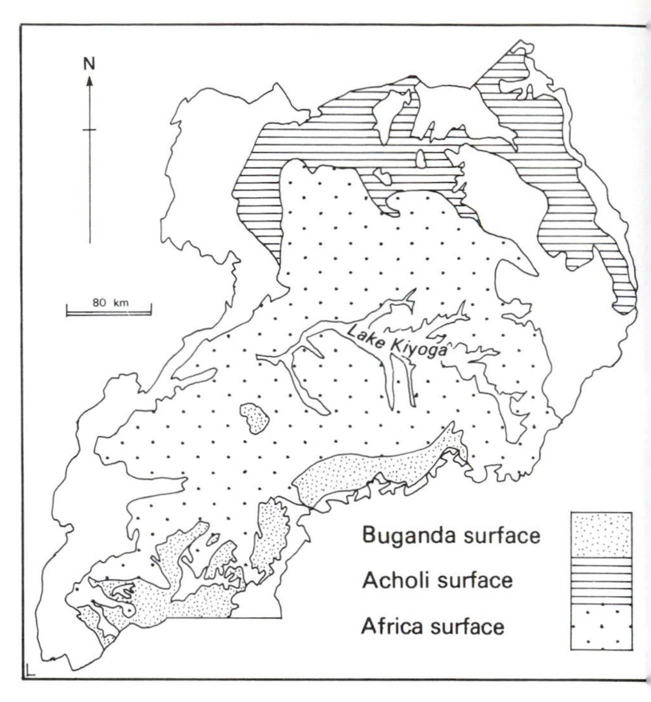


*Figure 3.16.* Surfaces in Uganda (After Ollier, 1981.)

### *The inter-rift plateau and Lake Victoria*

Between the eastern and western rifts in Uganda, Tanzania and partly in Kenya the general elevation of the land is between 1000 and 1200 m above sea level. The north-central part of this upland plateau consists of the waters of Lake Victoria (1135 m) which cover about 70 000 km². The underlying rocks of the inter-rift plateau are of the basement complex which in places has been granitised. Limited areas of Karroo strata crop out but volcanic lavas which began to accumulate in the Tertiary are more common and eruptions continue to occur in the eastern rift.

The Karamoja plain (deeply-weathered African surface) occupies most of Uganda north of Lake Victoria but eastwards the elevation of the land rises into the Cherangami, suk and Chermorungi ranges of western Kenya. Mount Elgon (4821 m), on the border between these two countries, is volcanic and has been glaciated. In northern Uganda, the Acholi surface is cut across fresh rock (Figure 3.16). South of the lake, the Serengeti National Park and the plains around Tabora remain as wildlife reserves or as unproductive tse-tse infected woodlands. Tors and inselbergs rise above the general level of the plains which rise in altitude to the rim of the rift overlooking the southern end of Lake Tanganyika or to the southern highlands near Mbeya and Njombe.

The mineral wealth of this region consists of gold which is mined south of Lake Victoria near Mpanda and Mbeya; diamonds which are mined south of Shinyanga; lead and copper at Mpanda and tin which is obtained from the uplands of Rwanda.

The drainage of the northern and central areas of this province was formerly westwards, initiated on the western slopes of the uplift which took place before rifting occurred. The tectonic activity associated with rifting obstructed the westward-flowing Kagera and Katonga rivers, which were forced to reverse their direction of flow. The central part of this inter-rift plateau sagged as the edges were uplifted and Lake Victoria came into being. The water accumulated until it could escape from the basin at the lowest point, Owen Falls. It spilled over into the valley of the Kafu which was also affected by the uplift on the rim of the rift causing it, too, to form an extensive shallow, ria-like lake until it also overflowed and eventually descended into Lake Moboto over the Kabarega (Murchison) Falls.

*Madagascar*

The large Indian Ocean island of Madagascar has an area of 370 000 km$^2$ and is 1600 km from north to south and 576 km from east to west. The central part of Madagascar is a plateau of 800 m, tilted slightly west; to the east is steeply sloping land leading to a narrow coastal plain; in the west sedimentary rocks at lower elevations form escarpments and plateaux, with intervening riverine plains.

Two-thirds of the island is made up of a plateau formed from the basement complex rocks similar to the neighbouring continent of Africa. Some Upper Pre-Cambrian, Lower Palaeozoic sedimentary rocks are incorporated into the uplifted block which has also been subjected to granitisation. Capping the plateau towards the eastern side are volcanic rocks which include Cretaceous basalts and Quaternary to Recent basalts, andesites and trachytes. These extrusive rocks form the highest parts of the island at 1200–1500 m above sea level, but another large volcanic mass occurs near Diego Suarez on the northern tip of the island. Quaternary vulcanicity is particularly well developed near Lake Itasy which has formed because of a lava dam. Explosive craters, ash domes and extrusion domes occur. Gabbro laccoliths are reported from near Morafenobe and exfoliation of granite occurs near Ambalavao in part of the Andringitra massif which reaches the height of 2658 m in Mount Boby, the second highest peak in Madagascar. According to King, the plateau surface in Madagascar is equivalent to the African cycle which was cut across Pre-Cambrian and later rocks. A late Tertiary plantation cut a lower surface on the sedimentary strata of the western part of the island. Extensive laterisation took place on the African surface.

In western Madagascar, the basement complex is overlain with rocks which include a sequence from Karroo to Tertiary. This sedimentary sequence includes Eocene limestones with karstic plateaux, sinks, dolines, caves and gorges in the Mahafaly plateau in the south-west and the Ankarana plateau in the west of the island.

The northern and south-western coasts of Madagascar have coral growing upon them but the long straight east coast is a sand bar and lagoon coast with extensive areas of dunes at Mandrare river mouth on the south coast. In the lee of the islands of Sainte Marie a succession of beach ridges has built out into the sea but the southern side appears to be in the process of being cut back.

Economic minerals include coal from the Karoo sequence in the south-west of the island, mica from the south-east and beryl near Fianarantsoa.

## South African plateaux

South of the Congo basin and west of the rift valley system, the whole of southern Africa has a very similar geological history and geomorphological development. The exception is in the extreme south of Cape Province where the Cape ranges and coastal plains comprise a very different landscape and development history. The core of south Africa consists of the saucer-shaped segment of the basement complex which has been strongly uplifted in the south and around its edges with a less elevated central portion. Large integrated rivers such as the Zambesi and Limpopo drain the north and central areas and the Orange the southern part of this tectonically raised continental division. There is a large central area centred on the Kalahari desert which has an internal drainage system. The geomorphological provinces identified in this area are shown in Figure 3.17 and are:

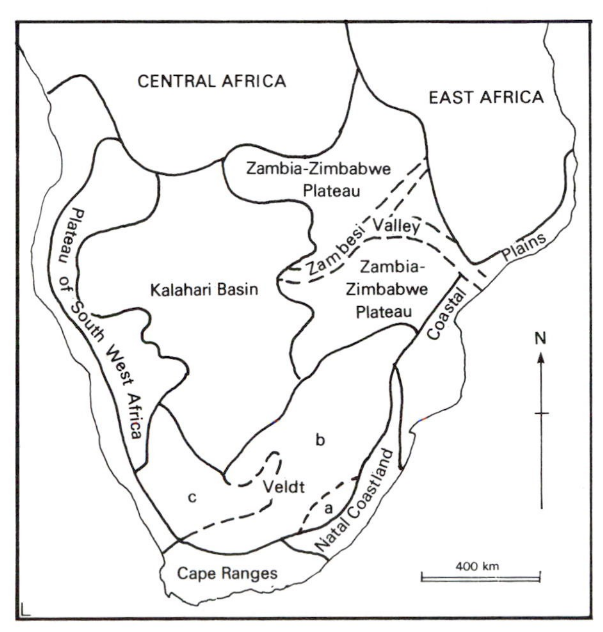


*Figure 3.17.* The provinces of southern Africa (*a*: High Veldt, *b*: Middle Veldt. *c*: Low Veldt).

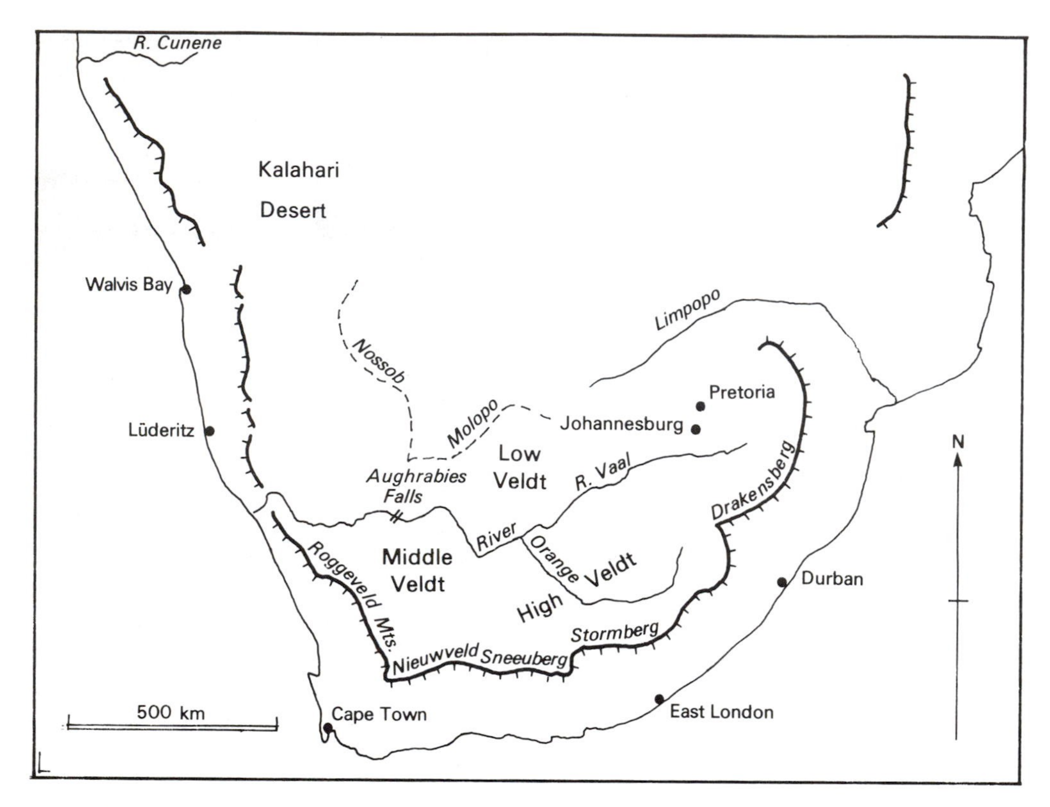


*Figure 3.18.* The great escarpment.

The Great Escarpment
The Zambia–Zimbabwe plateaux
Zambesi valley
High and Middle Veldt
The Orange river
The south-west African plateaux
The coastal margin of Southern Africa
The Natal coastlands
The Cape ranges
The Kalahari basin

*The Great Escarpment*

This major physiographic feature lies between 50 and 250 km inland from the coast and extends from the Zambesi river in Zimbabwe, around the southern end of Africa to form the western edge of the plateaux in Namibia and Angola (Figure 3.18). It varies in height considerably with the greatest elevations occurring at the Drakensberg (3299 m) (Figure 3.19); at other places along its length the escarpment is known as the Stormberg, Sneeuberg and Roggeveldtberg.

After South America was rifted away from the west coast of Africa, India from the east coast, and Africa itself was separated from Antarctica, the new continental margin flexed upwards as isostatic adjustment took place around the edge of the new smaller continent. Consequently, any erosion surface present on Gondwana prior to rifting would be uplifted. Thus the highest parts of the present African landscape belong to the Gondwana surface, identified by King at many places around the edge of the Great Escarpment. Separation of the land masses set in motion the cutting of a lower erosion surface, the African surface, this too is extensively developed on the slightly less elevated parts of the Great Escarpment (Figure 3.20).

Erosion also took place on the continental edges, and the Great Escarpment was gradually eroded back, so that it is now 50 to 250 km inland. The Great Escarpment is a watershed with many short streams with precipitate courses flowing directly to the sea. Inland the drainage is well integrated into large river systems, such as the Orange or Limpopo, which often almost completely traverse the width of the continent. In

*Figure 3.19.* Drakensberg escarpment. (By courtesy of H.J.R. Henderson.)

contrast to the scarp front streams, which cannot be older than Cretaceous, these plateaux rivers are of much greater antiquity.

### *The Zambia–Zimbabwe plateaux*

The north-east part of the south African major physiographic division comprises the two elevated areas of Zambia and Zimbabwe which take the form of plateaux at 1250 m and 1500 m respectively. These plateaux are separated by the Zambesi valley which forms the international boundary between the two countries. Most of the plateau surface is drained by tributaries of the Zambesi, particularly the Kafue in western Zambia and the Luangwa in eastern Zambia. An area of internal drainage into Lake Bangweulu occurs in northern Zambia where the largest stream is the Chambezi. Zimbabwe is drained mainly by tributaries of the Zambesi but the south-eastern part of the country drains to the Limpopo or Save rivers (Figure 3.21).

The basement complex crops out over much of this region and has been granitised by a series of intrusions with metamorphic aureoles surrounding them. Limited areas of Karroo sedimentary rocks crop out in the west of the province, including coal mined at Wankie and the Bataka basalt flows which may be seen in the Zambesi gorge below Victoria Falls. In Zambia, outcrops of Upper Pre-Cambrian rocks include sandstone, shale and limestone in which copper ores occur. A major geological feature of Zimbabwe is the great dyke which extends for 500 km from the Umkmes ridge, north-west of Harare to near Belingue. At least four intrusions of basic volcanic magma took place in the basement complex; each one cooled sufficiently slowly for separation of mineral constituents to take place, concentrating the economically valuable minerals into ores of workable quality at the base of each flow. Chromite is the major mineral associated with the Great Dyke itself, but gold and asbestos occur in the adjacent metamorphic basement rocks.

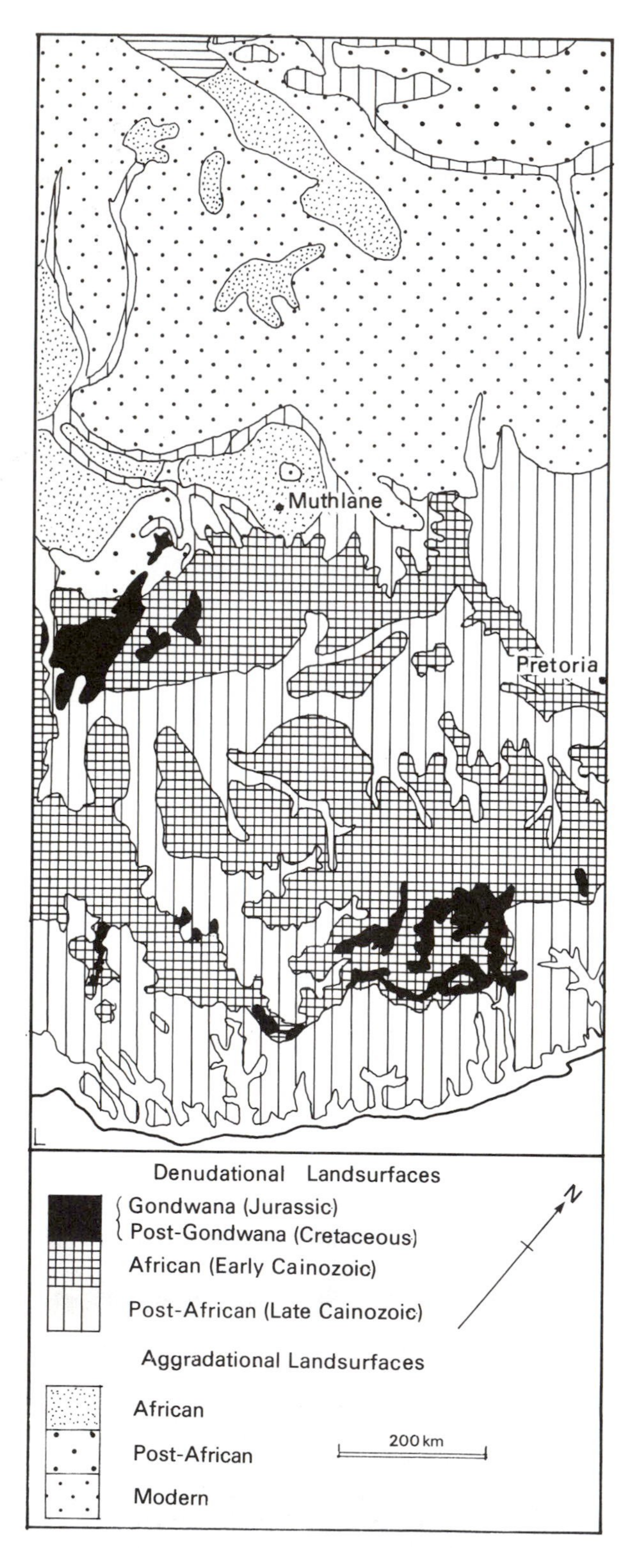


*Figure 3.20.* Surfaces in Basutoland.

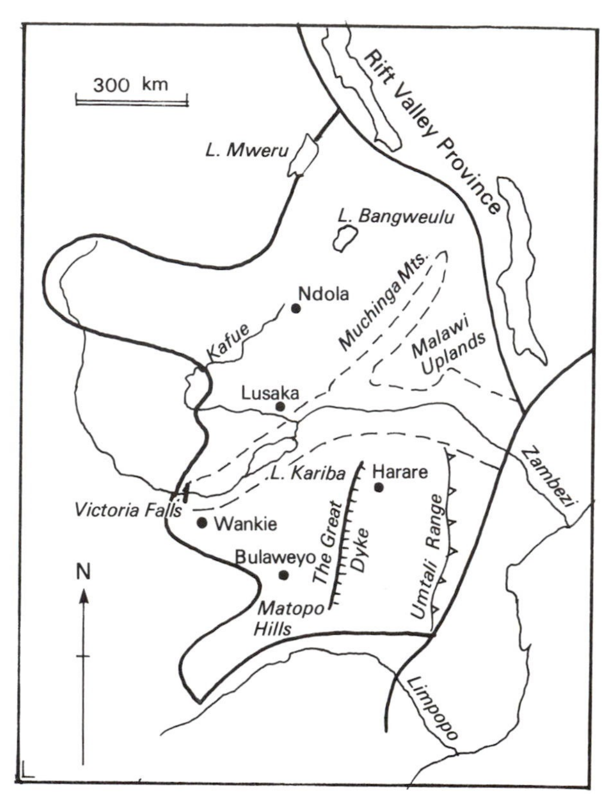


*Figure 3.21.* The Zambia–Zimbabwe plateau.

Parts of the Zimbabwe plateau are attributed to King's African surface as are the Zambian highlands, but the Zambian plateau is part of the end-Tertiary planation surface. On both surfaces valleys are poorly defined and the plains have a scatter of inselbergs and tors such as the Matopo hills, south of Bulawayo. These major surfaces are interrupted by faulting associated with the Luangua trough which extends westwards and far as the Okovango inland delta. This trough is 3400–1000 m below the plateau surface and it contains Karroo beds including the Bataka basalts. Following a period of erosion, Cretaceous sediments accumulated in the lower Zambesi valley and Luangua trough.

## *Zambesi valley*

The Zambesi is 3000 km in length and after rising in north-west Zambia flows across the Kasisi plains and plunges 110 m over the Victoria Falls into the zig-zag gorge formed by the river exploiting weaknesses in the basalts caused by faulting in two directions (Figures 3.22, 3.23). It has been suggested that the waters of the

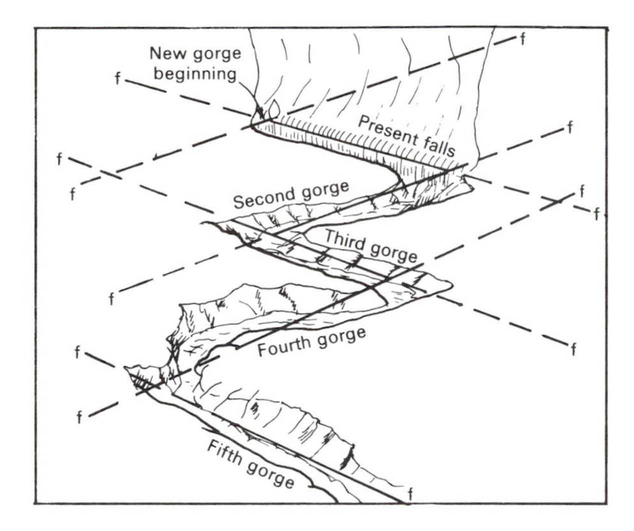


*Figure 3.22.* Diagram of the Zambesi Gorge and Victoria Falls showing fault lines (f) running in two directions.

Zambesi may have contributed to an inland drainage basin based upon the Kalahari. Tectonic activity after separation of the fragments of Gondwana provided a short, steep course to the sea from the Kalahari basin through Mozambique. Erosion has cut back along the rivers to produce the gorges which provide suitable sites for dams, as at Cabora Bassa and Kariba.

### *High and Middle Veldt*

These terms are used in purely an altitudinal sense to distinguish the plateau surfaces above 1200 m and those which lie between 600 m and 1200 m. The High Veldt is the extensive area of high land lying immediately north of the Great Escarpment in Cape Province and extending northwards through the Orange Free State into Transvaal. The most elevated parts are the Lesotho highlands, attributed to the Gondwana erosion surface,

*Figure 3.23.* The Victoria Falls. (By courtesy of H.J.R. Henderson.)

which rise to over 3000 m above sea level. Although these highlands are often snow-covered during winter, they are not high enough for permanent snows, but nivation hollows occur on south-facing slopes of the central ridge of Lesotho and the Witberge of eastern Cape Province. These features were formed during a colder climatic episode and are essentially periglacial rather than glacial features.

The Middle Veldt which lies north and west of the High Veldt in west-central Cape Province has an ill-defined boundary with the High Veldt, except where the Kaap plateau, part of the African surface at 1800 m, rises abruptly from the plateau surface north of the Vaal river. In central Transvaal an area known as the Bushveldt occurs north of Pretoria. This is a lopolith, where there have been successive plutonic intrusions of gabbro and red granite into the limestone country rock. Subsequent volcanic activity introduced syenite and mineral wealth. The Bushveldt complex is almost 500 km long in an east–west direction and the igneous complex passes 10 000 m into the Earth's crust in a large saucer-shaped structure. Erosion has removed the superincumbent rocks to reveal this complex structure at the surface.

Rocks of the Karroo system cover the greater part of the High Veldt and the southern part of the Middle Veldt. The Karroo system has as its basal member the Dwyka tillite, which reaches a thickness of up to 750 m in the south but decreases to less than 30 m in the Transvaal. These stony mudstones originated as glacial drifts during the Carboniferous when Gondwana was situated at the South Pole. They are succeeded by sandstones and shales of the Stormberg series and the sequence is completed by the basaltic lavas which, in the Lesotho highlands, reach a thickness of 1500 m. These Karroo rocks are mostly level bedded, resting on the uplifted basement complex and are responsible for the monotonous plains over much of the High and Middle Veldts. It is thought that the Karroo rocks were the erosion products from the Cape fold mountains.

Wherever the Karroo rocks are eroded away, as in the northern part of the Transvaal, the basement rocks of Upper Pre-Cambrian and Archean age are revealed to give a more varied landscape. The quartzites in the Bushveldt complex form escarpments upon which drainage has been superimposed from the Karroo beds. Mineralisation accompanied the volcanic intrusions and copper, antimony, platinum, gold, asbestos, tin, and chromite occur. Iron ore is mined on the northern limb of the Bushveldt basin and diamond mining occurs near Pretoria where the world's largest diamond, the Cullinan (0.5 kg), was found. The Upper Pre-Cambrian limestones provide valuable acquifers in central Transvaal.

### *The Orange river*

The drainage of much of the High and Middle Veldt is collected into the Orange river. This river, which has a length of 1100 km, rises in the Malauti mountains of Lesotho and proceeds to flow westwards across the continent to enter the south Atlantic on the border between the Republic of South Africa and Namibia. Its longest tributary, the Vaal, also rises on the western slopes of the Great Escarpment 200 km east of Johannesberg. As it crosses the High and Low Veldt, the Orange has low gradients and it steadily loses water through evaporation and to irrigation schemes as it passes into increasingly arid regions of the southern Kalahari. After passing over the 97 m Aughrabies Falls (Figure 3.24) the river leaves the late Tertiary land surface and enters the part of its course eroded in the Pleistocene erosion cycle. The river does not always have sufficient water to reach the sea during the dry season.

### *The south-west African plateaux*

North of the Orange river, the Great Escarpment continues through Namibia and Angola where it forms the western edge of the south-west African plateaux. From south to north these are: the Namaqualand plateau; the Khomas highlands; the Damaraland plain; the Chela–Otavi highlands; and the Bie plateau. The landforms all result from the uplifting of the basement complex which forms the surface geology through most of these upland areas.

The Namaqualand plateau is developed on granites and gneisses, the weathered debris of which is said to have contributed the sands which now cover much of the Kalahari. The Khomas highlands near Windhoek, which have peaks up to 2500 m, are much more rugged, formed from Pre-Cambrian metamorphic rocks. The continuation of the plateaux northwards in the Chela–Otavi highlands include Upper Pre-Cambrian dolomites. Copper is mined near Windhoek and at Tsumeb on the Damarland plateau where lead and zinc are mined also.

*Figure 3.24.* The Aughrabies Falls. (By courtesy of H.J.R. Henderson.)

Rainfall amounts increase northwards on these plateaux but it is not until the Bie plateau is reached that sufficient rain is received to provide for permanent streams such as the Cunene, and others which drain towards the inland Okovango delta.

### *The coastal margin of southern Africa*

The Angolan coastal plains are dry, with rainfall only reaching 500 mm on the rising land near the Great Escarpment. The northern coastal lowland is crossed by many short streams originating on the western slopes of the Bie plateau. At Lobito, and other places on the south Angolan coast sandspits have developed from the sediment load brought down by these streams. The Cunene is the last major stream to reach the Atlantic on this coast until the mouth of the Orange river is reached, and the climatic régime becomes increasingly arid for 2000 km.

Between Walvis Bay and Luderitz a series of seif dunes occurs in a sand desert which extends inland for 150 km. The dunes have an average height of between 50 and 100 m but many exceed 250 m. The direction of the most significant wind affecting these dunes is from the south-west which propels the sand northwards as far as the deep wadi of the Kuiseb which intercepts and arrests the sand. Between the dune-covered coastal lowlands and the Great Escarpment, a dissected plateau formed from Karroo sandstones occurs called the Kookoveld. It is an area of flat-topped ridges dissected under an arid environment. Altogether the Namib desert extends for 2000 km along the west African coast. Many of the beaches and raised beaches in this desolate area from Conception Bay to Cape Hondikhe are valuable resources for the diamonds they contain.

### *The Natal coastlands*

This province lies between the Great Escarpment and the Indian Ocean. It is distinctive for its geological structure which is monoclinal. The Karroo sedimentary rocks, including the Dwyka tillite are downfolded towards the coast. Further north, in Mozambique, the Jurassic rocks are brought in by this downfold, but in the Natal coastlands this occurs offshore on the narrow continental shelf. The streams of this area are short, originating below the Great Escarpment.

### *The Cape ranges*

The South African coast from Great Paternoster Point, 120 km north of Cape Town, around to East London is influenced by the folded structures of the Cape ranges which meet the coast at an angle and provide a series of headlands with rocky cliffs and sandy bays. The famous Table Mountain overlooking Cape Town has a synclinal structure from one of the folds. North of East London the coast has a smooth line and the slopes up to the Great Escarpment completely fill the area between the coasts.

### *The Kalahari basin*

The Kalahari is less arid than the Sahara with a rainfall of between 150 and 1000 mm, which is sufficient to encourage a growth of grasses and trees, especially in the north. The Kalahari desert lies north of the Orange river, west of the Zambian–Rhodesian uplands and east of the Great Escarpment and plateaux of Namibia and Angola. The surface of the Kalahari has an elevation of about 1000 m above sea level, it is sand-covered, probably the most extensive sand surface in the world. Uplift of the basement complex in the surrounding areas left the Kalahari as a large downwarp in the African plateau surface. Upon the Gondwana surface terrestrial deposits of Cretaceous age were laid down which are the equivalent of the continental terminal deposits of northern Africa. These were subject to planation at the end of the Tertiary but the sands continued to accumulate into the Quaternary and are 100–150 m thick. The sands were derived from the Karroo Stromberg sandstones which have been reworked by wind and water (Figure 3.25).

Surface water is absent for much of the Kalahari because of these sands, but there are three internal drainage basins: the Etosha pan, the Okovango delta and Malopo–Nassab system which formerly joined the Orange river. The Etosha pan lies in Namibia, in the extreme north-west of the Kalahari. It is the former inland delta of the Cunene before river capture took the waters of this river over the edges of the Great Escarpment at the Ruacana Falls. The floor of the pan is described as a partly saline clay plain. Other streams, like the Gubango also rise in the Bie plateau but flow with a south-easterly course to the Okovango delta which is situated in the north of Botswana. The delta is an extensive swampy area (comparable in size to the Nile delta), parts of which are permanently flooded. The abrupt south-east margin of the delta is attributed to faulting as this is a seismic zone. The northern distributaries of the Okovango and Cuando may have supplied water to the Zambesi, but tectonic adjustments appear to have closed this channel. In former pluvial times flow continued from the Okovango delta to the Makarikari pans where formerly a 34 000 km$^2$ lake existed.

The Malopo–Nassab system drains the southern part of the Kalahari basin and formerly joined the Orange river but rainfall is no longer sufficient to maintain flow, which becomes lost in a dune field. Pictures from satellites have given an overall view of the sand dune distribution in the Kalahari. Between the Etosha pan and the Okovango delta the dunes have an east–west orientation but between Windhoek and the Orange river the direction is north-west–south-east. The dunes in the north of the Kalahari are much subdued and covered with grass and trees but the existence of former arid conditions, much extended to the north, is confirmed by the dunes extending right into the Congo basin.

## Oceanic basins

The submarine provinces of the African lithospheric plate result from sea-floor spreading which has taken place around the continent since the disruption of Gondwana. The north Atlantic opened first with rifting occurring between 180 and 200 million years ago and creation of new sea floor is apparent from 160–170 million years ago. South America and Africa split apart in the early Cretaceous, 120-130 million years ago, and separation from Antarctica took place at the same time to form the south-east coast of Africa. 20 million years ago, the Arabian peninsula was joined to Africa, but rifting occurred and new sea floor is being created in the Red Sea.

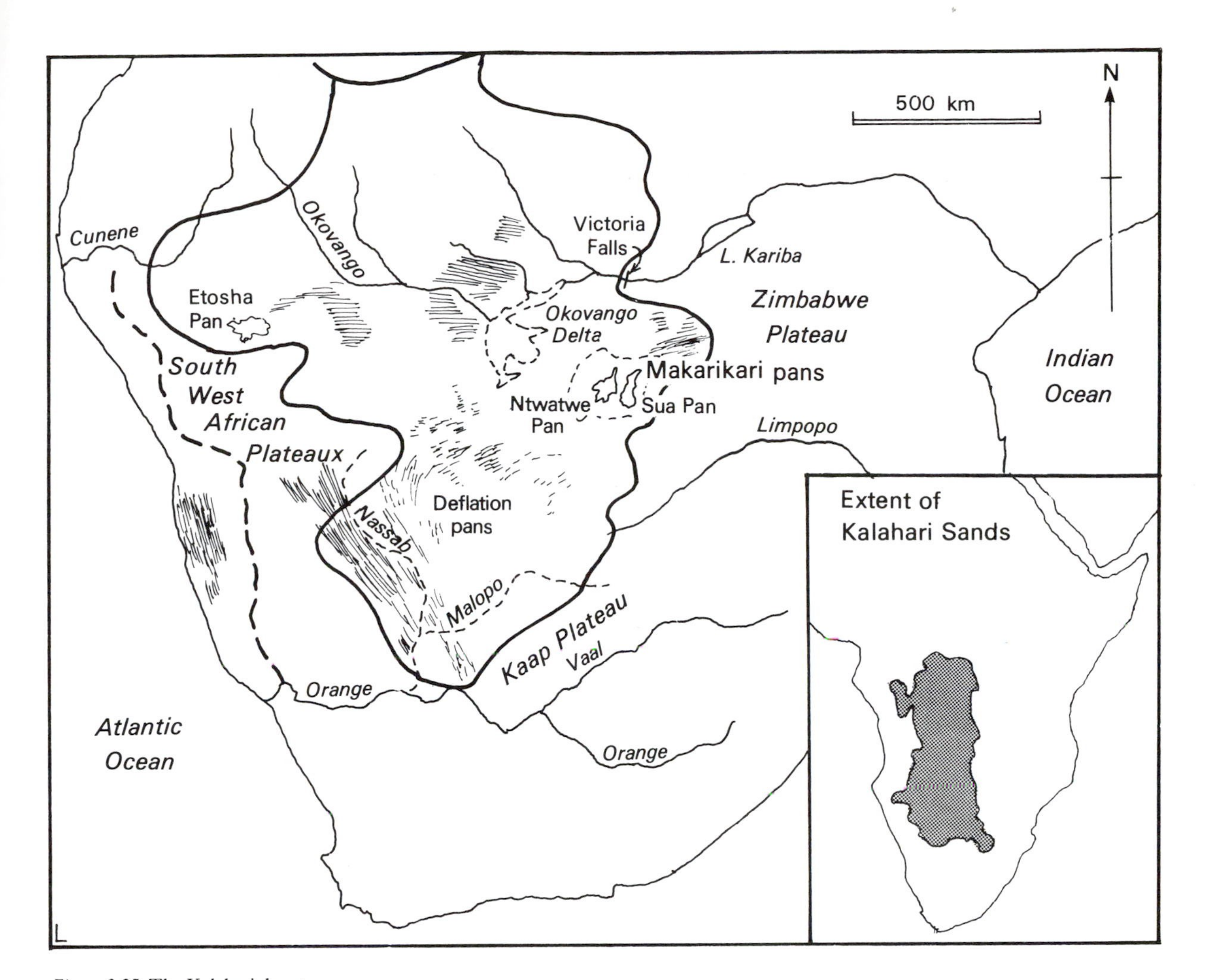


*Figure 3.25.* The Kalahari desert.

*The Canary basin*

The north-west oceanic province of the African plate lies between the mid-Atlantic ridge and the African coast extending from the Azores to the Cape Verde Islands. There is a narrow continental shelf along the Moroccan and Mauritanian coast, but a wider terrace lies at greater depth. On this terrace are situated the Canary and Cape Verde Islands, some of the oldest oceanic islands of the Atlantic. West of this terrace, the abyssal plain lies 6000 to 7000 m below sea level, and at its greatest width, approximately along the Tropic of Cancer, the Canary basin is 2000 km wide before the sea floor rises on to the mid-Atlantic ridge. Numerous seamounts occur which rise from the sea floor to within 250 or 300 m of the sea surface, but clusters of volcanic seamounts, such as the Canary Islands (Figure 3.26) are sufficiently high to break the sea surface to form islands, and retain residual volcanic activity.

*The south-east Atlantic basin*

Between the mid-Atlantic ridge from Ascension (859 m) to Tristan da Cunha (2060 m) and the African coast lies the south-east Atlantic basin. Its floor is an extensive abyssal plain, 6000 m below the sea, but several seamounts occur including St Helena, a 20 million-year-old island, 700 km from the mid-Atlantic ridge.

*Figure 3.26.* Oceanic volcanic island: Tenerife.

Tristan da Cunha is a central volcano with many secondary cones. Its lava slopes are up to 20° with cliffs along the coast. A narrow continental shelf lies along the coast of Cameroon, Gabon and Angola which is broken by one of the World's largest submarine canyons, associated with the Congo river. It extends for about 1000 km across the continental shelf and the continental slope. The southern boundary of this province is the Walvis ridge, a linear range of volcanic material from Tristan da Cunha to the African coast near the mouth of the Cunene river. It was extruded from a stationary 'hot spot' as Africa and South America moved apart, thus leaving a trail indicating the continental movement during the past 120 million years.

### *The Cape Verde basin*

Between the Cape Verde Islands, the mid-Atlantic ridge and the Sierra Leone rise, the abyssal plain of the Cape Verde basin is 6000 to 7000 m deep. The African continental shelf is wider than further north and between Banjul and Freetown extends as much as 500 km out into the Atlantic. The western margin of this province is the mid-Atlantic ridge which only breaks the surface at St Paul's Rocks just north of the Equator.

### *The Sierra Leone and Guinea basins*

A very strong transform fault, the Romanche fracture zone, offsets the mid-Atlantic spreading centre by 1500 km on the southern margin of the Sierre Leone basin. One of the deepest parts of the basin is associated with this fracture zone, the Romanche gap (−7758 m). From Freetown to Accra there is a narrow continental shelf, but a large fan of sedimentary material lies in the Gulf of Guinea derived from the Niger river. The volcanic islands of Annobon, Sao Tomé, Principé and Fernando Poo protrude through this sedimentary material.

### *The Cape and Agulhas basins*

The Agulhas bank extends the African continent some 400 km south of the Cape of Good Hope in a continental shelf which is only some 150–170 m below the sea surface. The abyssal floor of the Cape basin extends south of the Walvis ridge, around the Cape and east as far as Port Elizabeth where the basin is known as the Agulhas basin. Although the abyssal floor is about 6000 m deep, seamounts rise almost to the surface in many places including Discovery (−390) and Meteor (−560). An extensive undersea plateau called Cape Rise (or Agulhas plateau) lies 2300 m below the surface south of Port Elizabeth, but the eastern margin of this basin is the Mozambique plateau which extends from Maputo southwards, only 1200 m below the sea.

### *The Natal basin*

This deep sea basin extends from the Atlantic–Indian mid-oceanic ridge northwards between Africa and Madagascar. Its western limit is the Mozambique plateau and its eastern margin is formed by the undersea Madagascar plateau which at one place is only 18 m below the sea surface. In the south of Natal Basin is part of the abyssal plain but the basin becomes shallower northwards between Madagascar and the African mainland. The coast of Natal and Mozambique is characterized by a narrow continental shelf, the west coast of Madagascar, however, has a somewhat wider continental shelf.

### *The Mauritius basin*

This deep sea basin lies between Madagascar and the mid-Indian Ocean ridge. The east coast of Madagascar is formed by a large fault scarp and the land surface descends rapidly to abyssal depths close inshore. On the eastern margin, the volcanic islands of Réunion and Mauritius lie at the southern end of an extensive undersea plateau over which shallow depths of water occur; only 20 m on the Saya de Malha bank. The Seychelle Islands at the northern end of this underwater plateau have much older rocks (Palaeozoic) and are considered to be a fragment of continental crust, perhaps carried north by the movement of India towards Asia. It has been suggested that there may be deep sea trenches east of Mauritius and Réunion and west of the Amirante Islands, but oceanic crust is not being destroyed actively in them at present.

### *The Somali basin*

Off the Somali and Kenya coasts there is a very narrow continental shelf and an extensive abyssal plain continues across to the mid-Indian and Carlsberg ridge. Few seamounts occur, the most significant of which are the Aldabra Islands and the Comore Islands north-west of Madagascar.

# 4

# The Americas

The Americas consist of fragments of the former continents of Laurasia and Gondwana, rifted apart in the late Mesozoic and moved westwards away from Europe and Africa. The two separate lithospheric plates of North and South America only became linked together in the late Cretaceous by the orogenic activity which produced the Central American mountain chains (Figure 4.1). Thus, the Americas fall naturally into three parts and it is convenient to discuss their geomorphology under the headings of North, Central and South America.

Information about landforms in the Americas is uneven in amount. Much less research has been done and little has been written about Central and South America compared with North America. This disparity is reflected in the text which follows.

## NORTH AMERICA

The work of American geomorphologists has made the geomorphology of North America widely known, and examples from North America have illustrated geomorphological textbooks to the exclusion of comparable examples from other parts of the world. The area of North America is approximately 21.5 million km$^2$, and includes the countries of Canada, USA, Mexico and Greenland. It extends from 25°N latitude to beyond 80°N (over 6000 km) and from longitude 60°W to 170°W, although the main part of the USA lies between 70°W and 125°W, a distance of 4200 km.

The eastern margin of the North American lithospheric plate is a constructive margin, the mid-Atlantic

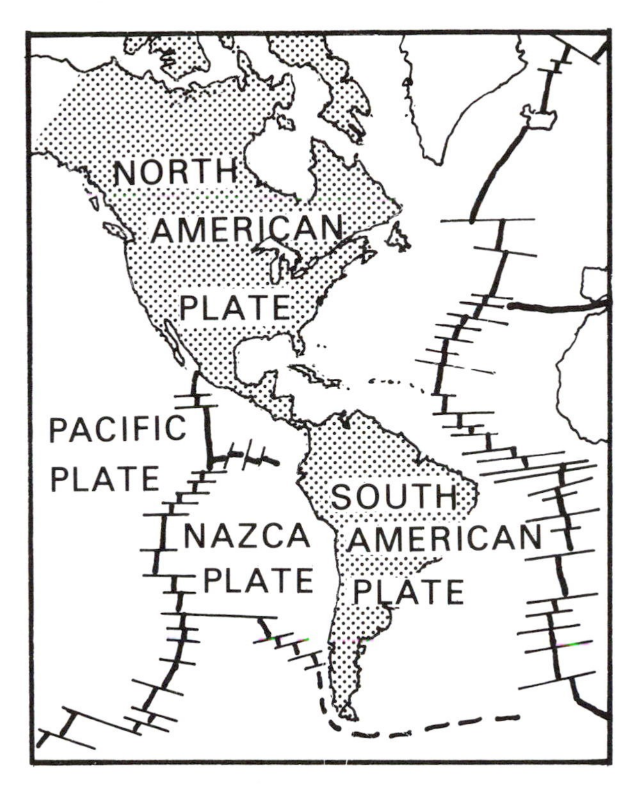


*Figure 4.1.* The North and South American plate.

ridge, extending from approximately 20°N to the Azores and then to Iceland. The northern margin lies within the Arctic Ocean and the western margin of the continental plate has overridden the oceanic floor as far as the east Pacific spreading centre which is strongly offset in the San Andreas transform fault. Only two small parts of the eastern Pacific floor remain, the Cocos and Farallon plates, the eastern margins of which are marked by subduction zones beneath the continental plate.

It has been estimated by Butzer that young mountain belts occupy the largest percentage of the landsurface in North America with 31%, followed by shield areas with 25%. Volcanic plains such as the Snake river plains make up 4% and depositional plains 18% of the land area. Erosional plains cover 18% and old mountain belts only amount to 5%. The mean elevation of the North American landmass is 610 m above sea level.

## Geological evolution

North America has developed from the nucleus of the Canadian shield which contains some of the oldest (2400 million years) rocks on Earth. These rocks crop out north of the Great Lakes and were probably the initial cratons from which the shield was formed. Orogenic activity around these cratons gradually extended the shield during the Archean and most of it has been granitised. Towards the end of the Archean, limited amounts of basaltic and andesitic lavas poured on to the sea floor and sediments, rich in iron, accumulated. The latter now form the iron ores of Minnesota. The succeeding Proterozoic rocks rest unconformably upon the Archean and include metamorphic and sedimentary deposits. Some have been intruded by granites, but considerable areas of the Proterozoic rocks were not metamorphosed. The conglomerates and mudstones of the Huronian rocks north of the Great Lakes include a tillite from a Pre-Cambrian glaciation. The youngest (Proterozoic) rocks occur along the south-eastern margin of the shield from the Great Lakes to Labrador, consisting of a belt of folded, metamorphosed Proterozoic rocks (Grenvillian) which were metamorphosed about 1000 million years ago.

East North America has a geological history closely related to that of Western Europe, and the suggestion has been made that a proto-Atlantic Ocean called 'Iapetus' had opened by late Proterozoic times and the Torridonian sandstones of northern Scotland were eroded from the Canadian Shield and laid down on the northern side of this sea. Sedimentation continued into the Lower Palaeozoic with deposition spreading on to the peneplained surface of the shield. When this ocean closed in the late Ordovician, rocks with a faunal composition from the 'American' side were attached to the 'European' side in the Taconic orogeny, and there was strong thrusting to the west along the Hudson valley and in Tennessee. The mountains produced were eroded during the Silurian until the Acadian orogeny, when the Atlantic uplands of Nova Scotia and the central highlands of New Brunswick were raised.

During Lower Carboniferous (Mississippian) times, sedimentation along the Appalachian geosyncline brought the land to sea level, and in the Upper Carboniferous (Pennsylvanian), extensive swamps spread from which the coal measures developed. West of the Appalachians, the shield rocks had been reduced to a plain of low relief before the end of the Pre-Cambrian and both Lower and Upper Carboniferous rocks were laid down when shallow epicontinental seas transgressed. The last mountain building episode along the Appalachians was the Allegheny orogeny which deformed these Carbonniferous strata. Throughout this period, North America and Western Europe were in close proximity to each

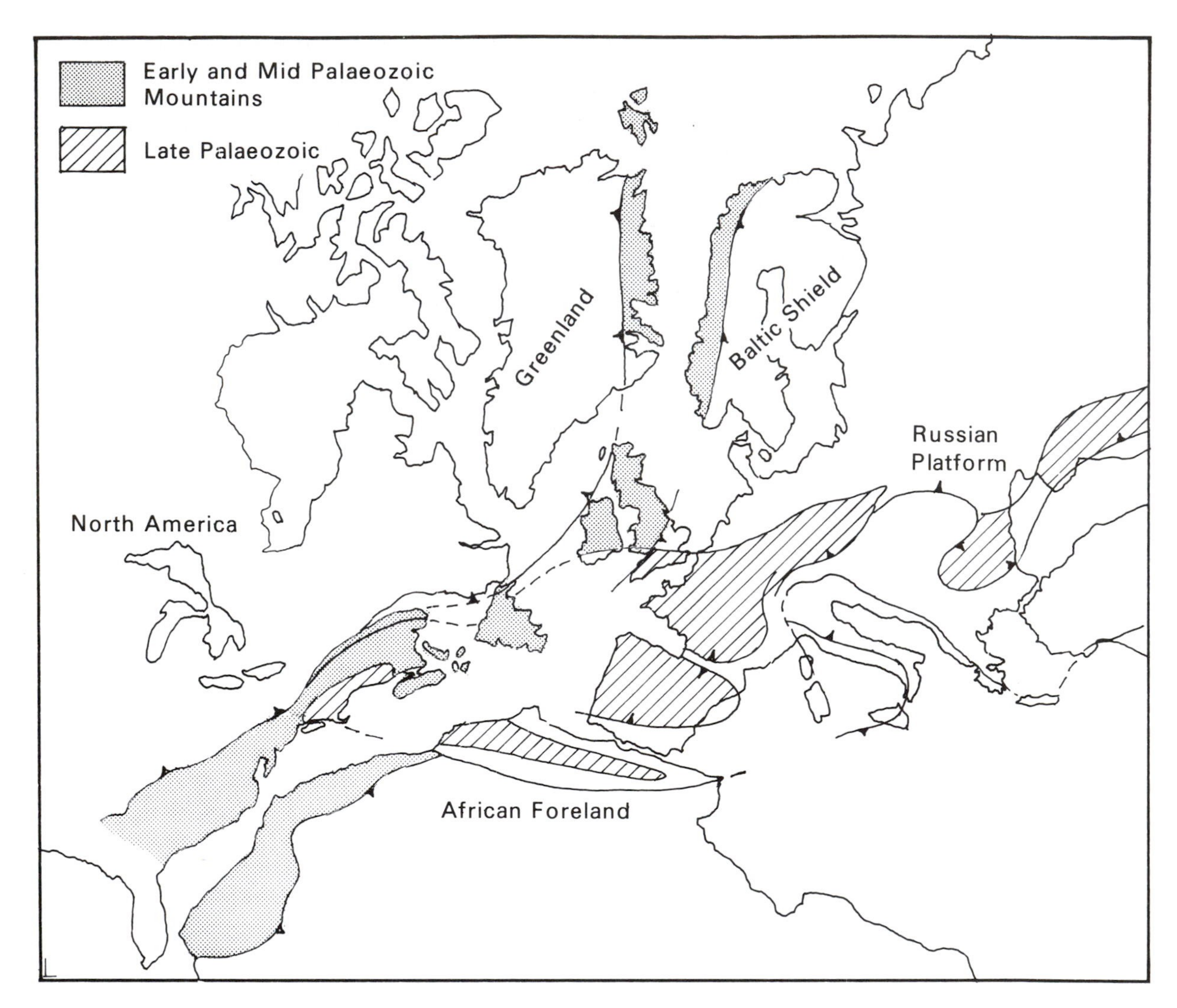


*Figure 4.2.* Relationship between North America, Europe and North Africa in the Palaeozoic.

other and the relationship of the rocks can be seen in Figure 4.2. The present Atlantic began to open early in the Cretaceous, and a geosynclinal couplet has been laid down on the trailing edge of the American plate. The miogeosyncline consisting of Cretaceous and Tertiary sediments dipping gently seawards with the eugeosyncline deposits banked against the continental shelf (see Figure 2.19). In the far north, the Franklinian geosyncline received sediment until the Devonian after which parts of Parry and Ellesmere Islands were uplifted in an early Carboniferous orogeny. Sedimentation then continued in the Sverdrup basin, the contents of which were folded to produce the north-western parts of these islands. The origin of this sediment is not known for sure, but it has been suggested that Greenland might have been the source.

The western margin of the Canadian shield was also the site of sedimentation and some vulcanicity from Lower Palaeozoic times, but in the middle Jurassic, the Columbian orogeny folded and metamorphosed the coastal ranges. Batholiths of granite were intruded beneath British Columbia, the Sierra Nevada and Lower California. Sedimentation continued east of this new range of mountains along the line of the Rocky mountains, which were uplifted in the Laramide (Alpine) orogeny as a result of the westerly movement of

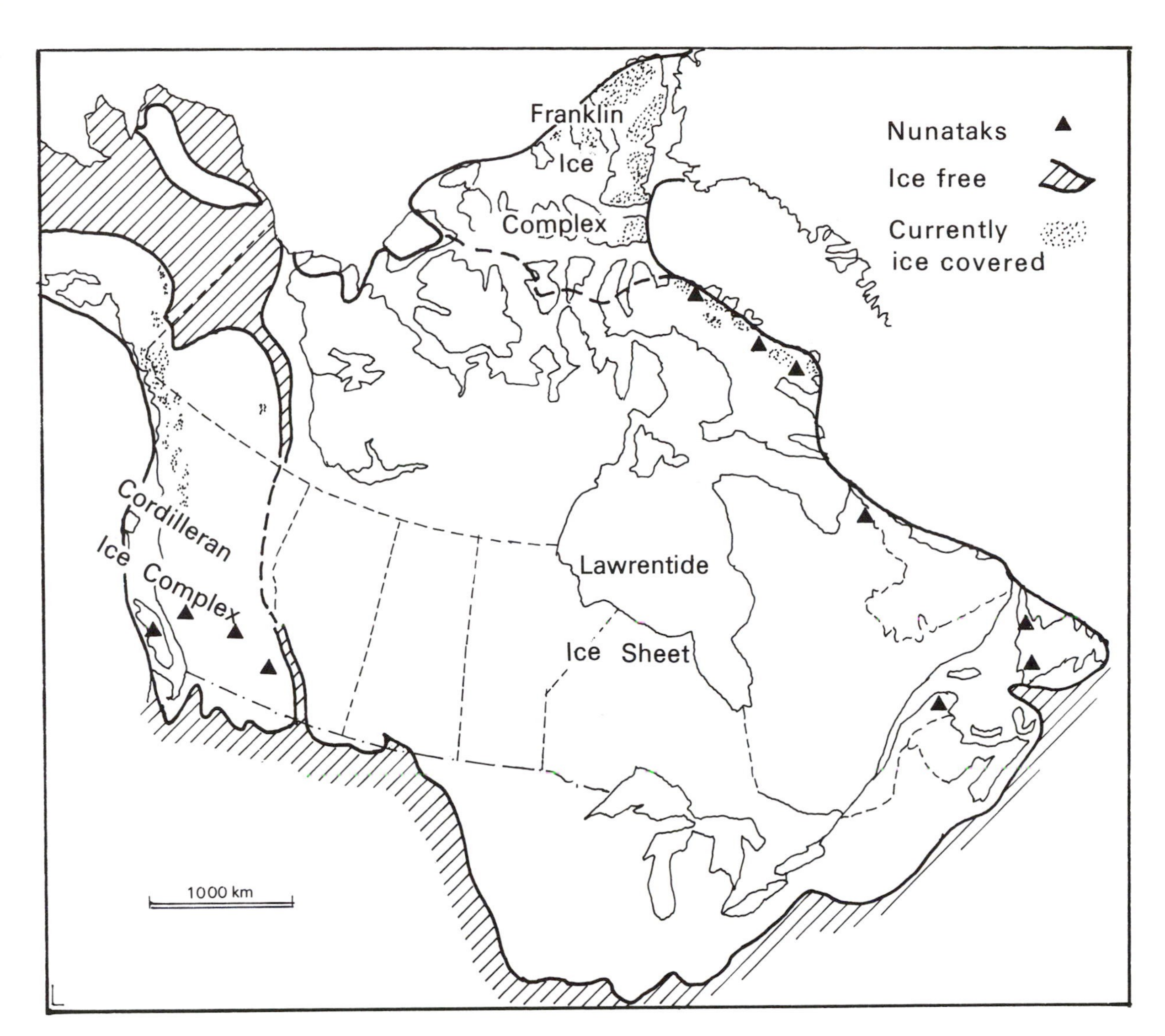


*Figure 4.3.* The extent of the Pleistocene ice caps and the present day glaciers in North America.

the North American plate. The intervening plateaux and the basin and range province were uplifted, faulted and subjected to erosion. Basaltic flows occurred, particularly on the Columbia and Colorado plateaux. Erosion of the emergent Rocky mountains provided a large amount of sediment which was deposited on the western part of the central lowlands during the Tertiary to form the great plains.

Virtually all of Canada and several of the northern states of the USA were glaciated during the Pleistocene. Evidence of erosional features and glacial deposits have enabled geomorphologists to reconstruct a history of successive glacial advance and retreat, but evidence of the earlier Kansan, Nebraskan and Illinoian glaciations has largely been erased by the last major glaciation, the Wisconsin. The main Lawrentian ice dome was centred upon the Hudson Bay area of the Canadian shield with smaller ice caps on the western cordilleras and upon the islands in the far north (Figure 4.3). Small ice caps only remain in Canada but most of Greenland lies beneath an ice sheet which attains a thickness of 3400 m.

The various geological deposits, their structure and subsequent events of erosion and deposition enables the geomorphologist to divide the North American lithospheric plate into seven major sub-divisions based on terrestrial land forms, and one on submarine landforms:

The Canadian archipelago and Greenland
The Canadian shield
The Appalachian highlands
The central lowlands
The great plains
The western cordilleras
The Atlantic and Gulf coastal plain
Oceanic landforms of the North American plate

The following pages give a brief account of the provinces included within these major sub-divisions.

## The Canadian archipelago and Greenland

Between the mainland of Canada and Greenland a series of islands lie mainly within the Arctic Circle, extending to beyond 80°N. The largest are Baffin, 344 000 km$^2$; Victoria, 128 000 km$^2$; and Ellesmere, 120 000 km$^2$. Greenland too, lies mostly within the Arctic Circle, but unlike the Canadian archipelago it is actively being glaciated. Ice covers most of Greenland but the Canadian Arctic islands are free of ice except for small ice caps on the mountains adjacent to Baffin Bay.

The northern margin of the Canadian shield occurs within these lands. Folded sedimentary rocks mark the edge of the Pre-Cambrian rocks, which roughly corresponds with the line of Parry Channel. The islands to the south are formed from relatively undisturbed limestone rocks overlying the basement complex, whereas north of the line, Cambrian, Devonian and Carboniferous rocks have been laid down in a synclinal structure, folded and uplifted in the Caledonian orogeny to give the uplands of Melville, Bathurst and central Ellesmere Islands. Sedimentation then continued from the Carboniferous to the Cretaceous in the Sverdrup basin.

The origin of the sediments which infilled these northern basins is in doubt, but there is evidence that the detritus came from the north and that land existed where there is now Arctic Ocean. One possible interpretation is that Greenland has moved in a clockwise direction around Canada as the Atlantic opened. It has been suggested that the fold mountains across the north of Greenland correspond to those of the northern Canadian islands.

A feature which affects all the lands of Arctic Canada is the presence of permafrost. In the North-West Territories it occurs north of a line from the Mackenzie delta to just south of Hudson Bay, but in Ungava only the northernmost peninsulas have continuous permafrost. Discontinuous patches occur as far south as northern Alberta, Saskatchewan, Manitoba and Ontario. In Quebec, discontinuous permafrost extends south to a line from James Bay to Goose Bay on the Labrador coast. The thickness of permafrost increases from 1 m on the southern boundary to more than 400 m in the Tundra region.

Within the Canadian Archipelago–Greenland major sub-division the following provinces are recognised:

The northern mountains
The Arctic lowlands
Greenland

### *The northern mountains*

Ellesmere and Axel Heiberg Islands comprise the northern mountains province. These islands, formed from Lower Palaeozoic rocks, were folded and uplifted in an orogeny early in Carboniferous times. The resulting mountains and plateaux rise to over 3000 m above sea level including ranges such as the Challenger mountains and the United States range of northern Ellesmere. The islands are situated at the head of Baffin Bay, the waters of which provide sufficient moisture to maintain ice caps on the plateau surface, particularly along the eastern side of Ellesmere Islands. Elsewhere in this province, Pleistocene glaciation has affected landforms. The coasts of Ellesmere and Axel Heiberg Islands were deeply trenched by fjords, and the relationships between land and sea have been complicated by the eustatic rise in sea level and between 50 and 100 m of isostatic rebound since the Pleistocene. Glacial deposits are reported to be limited, presumably as the islands were a centre for ice dispersal.

### *The Arctic lowlands*

The southern parts of the Arctic Archipelago, including Banks, Victoria, Prince of Wales and Somerset Islands, together with the northern coastal lowlands and Southampton Island, comprise this province. It is an area of low relief, mostly less than 500 m above sea level, where level-bedded Lower Palaeozoic rocks overlie the basement complex. The region has been subjected to fault-

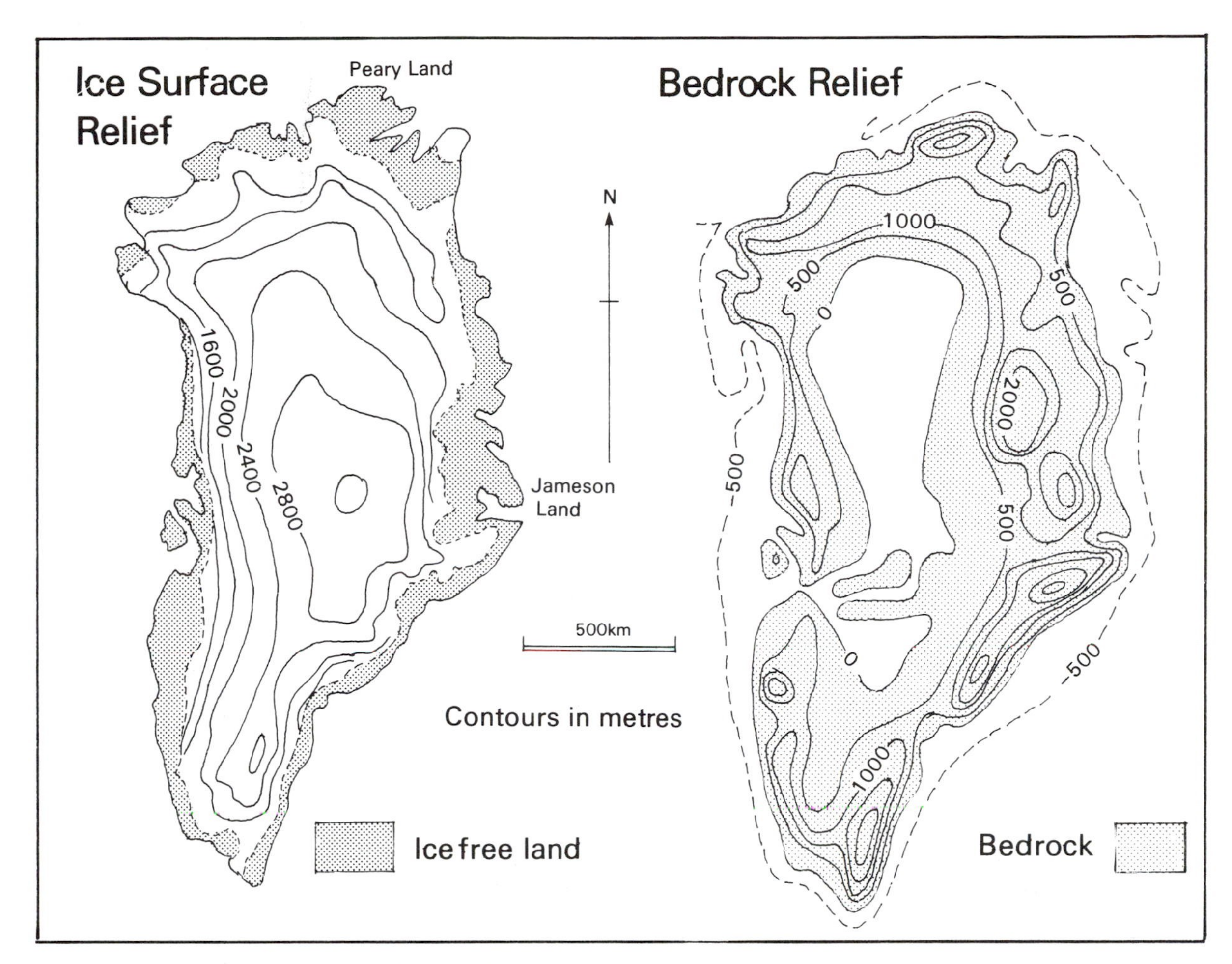


*Figure 4.4.* Surface relief and bedrock relief of Greenland.

ing and the Parry Channel is a major structural feature across the region from east to west, with many of the smaller channels between islands also formed by faulting. Tertiary drainage across this area was to the Arctic Ocean, so fluvial erosion also has played its part in landscape evolution. However, glaciation by Lawrentian ice in the Pleistocene extended north as far as Parry Channel. Evidence for this ice cover remains throughout the province, except for Banks Island which appears to have remained unglaciated. Glacial till is widespread on eastern Victoria Island with kames and end moraines on the western part of the Island. Southampton Island is a low plateau of Ordovician limestone with raised beaches left by former strand lines as isostatic readjustment has taken place. Most of this province has experienced between 10 and 200 m of post-glacial crustal uplift. Beaches of former proglacial lakes are present on the coastal lowland of Keewatin. The climate of the Arctic Lowlands is much drier than that of the Northern Mountains and in many places precipitation is sufficiently low for cold desert conditions to prevail.

### Greenland

The large island of Greenland, with an area of 1.25 million km$^2$, lies mostly within the Arctic Circle. It is 2650 km from north to south and its breadth is 1120 km at its widest part. Only about 80 000 km$^2$ around the coast is ice-free (Figure 4.4).

Most of Greenland is underlain by the basement complex, structurally a continuation of the Canadian shield. However, the arm of the Atlantic extending towards Baffin Bay resulted as sea-floor material was introduced north-east of Labrador along a now inactive spreading centre. Caledonian fold mountains occur along the east coast of Greenland and a second range of fold mountains runs east–west across the northern extremity of the country. Palaeographical reconstruction suggests that Greenland may have been located further north and west of its present position during Lower Palaeozoic times, in which case the fold mountain belts can be matched with those in the northern mountains province of Canada and the Caledonian province of Europe.

Geophysical examination of the Greenland ice cap has shown that it is over 3400 m thick and that the rock surface is below sea level in the centre. If the ice were removed from Greenland, the peripheral mountains would form a ring of islands surrounding an inner sea as the ice has depressed the earth's crust. Before glaciation, it is thought that Greenland had a similar topography to the Canadian shield.

The ice-free coasts of Greenland are indented with deep fjords, some of which contain glaciers which descend to the sea to supply icebergs which float southwards along Baffin Bay and into the north Atlantic.

## The Canadian shield

The Canadian (or Lawrentian) shield with its rigid Pre-Cambrian rocks is the nucleus of the North American continent. A large part of the shield is composed of granite and gneiss with very complex structures. After at least three episodes of mountain building and erosion, the Canadian shield was reduced to a peneplain by the end of the Pre-Cambrian. In a similar manner to that described for Africa, the surface of the shield was warped into a series of basins and ridges. Major basins are centred upon Hudson and Foxe Bays; smaller ones occur in Ungava Bay, the southern part of James Bay and between Victoria Island and the Boothia peninsula (Figure 4.5). Other basins have been identified between Lakes Michigan and Huron and in south-east Saskatchewan, where the basement complex is covered by later geological formations. Lower Palaeozoic rocks, including Ordovician limestones, remain preserved in these basins. The limestones have been eroded from the ridges which now provide plateau-like landforms or palaeoplains 400-600 m above sea level in Labrador, Quebec, east of Great Bear Lake and in Keewatin.

The shield probably remained as land for much of the Mesozoic, but the western margin became covered by Cretaceous and Tertiary sediments as debris from the western cordilleras was transported in an eastward direction and deposited. During the Pleistocene, the exposed shield was glaciated by the Lawrentian ice sheet which extended from the Rockies to the Labrador coast and from Parry Channel to beyond the Great Lakes in the south. Glacial features, and particularly those of deglaciation, are an important constituent in the landforms of the Canadian shield.

The Canadian shield, which extends 4000 miles east to west and 4000 miles north to south, can be subdivided into the following provinces:

The shield plateau
The Hudson Bay and Foxe Bay basins
The Baffin–Labrador rim
The Mackenzie plains
The Peace–Slave lowland
The Great Lakes
The St Lawrence lowlands
The Lakes peninsula

### *The shield plateau*

The geological history of the Canadian shield has been dominated by erosion since the end of Palaeozoic times. The Lower Palaeozoic sandstones and limestones tend to have been removed from the more elevated areas of the shield where a palaeoplain, called the Lawrentian peneplain, has developed between 400 and 600 m above sea level. This surface is particularly evident upon the Labrador–Quebec plateau, the Contwoyto plains of the Mackenzie district and in north-east Keewatin. It also extends northwards into the Arctic lowlands along the Boothia and Melville peninsulas. Areas of sandstone form plateaux south-west of Dubawnt lake and south of Lake Athabaska.

The whole extent of the Pre-Cambrian rocks was glaciated during the Pleistocene and the shield is scoured by former meltwater channels and covered with till, eskers, and sediments laid down in proglacial lakes. The glacial lakes on the Shield were not as extensive, nor as persistent, as the precursors of the Great Lakes.

Glaciation has left the shield with a confused drainage pattern, but it has been claimed that it is possible to recognise a pattern of former consequent streams developed upon the Lower Palaeozoic strata

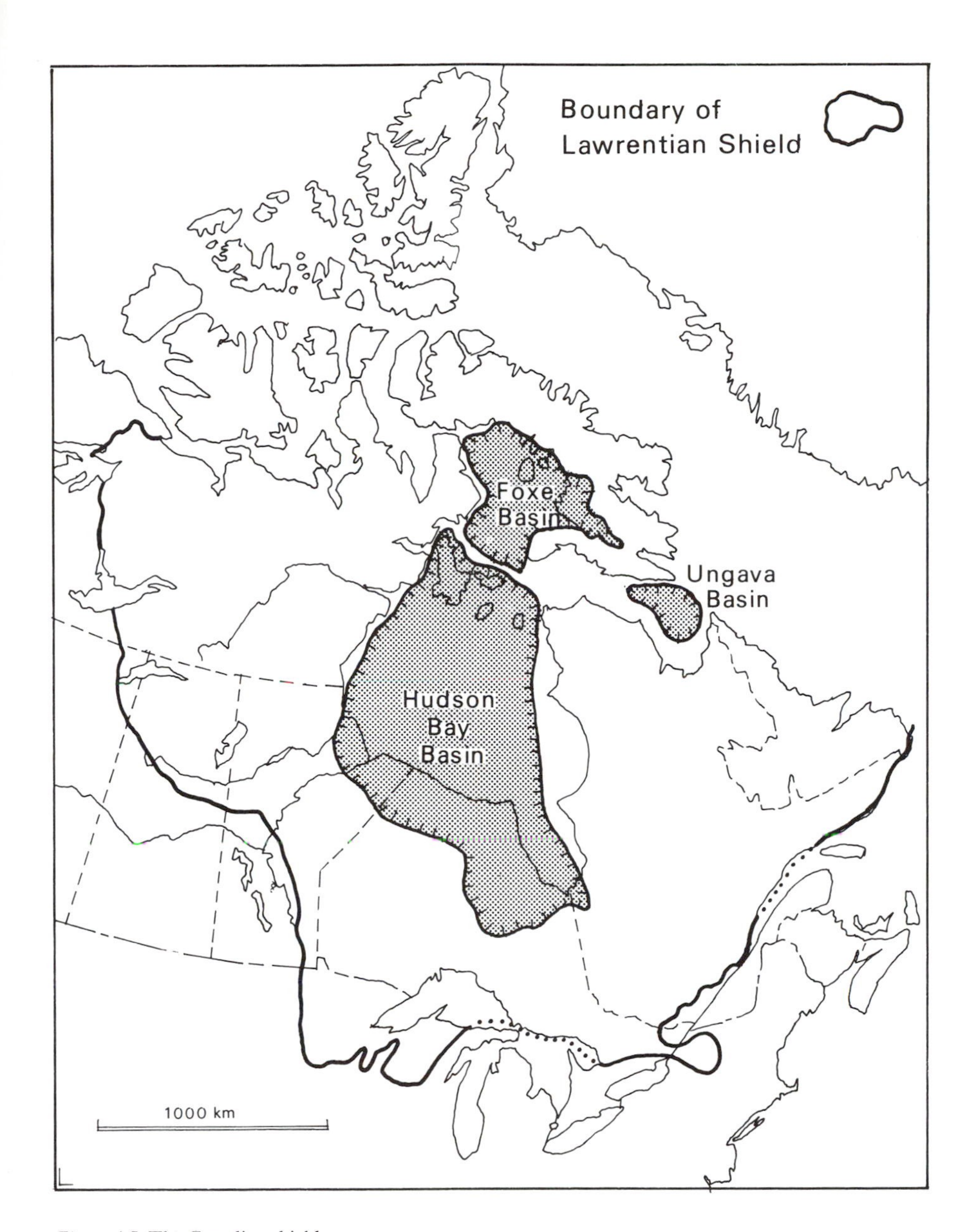


*Figure 4.5.* The Canadian shield.

and superimposed upon the Pre-Cambrian basement rocks. Preglacial drainage of the central part of the shield appears to have been towards the north-east across the Hudson Bay depression, but in the south-east, drainage was across the present line of the St Lawrence to the Atlantic; a trend which is preserved in certain rivers and gaps which cross the Gaspé peninsula from north-west to south-east.

*The Hudson Bay and Foxe Bay basins*

Two major basins merit consideration as provinces within the major division of the Canadian shield. The Hudson Bay basin extends for 1700 km from Southampton Island in the north to James Bay in the south. On the southern shore of Hudson Bay, the Ordovician limestones outcrop on a lowland up to 300 km wide and usually less than 150 m above sea level. The limestones dip towards the centre of the Bay where 1800 m of Lower Palaeozoic sedimentary rocks overlie the basement complex.

Although the Pleistocene glaciation probably began on the elevated rim of the shield, at its maximum extent, the ice sheet became centred upon the Hudson Bay basin which it completely filled. As deglaciation occurred, proglacial lakes occurred on the shield, one of which was located in the area of James Bay. Deposits of till were left by the ice, and sands and gravel laid down in eskers. Following the melting of the ice sheet, isostatic rebound of between 150 and 250 m has taken place, and is continuing at Churchill at a rate of 0.55 m per 100 years.

Drainage into Hudson Bay tends to be centripetal, following the dip of the Lower Palaeozoic strata, but Tertiary drainage crossed the basin to reach the Atlantic, and despite glaciation, evidence of a valley system has been identified on the floor of the Bay. The land around Hudson Bay is all subject to permafrost which limits soil permeability, even on normally permeable rock strata. Much of the land area is described as muskeg, a boggy terrain with vegetation composed of bog plants and stunted black spruce. Gravel ridges, former beaches, run roughly parallel to the present coastline.

The Foxe Bay basin lies further north between Baffin Island, Southampton Island and the Melville peninsula. This basin has a geological history similar to that of Hudson Bay and the Ordovician sediments extend from the basin on to the lowland of the western part of Baffin Island.

## *The Baffin–Labrador rim*

The Canadian shield ends abruptly along the north-east coast in the highlands of Labrador and Baffin Island. During the process of separation of North America from Europe, an extension of sea floor took place which resulted in the formation of an arm of the sea between Canada and Greenland which has been only partly infilled with later sediments. Uplift of the edge of the shield resulted in fast-flowing rivers and erosion during the Tertiary. Pleistocene glaciation followed and glaciers flowing out from the continental ice sheet turned the valleys into fjords. This geological history has resulted in a steep cliffed coast which rises 1500–1800 m from the sea to upland plateaux which still carry small ice caps. Some of the highest points of this northeastern rim of the shield may have protruded through the Pleistocene ice sheet as nunataks.

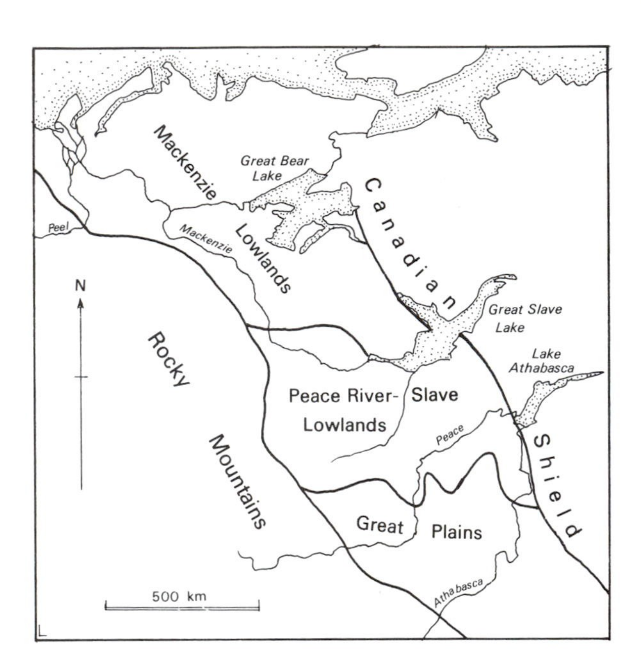


*Figure 4.6.* The Mackenzie-Peace lowlands.

## *The Mackenzie plains*

Land to the north of the Great Bear Lake is drained by the Mackenzie and Anderson rivers directly to the Arctic Ocean (Figure 4.6). The average elevation of the coastal plain is less than 300 m above sea level, but towards the south the land rises to between 300 and 600 m. The area is a drift-covered rolling plain with the Colville and Melville hills forming slightly higher ground. After leaving Richardson's mountains (part of the western cordilleras) the Peel river crosses an elevated plateau into which it is entrenched, and eventually joins the Mackenzie delta. The two rivers have constructed a delta between Richardson's mountains and the Caribou hills on the east bank of the Mackenzie; it is long (220 km) but not very wide (65 km). The whole of this region is underlain by permafrost and ground ice is often visible in river banks. Pingoes occur in the older parts of the Mackenzie delta, on the coastal plain and even beneath the sea where permafrost has been proven in the sediments of the sea bed.

### *The Peace–Slave lowland*

The lowland between the Canadian shield, the northern part of the high plains of Alberta and the Great Bear and Slave lakes is drained by the upper Mackenzie and its tributaries (Figure 4.6). The Peace–Slave lowland is generally poorly drained and large areas were formerly proglacial lakes during the process of deglaciation. Elsewhere, extensive till-covered, boggy areas with muskeg are common but where the Palaeozoic limestones crop out or are near the surface, the land is better drained. Scattered areas of permafrost occur.

### *The Great Lakes*

The southern margin of the exposed shield terminates in a series of fault scarps from the Strait of Belle Isle to the Gatineau hills in the vicinity of Ottowa. At its highest points, this scarp edge rises to between 500 and 1000 m above sea level, but is broken in several places where rivers in deep gorges have maintained their courses through to the St Lawrence.

The origin of the basins occupied by the Great Lakes is not known with certainty, but it is thought they had their beginning as lowland areas which were further deepened by repeated glaciation. As the lakes lie within the area covered by the Wisconsin glaciation, evidence of their earlier existence cannot be proven. When the Pleistocene ice finally melted and the margin of the ice sheet began to retreat northwards, the predecessors of the present lakes came into being. In order to avoid confusion with the present lakes, the predecessor of Lake Michigan was known as Lake Chicago; Lake Iroquois was the early Lake Ontario, and Lake Algonquin the predecessor of Lake Huron. At one stage, Lakes Superior, Huron and Michigan were confluent in Lake Nipissing. Lake Agassiz occupied a similar location in the Winnipeg basin, extending southwards into the Red River basin.

Drainage from Lake Chicago and Lake Agassiz took place southwards along the line of the present Illinois river to the Mississipi, but this outlet was abandoned about 10 000 years ago in favour of the Lake Nipissing–Ottawa River channel. These changes of outlet for the waters of the lakes resulted from isostatic uplift following glaciation, which has been greater in the north than in the south. The Nipissing outlet was abandoned about 5000 years ago, after which the Great Lakes drainage assumed its present form (Figure 4.7). Lake Superior is at 200 m above sea level and is over 300 m deep, Lakes Michigan and Huron are at 195 m, Lake Erie is at 190 m and Lake Ontario, 82 m above sea level.

The major drop in level between Lakes Erie and Ontario is at Niagara Falls which is about 50 m from the lip of the Falls to the water in the plunge pool below, where the lower Niagara river is about 40 m deep. Niagara Falls occurs where the waters cross the Lockport dolomite, one of the Lower Palaeozoic limestones which overlie the shield rocks, forming escarpments around the exposed Pre-Cambrian basement complex. Below the Falls, an 11 km gorge has been excavated as the Falls have retreated upstream. Recession is at a rate of 1.3 m per year during the last century; if this rate is used to calculate the time required to cut the gorge, it would have been 20 000–35 000 years. However, recent investigations have shown that a former gorge, now drift-plugged, exists from the whirlpool to Lake Ontario. Post-glacial activity by the Niagara river has been to remove drift rather than cut a completely new gorge. At the present rate of erosion at the Falls, Lake Erie will begin to drain in 25 000–30 000 years.

With the ice sheet damming the outlet of drainage waters, lakes were formed at different levels and at different times. The beaches left by these proglacial lakes have enabled a reconstruction of the sequence of development of the Great Lakes. Rochester, NY, is built upon a former beach level of Lake Ontario and the Hamilton Bar, 35 m above the present lake level, is another former beach. The Nipissing beach lies at 67 m above present lake level in the south but rises northwards, indicating that uplift has occurred. This uplift separated Lakes Superior and Huron at the Sault St Marie rapids.

The former lake margins can form significant bluffs in the landscape. Pressure of ice floes on the lake shores affects the beaches of the present lakes and redistribution of sandy material by waves has resulted in sandspits such as Point Pelee and Long Point in Lake Erie.

### *The St Lawrence lowlands*

The St Lawrence lowlands lie between the Canadian shield, the Adirondack mountains and the Canadian Appalachians. They have an elevation of less than 200 m above sea level and extend from west of Ottowa for 400 km to Quebec City.

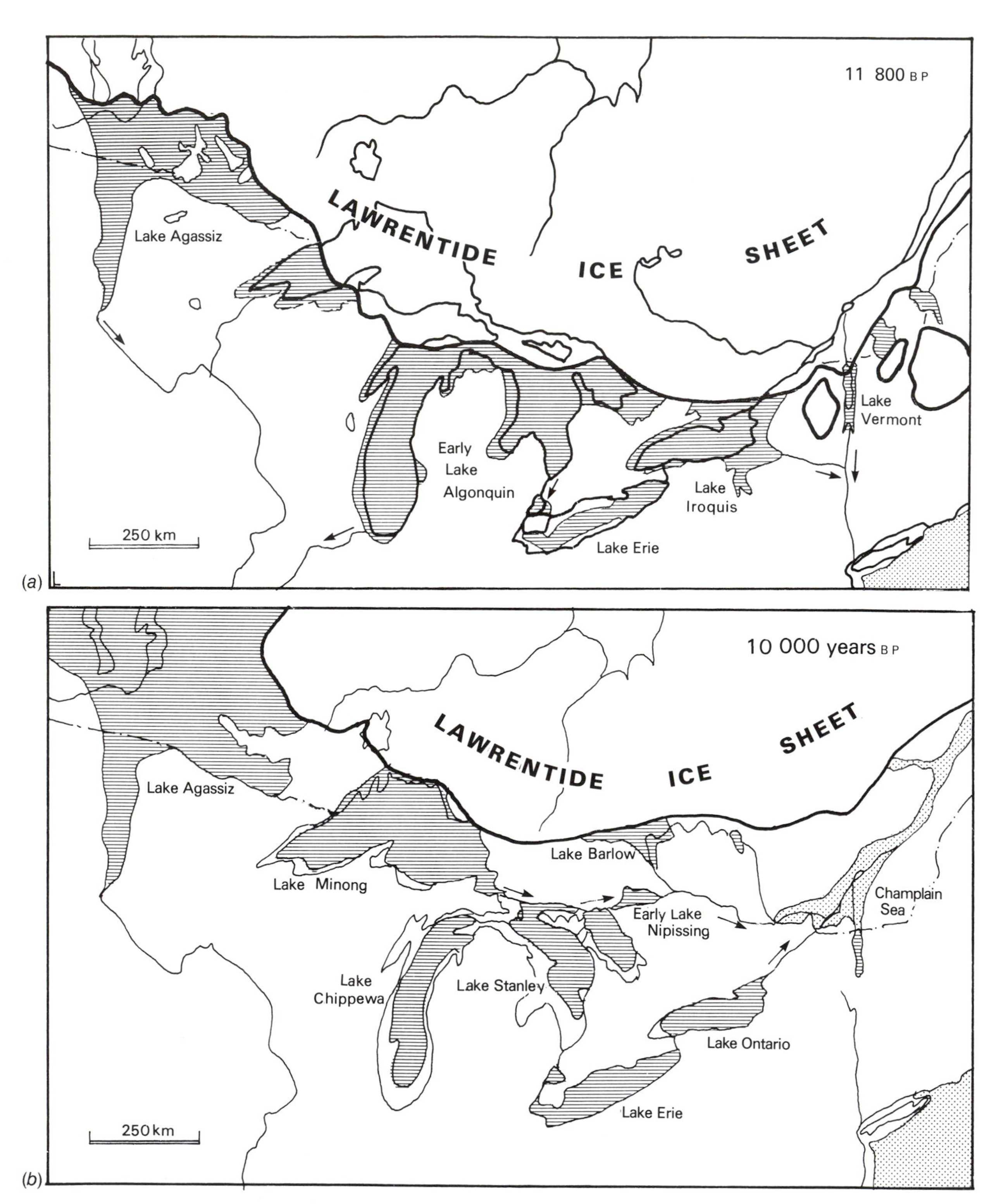


*Figure 4.7.(a)* and *(b)*. Development of the Great Lakes.

The basement complex here had been downfaulted and Lower Palaezoic sediments were deposited: Cambrian sandstones and Ordovician limestones. The lowlands were slightly affected by tectonic activity associated with the Appalachian geosyncline, but erosion kept this area at a low elevation, and although later rocks were deposited, they have since been eroded. Cretaceous intrusion of gabbro and syenite led to the formation of the Monteregian hills, one of which is Mount Royal in Montreal. Glacial deposits on the lowlands have been covered by estuarine sands, silts and clays deposited from the Champlain Sea. This 'sea' flooded the lowlands as the ice margin withdrew northwards, but as isostatic readjustment took place they once again became dry land. The structure and composition of some of these clay deposits is such that in certain circumstances the clays liquify and flow. More than 700 earth flows have occurred on these 'sensitive' clays in the St Lawrence and Lake St John lowlands. An example at Vianney in 1971 killed 30 people and 6.9 million m$^3$ of material moved out of a bowl-shaped hollow at a velocity of 26 kph into the Sageunay river. On the northern margin of this province, the retreating ice sheet left a major moraine about 20–30 km north-west of Quebec and Montreal.

### *The Lakes Peninsula*

The peninsula between Lakes Erie and Huron is a subdued scarpland formed by the Lower Palaezoic limestones and dolomites which overlie the basement complex. These escarpments dip south or south-westwards and are covered extensively by till and fluvioglacial deposits. After forming the isthmus between Lakes Erie and Ontario, the escarpment crosses the peninsula to the Cape Hurd peninsula and Manitoulin Island between Lake Huron and Georgian Bay. The escarpment swings around the western side of the Michigan basin to form the long peninsula between Green Bay and Lake Michigan. The northern margin of this province is the Black river limestone which crops out between Kingston and north of Lake Simcoe, where it forms a low escarpment overlooking the rocks of the Canadian shield.

Major fluvioglacial deposits occur in the 'horseshoe' moraines of the Lakes peninsula, but recessional moraines were also deposited parallel to the northern shore of Lake Ontario.

## The Appalachian highlands

The Appalachian highlands extend for over 2000 km with a north-east–south-west trend in the eastern USA. In the north they comprise most of the New England and Maritime provinces which lie between the North Atlantic and the St Lawrence river (Figure 4.8). South of New York, the Appalachian mountains are found at an increasing distance from the coast until in Georgia they are 400 km from the sea. Many ridges in the Appalachian highlands have a height of between 1000 m and 2000 m giving a uniform, level skyline with only the occasional mountain rising above the general level such as the famous Mount Monadnock (1052 m) in New Hampshire. The highest point of the Appalachian highlands is Mount Mitchell (2037 m).

The western slopes drain to the Ohio, Cumberland or Tennessee rivers, but drainage in the northern Appalachian highlands has been superimposed as it crosses many complex geological structures to reach the Atlantic. Further south many of the streams of the eastern slopes rise on the Blue ridge, cross the Piedmont province with rapids occurring at the eastern margin of the highlands where the rivers cross onto the younger sediments of the coastal plain.

The geological history of the Appalachian highlands is long and complex, beginning during Pre-Cambrian times. The distribution of the continental areas was quite different at that time, but it is assumed that the processes which shaped them were much the same as those active today. Examination of the geological deposits of the Proterozoic and Lower Palaeozoic in eastern USA and the north of the British Isles has led to the conclusion that there was a predecessor to the present Atlantic Ocean which has been called 'Iapetus'. In the late Pre-Cambrian to Ordovician times a spreading centre was established between the 'American' and 'British' sides of the ocean. Evidence for this lies in the type of sedimentation and different fossil assemblages present on either side of the sea. The sedimentary conditions present during the Cambrian were similar to those described as a miogeosyncline–eugeosyncline couplet. Iapetus was expanding during the Cambrian but by the end of that period of geological history deep water conditions had succeeded the shelf conditions which were responsible for extensive carbonate-rich rocks being laid down over the Pre-Cambrian basement complex. Iapetus gradually began to close during the

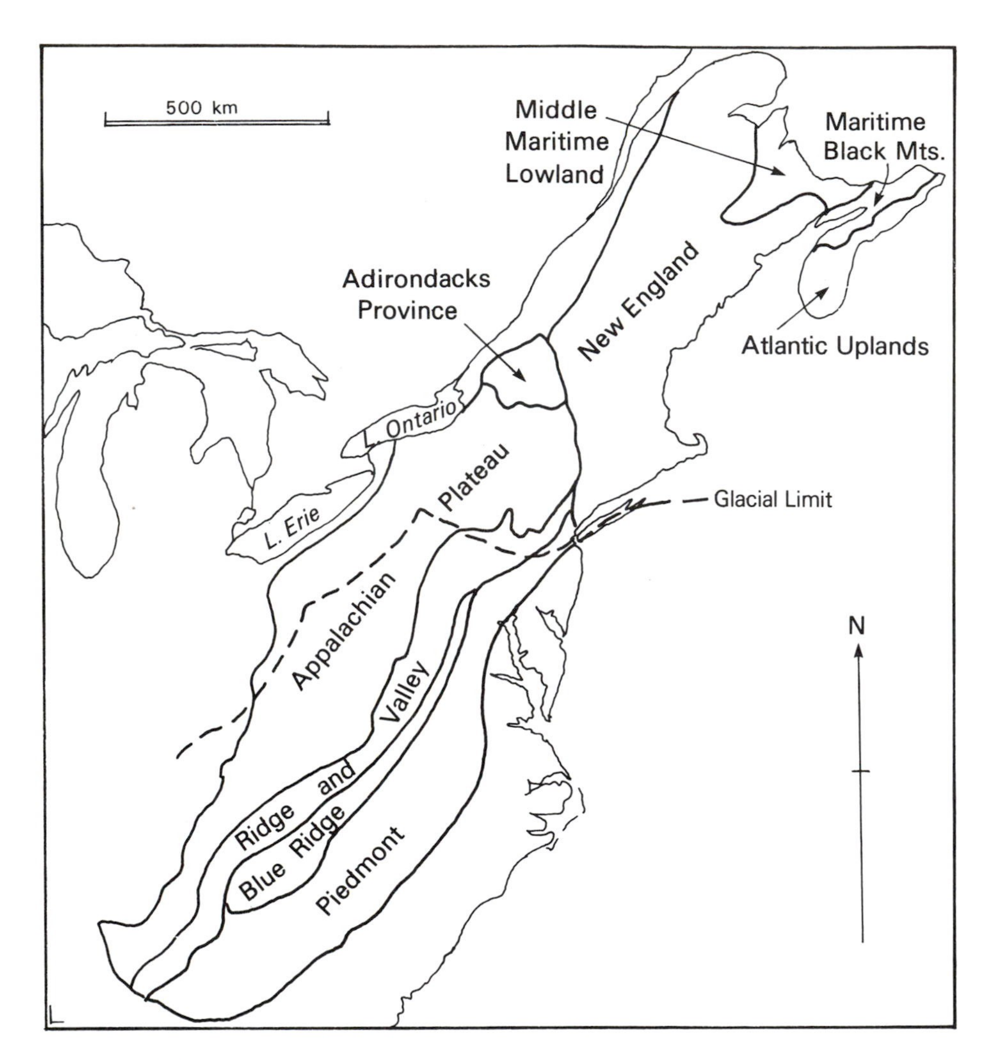


*Figure 4.8.* The Appalachian highlands.

early Ordovician and as subduction took place, volcanic activity associated with island arcs, similar to the western Pacific at the present time. The geosynclinal couplet was crushed between the advancing continental masses, granitisation occurred and the whole complex welded together in the earth movements known as Caledonian, or Taconic in America. Uplift took place as well as is shown by the presence of the greywacke sediments of the early Silurian (similar to the flysch deposits of the Alpine orogeny).

From Silurian to Jurassic times the Appalachian and European areas remained close and many common features are found in the coal measures of north-east USA and the British Isles. The Hercynian or Arcadian earth movements then affected the Devonian and Carboniferous sediments; the line of folding actually crossing that of the earlier Caledonian folding.

In the Jurassic, separation of North America from the European continent began as a spreading centre extended northwards along the north African coast, past the Iberian peninsula and west of Greenland. It was not until the beginning of the Tertiary that rifting occurred between Greenland and the British Isles accompanied by large outflows of basalt in Northern Ireland and eastern Greenland. As the Atlantic gradually reopened throughout the Tertiary a new geosynclinal couplet has developed along the east coast of the USA which formed the model for present concepts of sedimentation on trailing edges of continental plates.

Erosion occurred on the uplifted parts of the Appalachian mountains throughout the Mesozoic and the area was reduced to a peneplain. Uplifted again, this peneplain now forms the crests of the present mountains and is known as the Schooley peneplain. The present-day relief of the Appalachians results from further erosion below this datum line. Below the Schooley peneplain a second erosion surface has been cut, this is known as the Harrisburg surface, taking its name from the town in southern Pennsylvania.

Glaciation affected the northern parts of the Appalachian mountains so the regolith and soils are thin, but in the south weathering processes have produced a deeper regolith with more rounded landforms.

The provinces identified in the Appalachian mountains are listed below:

The Appalachian plateau
The ridge and valley province
The Blue ridge
The Piedmont province
The Adirondack mountains
The New England province
Newfoundland
The Ozark plateau
The Ouachita province

### *The Appalachian plateau*

This province extends from Lake Ontario southwards through Pennsylvania, West Virginia, Tennessee to Alabama. It consists of plateau landforms between 400 m and 1000 m above sea level which are drained and slightly dissected by streams flowing to the Ohio and Mississippi rivers. The rocks which form this plateau are mainly Upper Palaeozoic, Devonian and Carboniferous (Mississippian and Pennsylvanian) in age and include conglomerates, sandstones, shales and coals. The province boundaries are marked by outward-facing escarpments with a relief of 300 m on the western edge and 1000 m on the eastern side where the escarpment known as the Allegheny front overlooks the ridge and valley province. This escarpment may be traced from west of the Hudson river on the eastern side of the Catskill mountains with an elevation of 1000 m, further south it remains a significant escarpment but only 150–300 m high. The south-west boundary of this province occurs where the Palaeozoic rocks pass beneath the younger sediments of the Mississippi valley alluvium. The plateau nature of the eastern margin of this

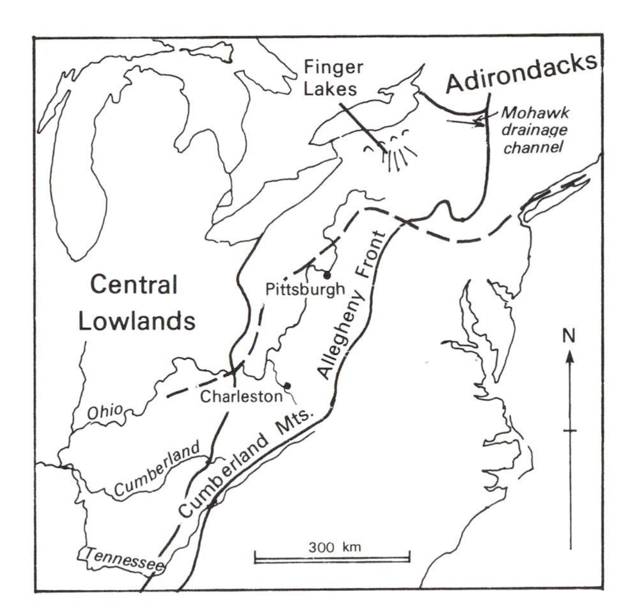


*Figure 4.9.* The Appalachian plateau.

province has been destroyed by erosion in the north, forming the Allegheny mountains and in the south the Cumberland mountains (Figure 4.9).

The northern section of this province has also been affected by glaciation where drifts of Kansan, Illinoian and Wisconsin stages have been identified. Immediately south of Lake Ontario, several valleys have been deepened by meltwater and glacial scouring; these are known as the Finger Lakes of which Lakes Seneca and Cayuga have basins which extend to below sea level. The boulder clay which is plastered over the area has been moulded into drumlins near Lake Ontario. The advance of the ice also disrupted the preglacial Teays river system and established the Ohio river in its present course. As the ice sheet melted, landforms were modified by glacial outwash and proglacial lakes. The northern boundary of the province lies along a major drainage channel from the Great Lakes which followed the Mohawk valley to join the Hudson river during the Iroqois stage about 12 000 BP.

### *The ridge and valley province*

The core of the Appalachian mountains is formed by the tightly folded lower Palaeozoic rocks of the ridge and valley province which extends 4800 km from New York State to Alabama. In the north it is 40 km wide, increasing to 160 km in the central section and narrowing again in the south to 80 km across. A complete

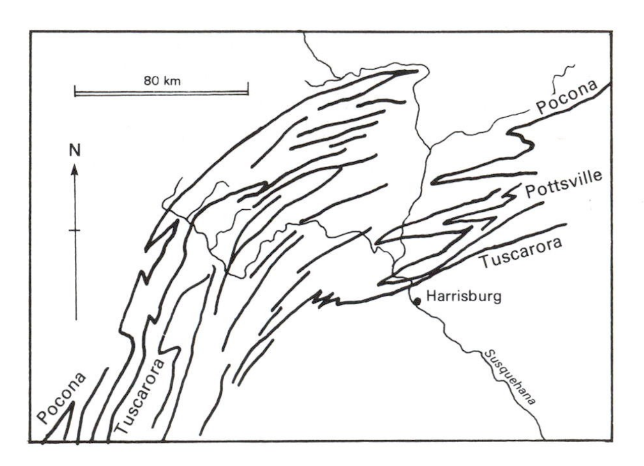


*Figure 4.10.* Pattern of sandstone ridges in the ridge and valley province.

succession of rocks from the Cambrian to the Permian formations is preserved in 13 000 m of sediments. After deposition, these rocks have been folded and overthrust to give anticlinal and synclinal structures which produce the striking ridge and valley topography. Silurian, Pennsylvanian and Missippian sandstones are the ridge-formers. On the ground these forested ridges give a uniform skyline interspersed with lowlands, but the pattern of pitching anticlines and synclines can most readily be seen from the air or on maps (Figure 4.10).

The uniformity of the summit levels led to the identification of the Schooley peneplain into which the Harrisburg erosion surface has been cut. Examination of the valley floors indicates a complex history of erosion as the presence of gravel terraces 15–20 m above the level of the flood plain of the Shenandoah river shows.

The drainage of the province was superimposed upon the folds, so there are many water gaps in the ridges. The alternating hard and soft rock strata allowed river capture to take place leaving many wind gaps and the area is noted for its trellis pattern of drainage developed on the lowlands between the resistant ridges.

Karstic features occur in the folded Ordovician limestones of the ridge and valley province. Cavern development has been related to terrace deposition in the valleys downstream and features such as sinkholes, a 30 m span natural bridge and a 300 m natural tunnel are geomorphological curiosities in the State of Virginia.

The eastern part of the ridge and valley province has fewer resistant sandstones and therefore fewer ridges. It is a lowland, extending 1900 km from the St Lawrence to Alabama which varies in width from 3 km in Virginia to 80 km at its widest.

### *The Blue ridge*

The Blue ridge includes some of the highest parts of the Appalachian mountains and extends 900 km from Pennsylvania to Georgia. There are several summits of over 2000 m south of the Roanoke river but the northern section of the Blue ridge is about 1300 m above sea level. In Pennsylvania it is a single ridge, this divides at the Maryland state border to form two ridges and south of the Roanoke river it becomes a complex mountain area 130 km wide culminating in an east-facing scarp of up to 800 m in height.

Lower Palaeozoic strata with varying degrees of metamorphism comprise the rocks of the Blue ridge. Unlike the Piedmont province to the east, there are no igneous intrusions. Thrusting has taken place westwards for 56 km in the Holston Mountain and Great Smokey Mountain thrusts, but elsewhere the younger strata dip away westwards from the Blue ridge.

It has been suggested that the accordant summit level of the Blue ridge could be an extension of the sub-Cretaceous fall line surface, but the evidence correlating these surfaces is not strong. A possible correlation with the Schooley peneplain of the Allegheny plateau has been suggested also. Even the Harrisburg peneplain, so well developed in the Piedmont province is poorly developed in the Blue ridge and difficult to trace as many of the rivers of the southern Blue ridge flow westwards and have had a different geomorphological history from those which flow eastwards. Later episodes of erosion have not penetrated into the more remote mountain areas.

North of the Roanoke river, the Appalachians are crossed by numerous rivers with no regard to the complex geological structures beneath. The James, Shenandoah and Potomac rivers cut across the Blue ridge in water gaps. Stream piracy by the Shenandoah south of Harpers Ferry where the Shenandoah is confluent with the Potomac has left three prominent wind gaps, known as Snicker's, Ashby and the Manassas gaps. The Ashby gap is about 30 m lower than Snicker's and Manassas is 30 m lower than Ashby, so it is argued that in the period between each river capture, the Shenandoah had lowered its bed.

### *The Piedmont province*

From the Hudson river at New York the Piedmont province extends 1600 km to central Alabama with a width which increases from 16 km in New Jersey to 200 km in South Carolina and Georgia. The eastern margin of the province is the fall line, marked by rapids or waterfalls on the rivers flowing to the Atlantic Ocean. The western margin is the foothills of the Blue ridge.

The geology of the Piedmont province is complex, consisting of the crumpled, metamorphosed sediments of Palaeozoic age. Granitic intrusions occur and faulting has preserved grabens containing Triassic rocks. Associated with the Triassic rocks are igneous intrusions, the most famous of which forms the Pallisades, a sill which crops out along the western bank of the Hudson river. This basaltic intrusion forms a cliff 80 km long and up to 300 m in height. When emplaced, it cooled sufficiently slowly for olivine to crystallise first and sink to the bottom, leaving plagioclase and pyroxene in the lower part and mostly plagioclase feldspar in the upper part of the sill. Contact metamorphism of the Triassic rocks below and above occurred. Other basaltic intrusions occur in New Jersey and southern Pennsylvania.

Many of the streams which drain the northern Piedmont province arise in the Appalachian plateau on the western side of the Appalachian mountains. These superimposed streams and the east coast streams which rise on the Blue ridge flow in roughly parallel courses across the Piedmont province to become extended consequents on the coastal plain sediments.

The relief of the Piedmont province is dominated by the Harrisburg surface which has a few residual monadnocks rising above it. This surface cuts across rocks of varying lithology and age to give a relief of between 60 m and 100 m in New Jersey rising to 600 m in Georgia. Some authorities identify an upland piedmont plateau in the southern parts of the province associated with granite domes which rise above the generally undulating erosion surface. Dissection of the landscape increases near the fall line along the eastern border of this province. Contemporary erosion is gradually working upstream from the coastal plain boundary cutting into the older rocks and producing the falls and rapids.

The Piedmont province may have been subjected to sub-aerial erosion during the Jurassic but in the Cretaceous a transgression of the sea occurred. Sediments were deposited which have been eroded back as far as the fall line to reveal the sub-Cretaceous surface. Erosion of Triassic sediments in New Jersey has provided areas of lowland which extend south-west from New York into New Jersey and southern Pennsylvania. The northern part of this province was affected by glaciation which left a mantle of drift deposits.

### *The Adirondack mountains*

Bounded on the east by the Hudson river and on the south by the Mohawk valley, the Adirondack mountains lie in the northern part of the State of New York. This is an anticlinal mountain area with a dome structure which rises to 1700 m above sea level and consists of a core of Pre-Cambrian rocks surrounded by Cambrian and Ordovician strata which form inward-facing escarpments. The western part of the Adirondacks is plateau-like at 300 m above sea level, but the eastern part is more dissected and mountainous (Mount Marcy 1700 m). Erosion surfaces have been identified and tentatively correlated with the Schooley and Harrisburg surfaces further south. Two terrace levels occur in the river valleys.

Drainage is roughly radial, but some integration has occurred as a result of river capture as streams have extended their catchments along the weaker rocks of the strike vales. It has been observed that the pattern of faulting in the Lower Palaeozoic rocks has influenced the stream pattern. Although subject to complete glaciation during the Pleistocene, the Adirondacks also have many late glacial corries on the higher peaks. As a result of glaciation, numerous glacial lakes, eskers and lacustrine deltas occur along the valleys.

### *The New England province*

The north-western boundary of the New England province follows the line of the St Lawrence lowlands where the Caledonian age fold mountains abut against the Canadian shield. These fold mountains continue southwards in the States of Vermont, New Hampshire and northern Maine and the Gaspé peninsula. (The continuation of this line of folding occurs in Newfoundland, the east coast of Greenland and western Europe.) The Caledonian folding was accompanied by westwards thrusting along the line of the Hudson valley and the southern margin of the St Lawrence lowlands. Uplift produced the Taconic, White and Green moun-

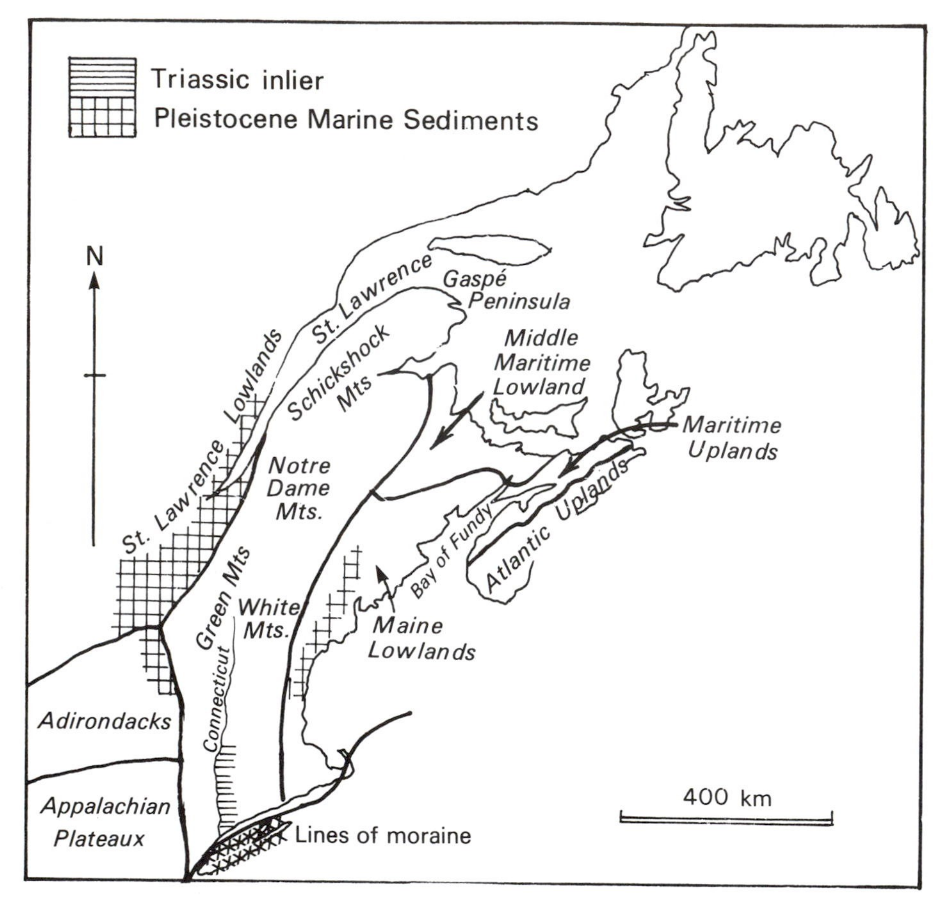


*Figure 4.11.* Geomorphological provinces of New England.

tains of New England and the Shickshock mountains of the Gaspé peninsula (Figure 4.11).

Erosion surfaces have been identified on these mountains; eleven different surfaces attributed to dates from Cretaceous to Pleistocene have been revealed by the projected profile method. Controversy exists about their origin, whether sub-aerial as a peneplain, pediment, or as a marine planation surface. The possibility of extending the Schooley and Harrisburg peneplain surface to the New England area has also been argued.

The rivers of New England include the Hudson and Connecticut, both of which cross the complex structure of the Appalachians assisted partly by the presence of grabens of Triassic rocks. However, it is difficult to conceive of their origin other than by superimposition at a time when the structure of the Lower Palaeozoic rocks could not influence their courses. In the Gaspé peninsula there is evidence of the original consequent drainage from the Canadian shield, passing at right angles to the St Lawrence and through the Shickshock mountains to the Atlantic. The presence of several wind gaps associated with river capture by the St Francis river of eastern Quebec indicates adjustments by rivers at a later stage in the evolution of the landscape.

Glaciation has further affected the geomorphology of the river valleys by plugging them with till, eskers or moraines or gouging them deeply to form lake basins. Fluvioglacial sands have been redistributed and formed into dunes south of Quebec. As the continental ice sheet melted, proglacial lakes formed in the major valleys, some of which acted as overspill channels. Lake Champlain is a small remnant of proglacial Lake Vermont which occupied the northern part of the Hudson valley when ice blocked its northern outlet. Isostatic readjustments subsequently altered the level of the land to retain the present lake.

The Atlantic uplands section of the New England province includes the Nova Scotia peninsula, the geology of which comprises metamorphic rocks and granite. These uplands have an elevation of 190–220 m with rocky hills which rise 30 m above the plateau-like surface. It is suggested that this is probably an exhumed sub-Triassic surface. Thick till occurs which in some places has been moulded at drumlins.

The Maritime uplands consist of a series of structurally controlled horsts and graben features including the Caledonian hills, Cobequid hills, the Truro plateau and the Cape Breton uplands. The Caledonian and Cobequid hills have Pre-Cambrian metamorphic and granitic rocks with sharply defined fault scarps. An upland surface occurs at 300–370 m and a lower surface at 245 m into which the rivers are deeply incised. The Cape Breton uplands have an upland surface at 475 m developed over granitic rocks. With high cliffed coasts this is reputed to be one of the most scenically beautiful areas of Canada.

The Maritime lowlands of New Brunswick are developed over Triassic rocks with igneous intrusions. These lowlands are mostly drift-covered and sea level changes have left a shore line 25–40 m higher than the present sea level.

The Bay of Fundy is noted for its exceptionally high tidal range, more than 14 m; in the upper reaches of the Bay are salt marshes, some of which were reclaimed by the early French settlers.

The Maine lowlands section is separated from the New England mountains by a break in elevation at about 150 m. Below this level the Carboniferous sedimentary rocks have been truncated to give a lowland, attributed either to Miocene plantations or to an extension of the exhumed Piedmont peninsula. Submergence of the area took place in the late Pleistocene leaving a grey silty clay dated 11 800 BP with a maximum thickness of 30 m. The coastal geomorphology is mostly structurally controlled with the effects of the post-glacial sea level rise apparent in the many embayments.

### *Newfoundland*

The north-east–south-west orientation of many physical features of this island result from the influence of the Caledonian structure. In the west, Long Range Mountain forms a dominant physiographic feature, it is formed of Pre-Cambrian metamorphic rocks with a tilted plateau surface at 600–800 m and steep stepped slope to the Gulf of St Lawrence. A limited area of limestone with karstic features occurs in the extreme north of the island.

The central part of Newfoundland is developed over slates and greywackes with granite intrusions. Surfaces on these rocks attain heights of 600 m in the west and like many upland surfaces they have been interpreted in different ways. Some people think of them as peneplains, tilted eastwards, sloping from 600 m in the west to 215–245 m in the east with a lower surface similarly inclined from 150–300 m to 105–120 m. South-east Newfoundland is underlain by metamorphic Pre-Cambrian sediments and intrusive rocks. The original folded structures are reflected in the physiography with anticlinal headlands and synclinal bays.

Newfoundland was completely covered by ice during the Pleistocene, probably by an independent ice cap, although there is some evidence that the Lawrentian ice sheet impinged on the northern peninsula. Elsewhere on the island, valley glaciers cut deep valleys in the western highlands and corrie glaciers occupied hollows in the plateau edges. Post-glacial eustatic rise in sea level has inundated the north coast at Hare Bay, but upward readjustments are still taking place.

### *The Ozark plateau*

The upland area south of St Louis and west of the Mississippi in the States of Missouri and western Arkansas is known as the Ozark plateau. It has many similarities with the plateaux which lie to the west of the Appalachian mountains. As in the Lexington–Nashville area, the Lower Palaeozoic rocks have been folded into a broad dome, the crest of which is formed by the Salem plateau which surrounds a limited outcrop (260 sq km) of the basement complex in the St Francois mountains. A variety of igneous and metamorphic rocks crop out, but one of the more interesting features is the exhumation of granite hills from beneath the overlying Lower Palaeozoic strata. The Salem plateau is formed by limestones and dolomites which overlie the Pre-Cambrian basement complex; its southern side is deeply dissected and has many prolific springs. Cherty limestones from the Mississippian (Carboniferous) underlie the Springfield plateau on the western flank of this province.

The southern margin of the Ozark plateau country is formed by the Boston mountains, a north-facing escarpment of Pennsylvanian rocks, which has been faulted and eroded into rugged topography. Within these uplands, three erosional surfaces have been identified, corresponding to the summit level of the Boston mountains, the Ozark surface on the Salem plateau (which extends northwards to pass beneath the glacial drifts in southern Illinois), and the Springfield surface. It is noted that many valleys have a small incision cut in the bottom of a much larger valley.

Despite the extensive outcrop of limestone, the Ozark plateau is not noted for its karstic features. Although there are many springs and caverns in southern Missouri; sinkholes are few in number. This may be because most of the surface features are choked with chert and clay debris from the weathering of the limestones.

### *The Ouachita province*

The Arkansas river is the northern boundary of the Ouachita province which continues the upland areas of the Ozark plateau southwards. Again the parallel is drawn between the features of the Appalachian mountains and these upland areas. The thrust-faulted, folded Ouachita mountains occupy a similar position to the ridge and valley province, although these mountains are not so high as the Appalachians, only reaching 800 m on the crests and 100–200 m in the valleys.

It has been suggested that the summit level of the Ouachita mountains can be carried southwards, decreasing in elevation until it passes beneath the sediments of the Coastal Plain. If this is correct the summit surface may be dated as an early Cretaceous peneplain. A second, lower surface in the Ouachita mountains has Eocene beds resting upon it which was probably cut early in the Tertiary and the valley floor of the Ouachita river has been labelled the Hot Springs surface.

## The central lowlands

One of the most extensive major divisions of North America is the central lowland plain, situated between the Appalachian mountains in the east and the high plains in the west. This large lowland, approximately 800 000 km$^2$ ranges in elevation from 150 m along the Mississippi to 300 m west of the Allegheny plateau and to about 600 m at the 'break of the plains' in the west (Figure 4.12).

In broad terms, the underlying geology is relatively simple, but in detail it is very complex. The surface of the basement complex with its granitised gneisses dips gently southwards beneath Lower Palaeozoic sedimentary strata. A slight warping into basins and ridges has resulted in escarpments in the overlying Ordovician limestones. Along the western margin of the central lowlands, Cretaceous limestones and shales crop out east of a Tertiary escarpment which forms the edge of the high plains.

Glaciation has played a major part in reshaping the geomorphological features of the central lowland plain. Ice from the Lawrentian ice sheet pushed almost to the line of the Ohio river in the east, and to northern Missouri and Kansas in the west. South of the glaciated areas, two sections of lowland were not glaciated; these are the Osage plains of southern Kansas and Oklahoma and the low plateau of Kentucky and Tennessee east of the Mississippi. An area in south-west Wisconsin also escaped glaciation and is known as the driftless area.

The present-day drainage of the central lowlands is largely dictated by the direction water flowed from the Pleistocene ice sheet as it wasted away. Northern parts drain either to the Arctic Ocean or across the Canadian shield to Hudson Bay and the St Lawrence. In the southern areas, streams followed valleys formed by outwash streams to join the Ohio or Missouri.

The central lowlands major physiographic division may be sub-divided into the following provinces:

The Manitoba lowlands
Area of younger drifts
Area of older drifts
Unglaciated areas

### *The Manitoba lowlands*

The lowland which surrounds Lakes Winnipeg and Manitoba lies across the boundary of the exposed Canadian shield with the Lower Palaeozoic limestones. Both the shield and the lower Palaeozoic rocks are drift covered, so landforms are almost entirely formed from Pleistocene and Recent materials. The lowlands of the Red River valley was the site of glacial Lake Agassiz which, at its maximum extent, covered 500 000 km$^2$. A major beach level was developed at 313 m above datum near the international border between Canada and the USA, but further north isostatic readjustment has raised its level by 30 m. Deltas were constructed where streams entered this lake, the largest of which, 55 km long and 50 m thick, was constructed by the Assiniboine

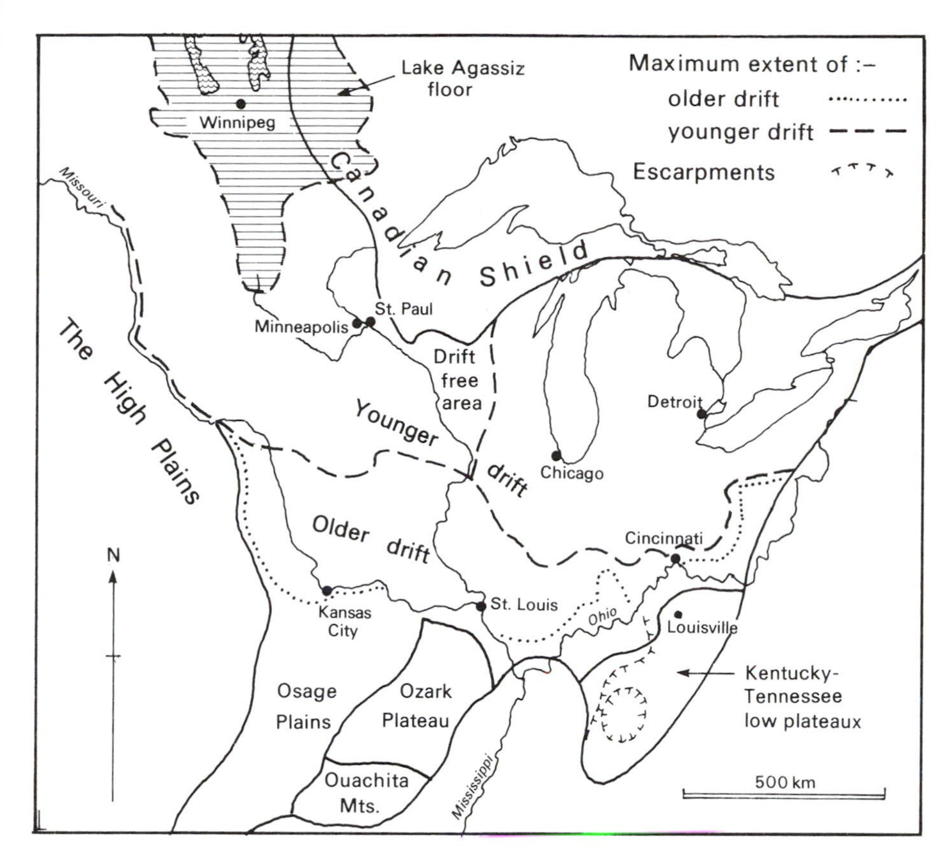


*Figure 4.12.* The central lowlands.

river near Brandon. The river has since incised into this delta, the surface sands of which have been blown into dunes by aeolian action. About 20 km south of the Assiniboine river, a meltwater channel was cut across the Pembina escarpment of Cretaceous shales to a depth of 150 m below the upland surface.

### *Area of younger drifts*

The younger drift landscapes occur on the glacial drifts of the Wisconsin stage of the Pleistocene which lie south of the Superior uplands and the Great Lakes. In this zone, the landforms are relatively 'fresh' with many lakes and a poorly organised drainage system. There are recessional moraines and beach levels associated with former water levels of the Great Lakes. Protruding from beneath the drifts are the glacially scoured cuestas developed from the Palaeozoic rocks overlying the basement complex.

South of this area of chaotic drainage is a second section of this province, the till plains. The topography in the till plain section is completely dominated by glacial deposition. Each of the glacial stages has helped to create these till plains which have a very flat topography, completely obliterating the previous preglacial landscape (Figure 4.13). Evidence of this former landscape has been traced by boreholes across several of the mid-west states. Before glaciation a river, called the Teays, rose in eastern Ohio and flowed westwards across Illinois, Indiana and Iowa to eventually join the Missouri. The Wabash river has revealed part of this former valley where it crosses the course of the former Teays river. Drainage in the till plain section is better integrated than in the area to the north, the major rivers following former outwash channels.

### *Area of older drifts*

At its greatest extent, glacial cover in the USA reached as far south as the line of the Ohio and Missouri rivers. The most extensive ice cover occurred in earlier glaciations and so considerable areas of older drift topography

*Figure 4.13.* Depositional drift landscape in Michigan.

have been exposed to the processes of erosion for a much longer time. Accordingly, landform features are more rounded and less angular than in the newer drift areas.

Glacial deposits from the earlier stages of glaciation occur as far south as Topeka in Kansas, southern Illinois, Indiana and possibly the extreme north of Kentucky. The till on these older drift areas is strongly dissected to reveal the sedimentary rocks beneath and the drainage pattern is in the process of adjusting to the bedrock structures. The glacial spillway channels occupied by rivers in the till plain section continue across this older drift landcape but additionally there are extensive deposits of loess on the interfluves.

*Unglaciated areas*

Three areas of the central lowlands remained unglaciated at the end of the Pleistocene, these are the Osage plains, the low plateau of Kentucky and a small area of south-west Wisconsin.

The Osage lowlands lie south of Kansas City, west of the Ouachita uplands and east of the great plains. The rocks of the area are mainly Carboniferous in age with some Tertiary outliers of the great plains rocks. Essentially the area has low escarpments of 200 m rising above the lowlands. Low hills also occur in this section, the Arbuckle and Wichita hills. A 300 m erosion surface has been identified on the Arbuckle hills, but the adjacent Wichita hills do not seem to have comparable evidence of planation.

The Kentucky low plateau lies between the slightly elevated Appalachian plateau of eastern Tennessee and Kentucky, the Mississippi lowlands and the lowland area covered by glacial deposits to the north in Ohio. The major structure of the area is a broad anticline, the long axis of which runs parallel to the Appalachians from northern Alabama to Ohio. Minor folding on this arching gives the Lower Palaeolzoic sediments a basin south-east of Nashville, surrounded by dissected plateaux. In the northern part of the province, the Muldraugh's Hill, Dripping Springs and Pottsville escarpments overlook the Lexington plain.

The Kentucky plateau is drained by the Licking, Kentucky, Green, Cumberland and Tennessee rivers which originate on the Appalachian highlands and flow across the province to the Mississippi. The presence of the limestones in the Lower Palaeozoic sequence permits the development of karstic landforms, including many caverns. The Dripping Springs–Chester escarpment has many thousands of sinkholes but is lacking in surface streams.

An erosion surface has been identified in the Lexington area at about 300 m where the landscape is only slightly dissected. However, near the major streams which are entenched 150 m below the surface there is considerable dissection. A lower erosion surface is invoked to account for the floor of the Nashville basin which it is thought was cut in early Pleistocene times.

The driftless area of south-west Wisconsin is interesting because this unglaciated enclave allows comparison of glacial and periglacial landscapes side by side. The area was protected from the advancing Lawrentian ice by low hills which deflected it. In the driftless area, landforms show signs of periglacial activity and mass wasting rather than ice scour or deposition.

## The great plains

The great plains extend from San Antonio in Texas to Central Alberta in Canada, forming an elevated area between 500 m and 1500 m above sea level. Their eastern margin is the 'break of the plains' escarpment and the western margin the foothills of the Rocky mountains (Figure 4.14).

The great plains are developed mainly over Tertiary deposits, derived from the erosion of the Rockies lying immediately to the west. A particular geological feature of the high plains is the Ogallala formation, a deposit of alluvial sands, gravels and silts, capped by an irregular but massive calcrete. On the surface of this formation, Pleistocene cover sands, terrace gravels and, locally, volcanic ash deposits occur. Erosion has removed the Ogallala formation from a wider area than it at present occupies and in the vicinity of Denver numerous erosion surfaces, pediments, have been identified. In

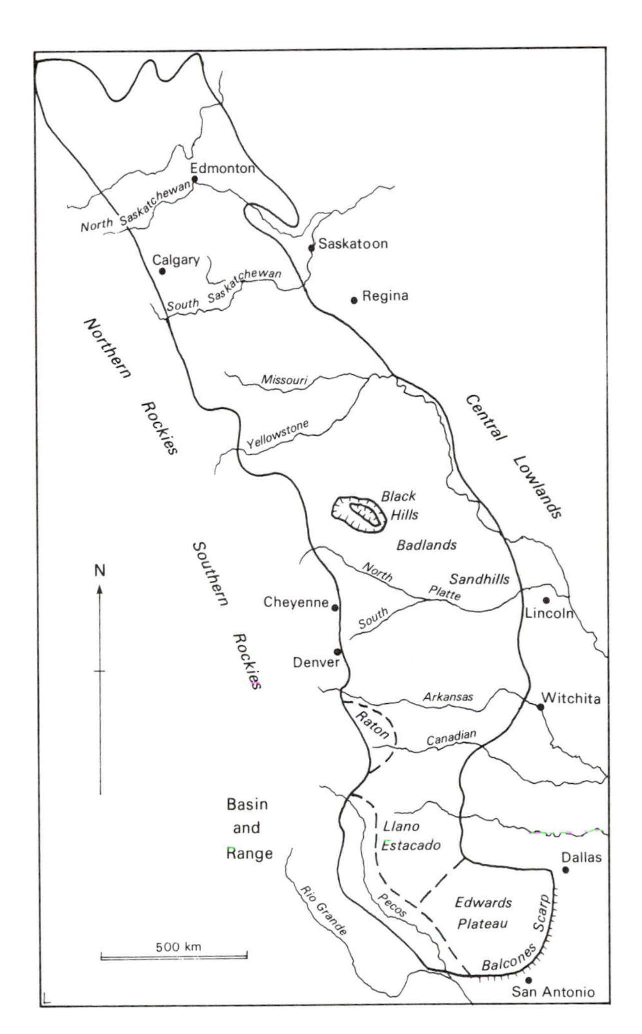


*Figure 4.14.* The great plains.

Nebraska, particularly, the surface sands have been worked into dunes during the Pleistocene and the finer constituents winnowed away and deposited further east as loess.

Past aeolian action can be identified in almost all parts of the great plains, reflecting a period of aridity during the Pleistocene when rainfall was insufficient for vegetation to form a stabilising cover. At several places on the great plains a strong north-west–south-east alignment of drainage occurs which is attributed to stream development along interdunal depressions. In Texas there are many enclosed depressions, some of which were formed by deflation of material by the wind, but others result from solution of the calcareous Ogallala formation. In the Canadian section badlands occur alongside the Red Deer river north-east of Calgary, and erosion has revealed the Cretaceous beds beneath the Tertiary sands and gravels.

Tertiary igneous activity occurred on the borders of the great plains on the Colorado–New Mexico state border in the Raton district where there has also been strong uplift. As a result, landforms in this area are strongly dissected by erosion into a series of mesas.

The southernmost section of the great plains is the Edwards plateau, which terminates eastwards in the Balcones escarpment, formed by Cretaceous sedimentary rocks.

Many large streams cross the great plains from the Rocky mountains to join the Missouri–Mississippi drainage system. These streams are incised into the plains and river terraces occur along their valleys at three or four levels above the present alluvium. The Pecos river in Texas has eroded its valley along the front of the easternmost mountains of the basin and range province and isolated the Llano Estacado as a plateau, 900 m above sea level. It has many areas of internal drainage with salinas lying about 30 m below the general surface of the plain. The Llano Estacado surface is composed of marls, loams and sands which contrast with the more southerly Edwards plateau, where the superficial materials have been stripped leaving the Cretaceous limestone cropping out at the surface. The edge of the Edwards plateau is marked by the Balcones scarp, north-west of San Antonio.

The final section of the great plains is the dome of the Black hills of Dakota. This inlier of crystalline and Palaeozoic rocks protrudes through the younger rocks of the great plains in an almost perfect structural dome. The hills, which occupy an elliptical area 200 km by 95 km and rise to 2400 m. They consist of crystalline rocks of the basement complex which crop out in the eastern half of the dome's core, but the western half is still covered by Palaeozoic limestone. Red shale has been eroded to form an encircling valley and outside it the Dakota sandstone forms an inward-facing escarpment.

## The western cordilleras

The western cordilleras of the USA and Canada are a large highland area lying parallel to the Pacific coast brought about by earth movements as the North American lithospheric plate moved westwards. Unlike the Andes which appear to have displaced the subduction zone in the Pacific trenches further westwards, the North American plate seems to have overridden the eastern part of the oceanic crust and to have become involved in one of the large transform faults of the spreading centre, the San Andreas fault. The tectonics of the western cordilleras are extremely complex and can only glossed over in a broad review of world geomorphology. The cordilleran belt has an active geological history which spans the period from Pre-Cambrian to Quaternary and according to Windley contains 'a tectonic collage of more than 50 fragments' including ophiolitic material, volcanic island arc materials, basalt and gabbro possibly associated with former rifting sites and other materials swept up against the continental margin. In simple terms, the western cordilleras comprise two major mountain chains, the Rockies and the West Coast mountains, with elevated plateaux between them (Figure 4.15)

The Rocky mountains
Yukon valley and uplands
Central plateaux and ranges of British Columbia
The Columbia-Snake plateau
The basin and range province
The Colorado plateau
The west coast mountains
The coast ranges
The Great Valley of California

### *The Rocky mountains*

The Rocky mountains form the eastern range of high peaks which lies parallel to the Pacific coast of North America. In Alaska, the mountains known as the Brooks range are part of the same system and in Mexico they are known as the Sierra Madre Oriental. The Rockies occupy a zone between 100 and 400 km wide and comprise igneous Proterozoic cores with sedimentary rocks from the Lower Palaeozoic and Mesozoic. These sedimentary rocks were uplifted and folded in late Mesozoic and early Tertiary time.

The highest mountains of the Arctic occur in the Brooks range of Alaska, with peaks of over 3000 m. The core of these mountains is composed of Lower Palaeozoic sediments which have been folded to form the highest land, but Mesozoic strata crop out on their flanks. In Tertiary times the area was peneplained, but subsequent uplift in the late Tertiary accompanied by thrusting to the north, gave the basis of the present relief. Four major episodes of glaciation have been

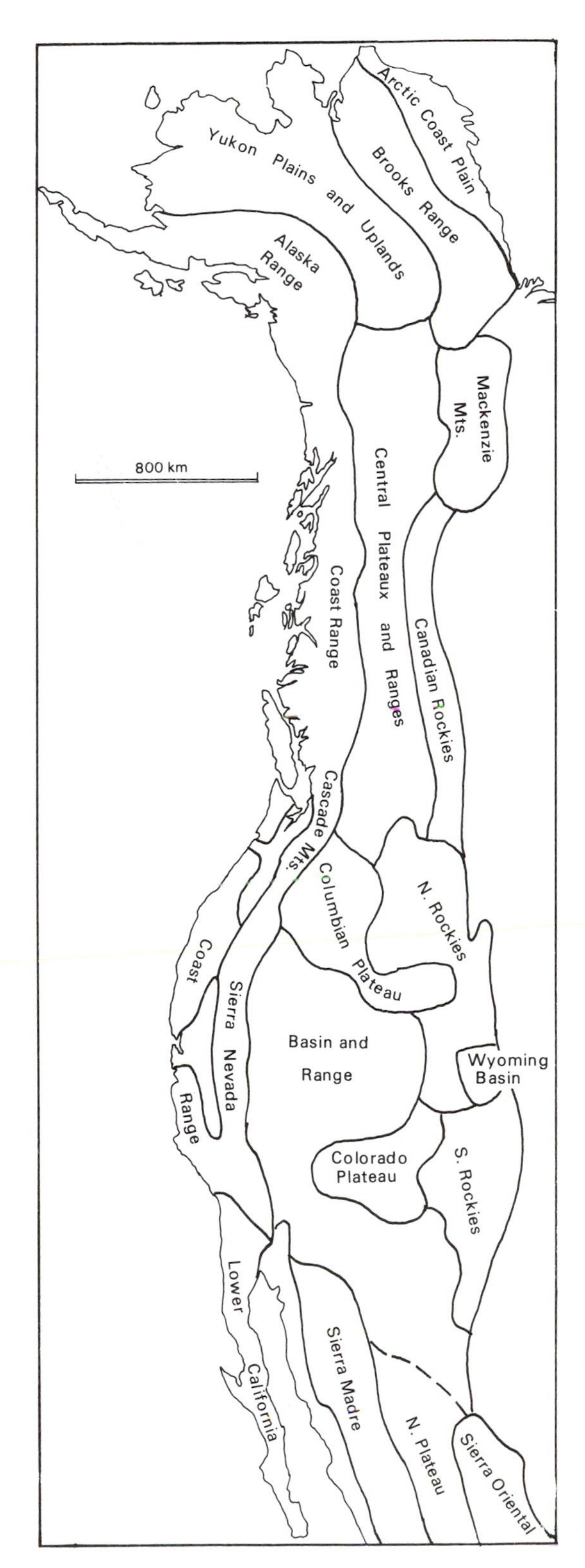


*Figure 4.15.* The western cordilleras.

identified in the Brooks range during the Pleistocene, and many small corrie glaciers remain to the present day.

The Canadian Rockies include the Mackenzie and Selwyn mountains of the Yukon, the Rocky mountains of Alberta and their foothills to the east (Figure 4.16). In the Canadian Rockies, Palaeozoic limestones are thrust eastwards over Mesozoic rocks, for example the limestones form the peaks in the Banff area and the younger rocks crop out in the valleys. The highest peak of the Canadian Rockies is Mount Robson (4000 m), and many peaks reach at least 3000 m. During the Pleistocene these mountains were glaciated. Mount Assinibione has many corries surrounding a horn peak which remains from this period of glacial erosion. The Columbian ice field lies on the continental divide at heights of between 3000–4000 m above sea level. It has an area of 325 km$^2$ and is estimated to be 100 m thick. Several valley glaciers emerge from the ice field including the Athabasca glacier which is 7.3 km long and 1.2 km wide (Figure 4.17). Ice movement in the glacier averages 70 m per year below the ice fall but only 5–10 m at the snout. Arcuate recessional moraines 3 to 6 m high mark the position of the glacier snout in 1919, 1945 and 1956. Lateral moraines 250 m above the snout indicate the extent of thinning and retreat in recent years. The western margin of the Canadian Rockies is marked by the Rocky Mountain trench. This is a straight, flat-bottomed valley 4–15 km wide with sides up to 1000 m deep. The trench has been glaciated and contains up to 1500 m of fluvioglacial deposits into which the modern streams like the Columbia have incised themselves. It probably represents the line where a new fragment has been welded onto the pre-existing continental cratons.

In the USA, the Rockies may be sub-divided into northern and southern sections. The northern Rockies include the Lewis and Bitter Root ranges of western Montana, the Big Horn and Wind River mountains of Wyoming and the Uinta and Wasatch mountains of Utah. The height of these mountains increases southwards until peaks of over 4000 m are reached in the Wind River mountains. The rocks which form the mountains are Palaeozoic sediments with intrusive granite cores, but the Wyoming basin contains Cretaceous and Tertiary deposits. The northern Rockies have a very complex imbricate faulted structure involving Palaeozoic and Mesozoic rocks, over which thrusting took place towards the east, the most spectacular example is the Lewis thrust, in which a normal

*Figure 4.16.* The Rocky mountain front, Canadian Rockies.

sequence of rocks from Pre-Cambrian to Cretaceous have been thrust eastwards for over 200 km. The surface over which the thrust is thought to have moved is known as Blackfoot surface, of early Tertiary age. Within the northern Rockies of western Montana there are many faulted basins, the structure of which is not known for certain. These Tertiary basins have an infill across which pediments have been cut or aggraded at two or three different levels, the uppermost of which has been tentatively related to a late Pliocene or early Pleistocene stage of erosion. Explanation of the drainage pattern, which is integrated from one basin to another with gorge-like sections between, is difficult.

The Wyoming and Big Horn basins are the largest of a number of elevated plains surrounded by ranges of the Rocky mountains. The Pre-Cambrian basement complex is upfaulted in the ranges such as the Uinta and Tenton mountains and Palaeozoic and Tertiary beds form hogsbacks or outward-facing escarpments around the basin edges. The rivers appear to be superimposed on these structures, possibly from Tertiary beds, since eroded away. The Devil's Gate on the Sweetwater river is an example of a narrow gorge cut in the older, harder strata. Dunes and badlands occur west of the town of Casper in the Wyoming basin.

The Yellowstone plateau which lies on the continental divide, is built of basaltic and rhyolitic lavas of Tertiary age. The site was once a topographic depression which was infilled with lava flows and covered with an ignimbrite deposit. The Tenton mountains form the southern boundary of the Yellowstone plateau. These were uplifted in the early Cretaceous Laramide orogeny, peneplained in the Pliocene and uplifted again so the peneplain now forms the summit surface. Lower surfaces were eroded during the Pleistocene. The Big Horn mountains have a similar geomorphological history. Glaciers extended down from the surrounding mountains onto the Yellowstone plateau, blocking the valley with moraine to form a predecessor of the present Yellowstone Lake. This glacial lake, 50 m

*Figure 4.17.* The Athabasca glacier, British Columbia.

higher than the present lake, drained southwards to the Snake river, but in post-glacial time the Yellowstone river has extended its headwaters to partially drain the lake. However, the outstanding features of the Yellowstone National Park are connected with residual geothermal activity. Over four thousand hot springs occur and geysers spout intermittently. Brightly coloured mineral springs and mud volcanoes add to the popular interest.

The southern Rocky mountains lie mainly within the States of Colorado and New Mexico, extending into southern Wyoming as far as the North Platte river. The southern Rockies are higher than the northern Rockies with many peaks over 4300 m; the chief ranges are the Front range, San Juan range and the Sangre de Cirsto range. Between these anticlinal ranges, broad downfolds underly the Park basins.

The Front range, immediately west of Denver, consists of a series of block-faulted Pre-Cambrian gneisses intruded with granites with Palaeozoic and Mesozoic rocks dipping away on the flanks. Pike's Peak (4301 m) is the top of a granite batholith. The geological and morphological history of the Front range illustrates the length and complexity of landform features in this province.

The southern Rockies were folded and uplifted in the early Cretaceous Larimide orogeny and overthrusting of 8 km to the east has been observed. The Eocene, Oligocene and Miocene were mainly characterised by erosion, resulting in the Flattop peneplain which now forms the summit of the upwarped mountains. In the Pliocene, further erosion produced the Rocky Mountain peneplain with a surface of moderate relief but renewed uplift at the end of the Pliocene led to removal of much of the Tertiary cover and the initiation of canyon cutting. Valley incision and terrace formation took place during the Pleistocene and on the foothills pedimentation occurred.

Other notable geomorphological features of the southern Rockies include the granite Sawatch mountains which were heavily glaciated to give well-developed alpine scenery. Moraines occur in most valleys which have been greatly deepened during the Quaternary. Six episodes of vulcanicity have been identified between the Miocene and the Pleistocene in the San Juan Tertiary volcanic district. A feature of the volcanic deposits overlying the Cretaceous shales of the San Juan mountains is their susceptibility to landslipping. The best known is the Slumgullion mud flow which blocked the valley to hold back Lake San Cristobal. On the western slopes of the Jemez mountains is the Valles caldera, a 24 km diameter crater. The volcano blew up in the early Pleistocene and following its collapse into a caldera, a ring of small volcanic cones has appeared.

### *Yukon valley and uplands*

The Yukon river drains central Alaska which is described as a rolling upland of less than 1000 m with discontinuous groups of higher mountains rising to 1800 m. Many parts of this province are mantled with loess and there are discontinuous patches of permafrost.

### *Central plateaux and ranges of British Columbia*

Between the Canadian Rockies and the coastal ranges from Alaska to the USA border there is a landscape of dissected high plateaux and mountain ranges. In the south is the Stikine plateau and further south the Fraser plateau, separated by the Skeena Mountains. The Columbia mountains include the Purcell. Selkirk and Monashee ranges which have peaks over 3000 m developed on Palaeozoic metamorphic and sedimentary rocks. This area has experienced some spectacular landslips such as on Turtle mountain in 1903 and the Hope slide of 1965, when 130 million tons of rock slid 1000 m into the bottom of the valley which it buried to a depth of 80 m.

### *The Columbia–Snake plateau*

This plateau lies between the northern Rocky mountains and the Cascade mountains. An intermontane area, it has an elevation of almost 1500 m in the south and less than 700 m in the north. The area is drained by the Snake and Columbia rivers and their tributaries.

The geology of the area is dominated by basaltic and andesitic lavas of Tertiary age which cover 390 000 km$^2$ in Washington, Oregon and Idaho. The lavas are up to 3000 m thick with individual flows of 30 m in thickness. Protruding through the lavas are Palaeozoic and Mesozoic sedimentary rocks which were part of a landscape submerged by the lava flows. A loessial cover has accumulated over the lavas to an estimated depth of 70 m, in parts the accumulation was sandy and dune areas occur. The northern section of the province is known as the Walla Walla plateau. A series of broad anticlinal ridges and synclinal valleys occurs in the western parts and a downwarp with sedimentary infill occupies the central section.

An area of great interest is the Channeled Scablands, south-west of Spokane. Meltwater from an ice-dammed lake, Lake Missoula, was suddenly released during deglaciation in the Pleistocene and the resulting flood poured across the plateau scouring away the silty cover of loess and cutting deep, broad channels in the basalts beneath (Figure 4.18). Individual valleys, for example the Grand Coulee, are 130 m deep and 8 km wide with steep walls which terminate upstream at dry waterfalls. Air photographs show anastomosing channels with giant ripple-marks.

In the Blue mountains district of the Columbia plateau, the lavas have been uplifted 1700 m above the plateau level, but the Snake river has maintained its course by cutting down as the land rose. The result is a canyon known as Hell's Canyon which is deeper than the Grand Canyon and at the bottom of which the lavas are seen to lie unconformably upon a rough granite surface.

The Snake river plains are a structural downwarp infilled with Quaternary lavas. The river is incised into the plains which descend from 2000 m in the east to 1100 m in the west near the town of Twin Falls. The cinder cones of the Craters of the Moon near Pocatello, Idaho stand above the plains in a wasteland of lava fields.

The southern section of the Columbia plateau is transitional, rising to the basin and range province. A pattern of faulting in the bedrock is developed which becomes more obvious in the Basin and Range province. The climate becomes more arid and it is an endoreic area in which Lakes Hervey (salt) and Malheur (fresh) receive the drainage. Desert conditions prevail on areas of pumice sand through which numerous volcanic cones protrude, the largest of these is the Newbery volcano (2600 m), a caldera structure, surrounded by precipitous slopes where the centre has collapsed. The south-east part of this section is com-

*Figure 4.18.* Dry Falls, Washington State. A former waterfall over basalts in the Channeled Scablands.

posed of acidic volcanic rocks which have been dissected to reveal granitic plutons.

Both Columbia and Snake rivers have had a complex history, influenced by the outpouring of lavas. The erosional history of the province is only sketchily known, although a pre-Tertiary origin for the drainage pattern has been suggested and there is evidence for one or more erosional surfaces.

### *The basin and range province*

This province lies between the Sierra Nevada of California and the mountains of Utah, Colorado and New Mexico. From north to south, it extends from the edge of the Snake river plateau to beyond the Mexican border, a distance of over 2000 km with an area of 780 000 km$^2$. Most of the basin and range province lies between 1500 and 2500 m above sea level, but the floors of the Bonneville and Lahontan basins lie between 1000 and 1500 m. Much of the area has endoreic drainage including streams like the Humboldt river of Nevada which flows into the Carson sink and the shorter streams which drain directly into the Great Salt Lake. The major exoreic stream is the Colorado, but this arises outside the province, which it enters as it passes into the Grand Canyon. The climate of the great basin is semi-arid or arid; included within its boundaries are the Mohave, Sonoran and Colorado deserts.

Sedimentation of Lower Palaeozoic rocks on the western margin of the shield was followed by Mesozoic folding accompanied by igneous activity. Block-faulting then took place during the Tertiary to give a landscape of horsts and graben; some movement continues to the present day. Detritus eroded from the upfaulted mountains has gradually infilled the basins. Even so, the floor of Death Valley lies below sea level (−86 m) but it still contains over 2700 m of alluvial infill overlying the solid strata. During the Pleistocene a

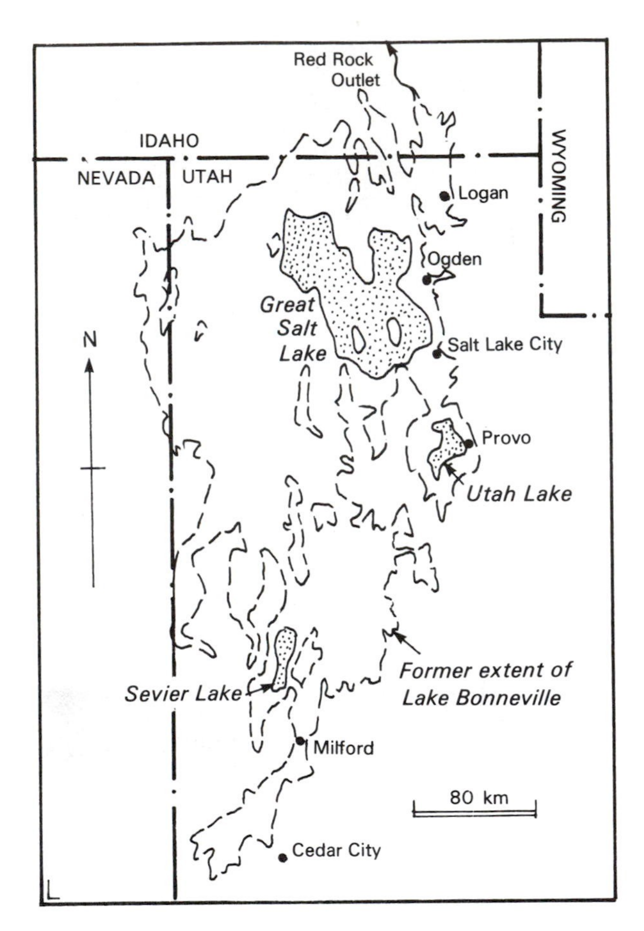


*Figure 4.19.* The Great Salt Lake and Lake Bonneville.

cooler, moister climate prevailed in this province and permitted the accumulation of water in the lower parts of the landscape, particularly in the Bonneville and Lahontan basins. Evidence of these former lakes is seen in the deltas, spits and bars associated with former strand lines. At its maximum extent Lake Bonneville was 52 000 km$^2$ and was over 300 m deep (Figures 4.19 and 4.20). Its shoreline occurs at about 1700 m and it drained northwards into the Snake river through the Red river pass, north of Salt Lake City. Tectonic warping since has altered the position of these strand lines by up to 100 m. Lower shorelines also occur at 1600 m and 1500 m above sea level; Salt Lake City stands on the 1600 m shoreline of the 'Provo' stage of Lake Bonneville. Similarly, Death Valley was occupied by a lake during the Pleistocene (Lake Manly) with a depth of 200 m. Tectonic movements have lowered the western shoreline by 7 m compared with the eastern one.

The basin and range province played an important part in the development of ideas about arid landforms. Many basins (or Bolsons) have centripetal drainage from the surrounding mountain ranges into a central playa which may periodically flood to become a saline lake. Below an abrupt mountain front there is a sharp break of slope to an upper eroded rock pediment which further downslope gives way to an agradational peripediment. Weathered debris is transported across these lower angled slopes to the basin centre by sheetwash from the occasional rain storms. Coarser debris is deposited near the mountain front, but clays and dissolved salts are carried to the centre of the playa. Individual valleys may have alluvial fans where they open onto the pediment or if several valleys are close together, their fans intersect to form a bajada.

The Californian section of the province is occupied by the Sonoran desert. The major drainage is by the Mojave river which rises on the San Bernardino mountains and proceeds to flow north-eastwards before it becomes lost in its own gravel bed. Typically, the landforms of the Sonoran desert are small mountain ranges with basins between containing playas. The Salton Sea occupies one of the deeper basins, 78 m below sea level, with its north-western shore formed by the San Andreas fault. The presence of the Sea resulted from flood water from the Colorado being diverted along an irrigation canal between 1904 and 1907. After leaving the Grand Canyon, the Colorado river cuts through four areas of uplift upon which it has been superimposed. Irrigation occurs alongside the river where water is available but elsewhere dunes and desert pavement soil occur on pedimented surfaces.

East of the Colorado, block-faulted mountains similar to the main part of the basin and range province occur, referred to as the Mexican highlands. The Rio Grande river follows a line of graben before it turns eastwards to form the international border between the USA and Mexico. Three valleyside pediments have been observed along the Rio Grande, the youngest dated 2500 BP. The Tularose basin, the lowest parts of which are 1300 m above sea level, was also occupied by a Pleistocene lake which left shorelines. The basin is floored by a playa where evaporation leaves an incrustation of gypsum. Wind has assembled the grains of

*Figure 4.20.* The Great Salt Lake, Utah.

gypsum sand into the dunes of the Whitesands National Park.

### *The Colorado plateau*

The Colorado river drains most of the area between the southern part of the Rockies and the basin and range province, most of which is included in the 40 000 km² of upland plateaux in the States of Utah, Colorado, Arizona and New Mexico.

The Pre-Cambrian platform of the Vishnu schists is revealed in bottom of the Grand Canyon, and Proterozoic sediments rest upon these Archean rocks. Cambrian, Devonian, Carboniferous and Permian sediments have accumulated to a depth of 2700 m, and the succession can be seen in the walls of the Grand Canyon. The presence of Mesozoic sediments in the synclinal Henry mountains north of the Grand Canyon indicates that sedimentation continued throughout the Triassic, Jurassic and Cretaceous but these rocks have since been eroded from parts of the plateau surface. Intrusive igneous rocks form stocks and laccoliths, for example Mounts Ellsworth and Hillers, but the greatest amount of volcanic activity occurred around the edge of the plateau after mid-Tertiary times. This is well illustrated by the San Francisco mountains north of Flagstaff where eruptions continued until the eleventh century.

Although referred to as the Colorado plateau province, the area is composed of many faulted blocks of country upon which plateaux are developed at different elevations. North of the Grand Canyon the Shivwits, Uinkaret, Kanab and Kaibab plateaux, developed on the Kaibab limestone, increase in height from west to east with the Kaibab plateau attaining 3000 m. North of these elevated plateau on the rim of the canyon, even higher plateaux occur on Mesozoic rocks. The Triassic Pink cliffs and Chocolate cliffs, Vermillion and White

*Figure 4.21.* The Grand Canyon, Arizona.

cliffs of the Jurassic rocks and the Grey cliffs of the Cretaceous are escarpments formed by strata dipping northwards. Each formation increases the elevation until the Paunsagunt and Markagunt plateaux are reached with elevations of 3700 m above sea level. Erosion of these strata has revealed the dissected topography seen in Bryce Canyon and Zion National Parks.

Glaciation has affected the higher peaks of the basin and range province. The 3700 m Grand Mesa in western Colorado has over 400 glacially-scoured rock basins with end moraines. Some glaciers descended to 2200 m above sea level during the Pleistocene.

The Uinta basin in the north and the Navajo basin in the east of the province are both synclinal structures in which Tertiary beds have been preserved. Erosion of the edges of plateaux formed on these rocks has revealed the brightly coloured rocks of the Painted desert where the silicified remains of trees can be seen in the Petrified Forest National Park. In this area the extension of pediments by scarp retreat is well exemplified.

The Colorado river rises in the southern Rocky mountains in the State of Colorado and flows southwest across the plateau. At Moab the river enters the uplifted plateau country where spectacular views of the canyons may be seen in Canyonlands National Park. A dam across the Colorado impounds water in Lake Powell, flooding Glen Canyon.

The Grand Canyon is most easily approached from the southern side across the Coconino plateau at an elevation of 2300–2500 m. In the canyon, erosion has given the almost horizontal strata a stepped appearance with vertical walls formed by the Tapeats sandstone (Cambrian), the Redwall limestone (Mississippian), the Supai sandstone (early Permian) and the Kaibab limestone (middle Permian). Extensive benches have developed on two of these cliff-forming strata, the Tonto platform on the Tapeats sandstone and the Esplanade on the Supai sandstone. The depth of the canyon varies from 1175 to 2000 m and its width is from 8 to 24 km across (Figures 4.21 and 4.22).

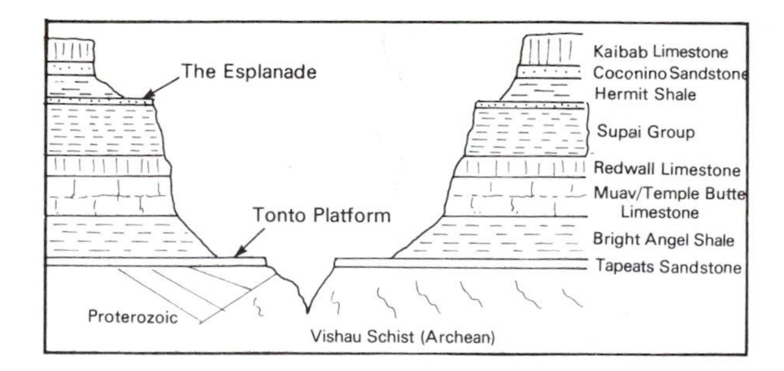


*Figure 4.22.* Section across the Grand Canyon.

Two other geomorphological features of this plateau are worthy of comment. North-west of Moab, as the Colorado enters the canyons, Cretaceous sandstones have been eroded to form a collection of natural arches (Figure 4.23). Fifty kilometres east of Flagstaff, Arizona, a meteorite landed about 22 000 BP producing a crater 200 m across and 200 m deep. The impact displaced over 300 million tons of rock and buckled the Triassic rocks around the edge of the crater. There is evidence of nickel-iron buried beneath the south-western rim which suggests the meteorite approached the earth's surface at an angle from the north-east (Figure 4.24).

*Figure 4.23.* Natural arches. Arches National Park, Utah.

### *The west coast mountains*

A double mountain range forms the western part of the western cordilleras of North America. The Sierra Nevada form the western limit of the elevated interior plateaux and the lower coast ranges lie immediately adjacent to the Pacific Ocean. Between these two mountain ranges, the Great Valley of California and the Willamette valley occupy the same structural depression. It passes below sea level on Puget Sound and continues between Vancouver Island and the mainland in the Strait of Georgia.

From the Aleutian peninsula a line of mountains can be traced through the Alaska ranges into the St Elias mountains. Uplift continued into the Pleistocene in the Alaska range and Mount McKinley (4188 m) is the highest peak in North America. These young mountains have ice caps upon them.

The Cascade mountains extend from British Columbia as far south as the Klamath mountains, a knot of mountainous country at the northern end of the Great Valley of California. The northern section of the Cascades is composed of Palaeozoic and Mesozoic rocks into which batholiths of granite have been intruded. Although volcanic peaks, Glacier Peak and Mount Baker, are present, the greater part of these mountains forms a plateau-like surface into which deep valleys have been incised. The northern Cascades were glaciated during the Pleistocene with ice extending as far south as the international border and just south of the Strait of Juan de Fuca on the northern slopes of the Olympic mountains of the coastal ranges. The ice pushed further south in the Willamette valley where end moraines were deposited.

South of Seattle, the Cascade mountains are formed from Tertiary volcanic rocks which make up a plateau upon which younger basalts and andesites have been deposited as shield and strato volcanic cones. The central Cascades rise to over 3000 m elevation and include the recently active Mount St Helens (2950 m). Other volcanic peaks in this section include Mount Rainier (4392 m) and Mount Good (3427 m). Crater

*Figure 4.24.* Meteor Crater, Arizona.

Lake, a National Park marked on most atlas maps, occupied a caldera formed by the collapse of Mount Mazama, an eruption of which spread ash widely over western Canada and the USA during the Pleistocene. Unfortunately, the attractive lake and the small cinder cone of Wizard Island in it have disappeared as a result of recent volcanic activity.

The southern section of the Cascade mountains is lower but the peaks consist of a line of volcanic peaks, dominated by Lassen Peak (3178 m) and Mount Shasta (4316 m). It is through this section of relatively low relief that the Klamath and Pitt rivers flow from the great basin to the Pacific Ocean across the major watershed of the Cascade mountains. Tertiary drainage of the area, identified by the alignment of gold-bearing gravels, was to the west, but only a few rivers, the Columbia, Pitt and Klamath, managed to maintain their westward courses through the Alpine orogeny.

Continuing the alignment of mountains south from the Klamath knot, the Sierra Nevada mountains of California are similar to the Cascades having a block-faulted, uplifted granite batholith structure. The Sierra Nevada have a sharp eastern slope to the basin and range province, but the western slopes are made up of folded Palaeozoic and Mesozoic strata which dip more gently beneath the alluvial infill of the Great Valley of California.

### *The coast ranges*

Parallel ranges of lower mountains, 1000 m to 1500 m, lie adjacent to the Pacific coast from the Strait of Juan de Fuca to San Francisco and beyond in the Santa Lucia and Diabolo ranges to another knot of mountains, the Tehachapi massif (2435 m) at the southern end of the Great Valley. The rocks of these ranges are Jurassic and Cretaceous limestone which were folded and faulted in Miocene times and subjected to granitic intrusions.

The San Andreas fault runs diagonally through the southern coast ranges from Tomales Bay on the coast north of San Francisco towards the Tehachapi mountains (Figure 4.25). This major dislocation of the earth's

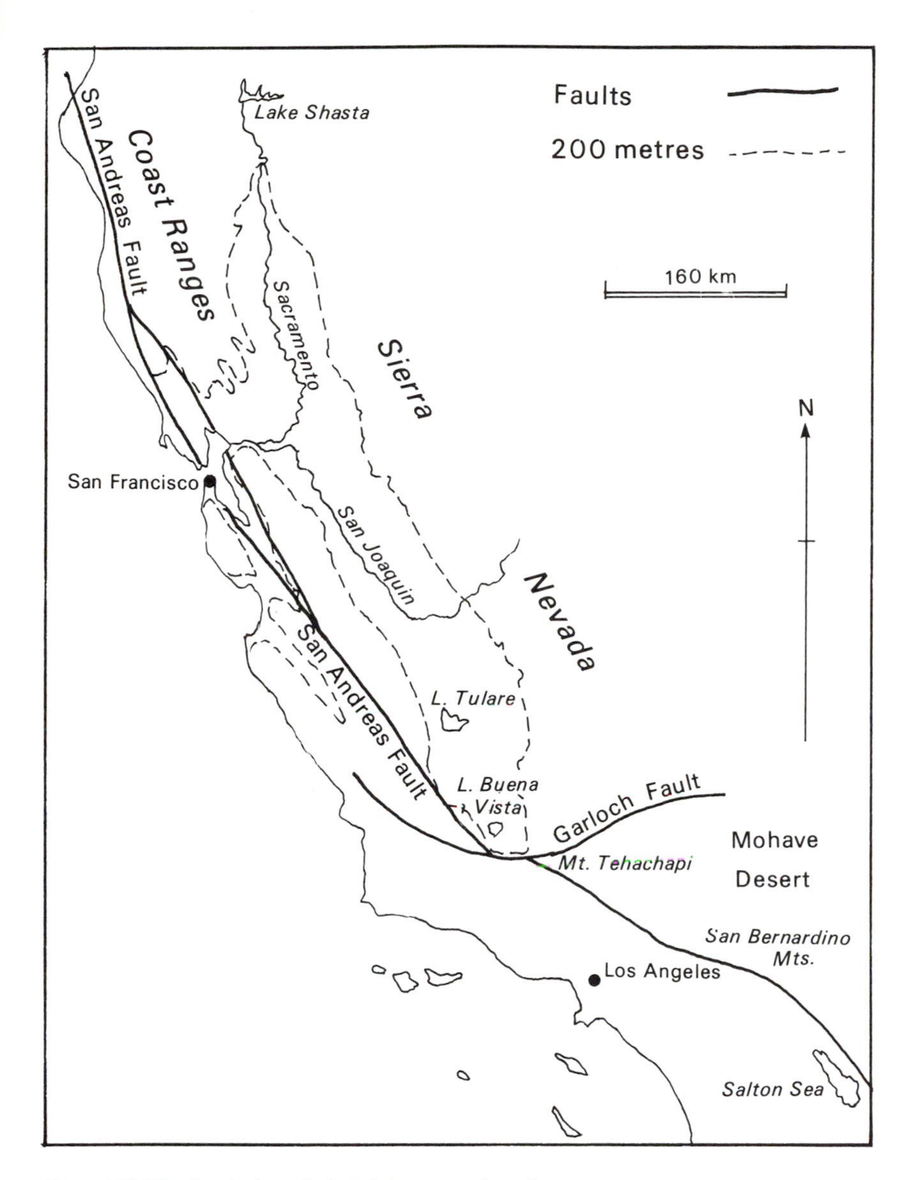


*Figure 4.25.* The San Andreas fault and the great valley of California.

crust is a transform fault on the eastern Pacific spreading centre. Along it, the western side is moving northwards relative to the eastern side at an average rate of up to 50 or 70 mm per year. After the large earthquake of 1906, which destroyed most of San Francisco, displacements of up to 6 m were found to have taken place.

*The Great Valley of California*

The Great Valley of California lies about 100 km inland from the Pacific Ocean and is 640 km long by 80 km wide. It lies between the Sierra Nevada mountains to the east and the Coast ranges to the west. The floor of the Great Valley is only about 150 m above sea level and is drained by the San Joachim and Sacramento rivers which pass through the Coast Ranges at San Francisco. It is structurally an asymmetric syncline into which an estimated 10 000 m thickness of erosion debris from the surrounding mountains has been deposited. The natural condition of much of the Great Valley is semi-desert with a rainfall of less than 250 mm, but irrigation is possible from the many mountain streams which flow into the basin. Two playas occur at the southern end of the Valley, but the rest of the southern Valley is exoreic, drained by the San Joachim river. The valley of the Sacramento river has a more humid climate, but both catchments have extensive alluvial fans extending into the valley floor from the surrounding mountains. Some recent uplift has given rise to the Kettleman hills in the San Joaquim valley and an igneous intrusion has left the Marysville Buttes, a laccolith and rhyolite plug which rises to about 600 m from the Sacramento plains.

## The Atlantic and Gulf coastal plains

These lowlands extend southwards from New York along the east coast of the USA, swing around the southern end of the Appalachian mountains and continue westwards beyond the Mississippi into Texas and Mexico. Throughout this whole area, the elevation is less than 150 m, and much of it is less than 35 m above sea level.

The geological structure is of seawards-dipping rocks from Cretaceous to Recent in age. The oldest beds crop out furthest inland where they may form escarpments or ridges, but the youngest sediments give rise to a featureless coastal plain which continues beneath the sea to form the continental shelf. The coastal plain is crossed by many extended consequent streams which rise in the Appalachian mountains. The inland margin of the coastal plain province occurs where these streams cross onto the younger sediments by rapids or falls at the 'fall line'. North of the Rappahannock river, the limit of tidal waters extends inland to the fall line. Marine terrace features have been identified along the Atlantic coastal plain.

### *Atlantic coast (north)*

The northern section of this province extends from Cape Cod, through Martha's Vineyard and Long Island to Cape Kennedy. The features of Cape Cod, Martha's Vineyard and Long Island are developed upon submerged ridges of the coastal plain sediments, capped by moraines. Longshore drift is an important feature of the coastal processes, maintaining the barrier beaches and contributing to the recurved spit of Cape Cod. It is deeply embayed by the Delaware and Chesapeake Bays and estuaries of other rivers. Barrier beaches extend along most of the coast, enclosing lagoons of various sizes. Pamlico and Albermarle Sounds, which are up to 50 km across, lie behind the barrier beaches of Cape Kennedy.

### *Atlantic coast (south)*

Between Cape Kennedy and Charleston, the pattern of geological outcrops which bring Cretaceous strata to the surface indicates an anticlinal flexure. Structurally, the Atlantic coast south is similar to the Atlantic coast north, but instead of the barrier beaches, there are many low coastal islands, the Sea Islands. It is thought that the reason for this difference is that the Atlantic coast south has had a slight subsidence of the coastline which has broken-up the former barrier beaches. Inland terrace features may be seen at 90 m, 71 m, 57 m, 33 m, 14 m, 7 m and 2.5 m above sea level parallel to the coast. Depressional features on these terraces have been a cause for controversy. Scattered widely over these terraces, the depressions have been attributed to solution or meteorite impact.

### *Florida*

The Florida peninsula extends for 600 km south of the main body of the USA forming the eastern limit of the Gulf of Mexico. It is an anticlinal flexure in limestones which forms an emergent low plateau with karstic features. These are most common north of Lake Okeechobee and are associated with the outcrop of the Ocala limestone. There are many lakes in subsidence hollows, and deep shaft-like ponors descend to over 30 m below the surface. South of Lake Okechobee, the organic swamps of the Everglades lie in a depression which originated in Pliocene times. Since then it has been partly infilled with Pleistocene marls and topped-up with swamp deposits. The Big Cypress Swamp of south-west Florida occurs on a saturated limestone plateau. Drier areas of grassland also occur, interspersed with wetter solution pockets. Lake Okeechobee, at the northern end of the depression in which the Everglades are situated is shallow, less than 7 m deep, although it is some 50 km across.

The east coast of Florida has a series of barrier beaches with sand dunes enclosing lagoons. The lagoons are backed by a low coastal rise upon which marine terrace features occur up to a height of 90 m. The southern tip of Florida ends in a string of islands called the Keys, extending 240 km south and westwards from Miami Beach. The eastern Keys are formed from dead Pleistocene coral, but the western Keys originated as shoals in the sheltered waters behind the reefs, consisting of the same limestone which lies beneath the Everglades.

### *The Gulf coast (east)*

Between Florida and the Mississippi valley, the Gulf coast is characterised by an increased thickness of the Cretaceous and Eocene formations which give a landscape of landward-facing escarpments. The marine terrace features noted in Florida continue westwards, and inland the Selma chalk is noted because it gives rise to the black vertisols formerly known as 'black cotton soils', valued for cotton production. The southern pine hills of Mississippi occur on Pliocene rocks which dip southwards from the cuesta crest at 170 m to the Pleistocene terraces at 60 m above sea level.

### *The Mississippi valley plain*

From the confluence of the Mississippi with the Ohio at Cairo in Illinois, the alluvial plain of the Mississippi extends 800 km to the delta which begins approximately at Baton Rouge. The boundaries of the alluvial plain are marked by prominent bluffs which diminish in height southwards towards the Gulf of Mexico. The trench, which the Mississippi has partly infilled with sediments, was probably cut by the river during the Pleistocene. The alluvial plain varies in width from 40 to 200 km, but it is interrupted by ridges of Cretaceous and Eocene rocks, covered by Pliocene gravels and loess. There is evidence of several changes of course by the Mississippi and Ohio rivers at the head of the alluvial plain, and four terrace levels occur above the present flood plain, known as the Prairie, Montgomery, Bentley and William terraces.

The Mississippi river itself has all the characteristics of a meandering stream with levées, backswamps and ox-bow lakes. Tributary valleys are deferred downstream as the bed and levées of the mainstream bed are above the level of the tributary valley. Backswamps occur in the Yazoo and Tensas basins between the levées and the floodplain margins. The Mississippi meanders within a relatively restricted meander belt and in its southern section has had several changes of course during the past 2000 years.

The classic birdsfoot delta of the Mississippi has a complex history with evidence of at least four major phases of deltaic construction. At different times, distributaries have built the western Teche and Lafourche deltas which overlap each other, followed by the eastern St Bernard delta and finally the present delta which dates from a diversion in the sixteenth century near New Orleans.

### *The Gulf coast (west)*

West of the delta of the Mississippi, the Gulf coast is characterised by barrier beaches and lagoons, frequently under attack from tropical storms which develop in the Gulf and then move northwards. The coastal barrier beaches are the first line of defence and are frequently swamped by waves driven by hurricane-force winds. Whilst they absorb the pounding, they are subject to considerable change.

A feature of the Gulf coast west of the Mississippi is the presence of salt domes or diapirs, resulting from salt flowing upwards under pressure through incompetent rocks to form hillocks rising 25–50 m above the coastal plain. These features, associated with oil reservoirs, also occur on the continental shelf offshore.

Along the coastal region of Louisiana and Texas relict beach ridges occur; they extend for 50 km and have a height of 3 m. Further inland the cuestas noted in the eastern Gulf coast again form the dominant character of the landscape.

## Oceanic landforms of the North American plate

The North American plate extends from the mid-Atlantic ridge westwards across the western Atlantic sea floor and the North American continent. A complex western boundary is partly trench, and partly overridden mid-oceanic Pacific ridge which passes overland as the San Andreas fault. So, land west of the San Andreas fault is not part of the North American plate, but belongs to the Pacific plate. In a north to south direction, the plate extends from the Caribbean to Greenland and the northern islands of Canada. So, the Canadian archipelago is the higher parts of a continental shelf some of which, notably at Hudson Bay, has been drowned by the sea.

The submarine landforms of the North American plate lie mainly below the Atlantic Ocean west of the mid-Atlantic ridge. There are four main structural basins on this part of the sea floor: the north Atlantic, Newfoundland, Labrador and Baffin Basins.

### *The North Atlantic basin*

This large area of abyssal plain between America and the mid-Atlantic ridge extends from the latitude of Barbados northwards to the Newfoundland rise. Numerous seamounts occur on the floor of the abyssal plain, Bermuda reaches above the sea, other submerged seamounts only reach to within 800–1000 m of the sea surface. Although rifting began 180 million years ago, sea floor only as old as 130 million years old can be traced.

From Florida northwards the continental shelf gradually becomes wider until it reaches its maximum width in the Grand Banks of Newfoundland where parts of the sea are only 4 m deep. The continental shelf and slope are furrowed by several submarine canyons. The Hudson canyon, discovered in 1891, was subsequently found to be 225 km in length. The origin of these canyons is uncertain, but they are thought to have been eroded by turbidity or suspension currents carrying sediment from the continental shelf down to the abyssal plain 6000 m below.

### *The north-west Atlantic basins*

The smaller Newfoundland, Labrador and Baffin basins are not so deep, about 4500 m in the deepest parts. Collectively these basins form an arm of the Atlantic which opened about 70 million years ago. It then stopped, however, and sea-floor spreading moved to the eastern side of Greenland and, beginning some 60 million years ago, developed the North Atlantic Ocean.

## CENTRAL AMERICA AND THE CARIBBEAN

The mainland and islands of Central America have a land area of about 2.6 million km². Mexico has the largest land area, with none of the Central American republics and Caribbean islands having an area of more than 150 000 km². The major part of this area lies south of the Gulf of Mexico and it will be appreciated readily that the division between North and Central America is an arbitrary one.

Central America was a late Mesozoic geological development: the link between the North and South American plates was formed in the late Cretaceous. The geomorphological unit is built around a stable area of basement complex which lies buried beneath Florida and the Gulf of Mexico. Palaeozoic sedimentation, folding and subsequent cratonisation brought an 'Atlantean' land mass into being which was eroded to low relief, downfaulted, and has remained almost entirely below the sea. The only parts of it which rise above are the Maya mountains of British Honduras, although it is thought to underlie the Yucatan peninsula and adjacent offshore areas. Other small areas may be present in the islands of Aruba, Curaçao and Bonaire off the north coast of South America.

The geomorphology of Central America may be considered to consist of two parallel mountain ranges with intermontane plateaux between them in Mexico. Lowland areas occur on both east and west coasts, related to structures further north in the USA. This pattern is abruptly cut off by the neovolcanic plateau in central Mexico, south of which there is a single complex of mountains which form the Isthmian ranges (Figure 4.26). The list of the regions is as follows:

Sierre Madre Oriental
Sierre Madre Occidental
Northern plateau of Mexico
Mesa Central
Neovolcanic plateau
Sierre Madre del Sur
Central American ranges
The Nicaraguan depression
Isthmian ranges
The Gulf coast plain
The Yucatan peninsula
The Gulf of California coastal plain
Lower California
The Caribbean arc

### *Sierre Madre Oriental*

This eastern range of mountains which forms the margin of the northern Plateau of Mexico is best described as a dissected mountain slope between the plateau and the Gulf of Mexico lowlands. The rocks from which it is formed are limestones of Lower Cretaceous and Triassic age which give rise to karst morphology. Small areas of sandstone, shale and Tertiary rhyolite lava occur, all of which are partly covered by unconsolidated fluvial and aeolian deposits. The

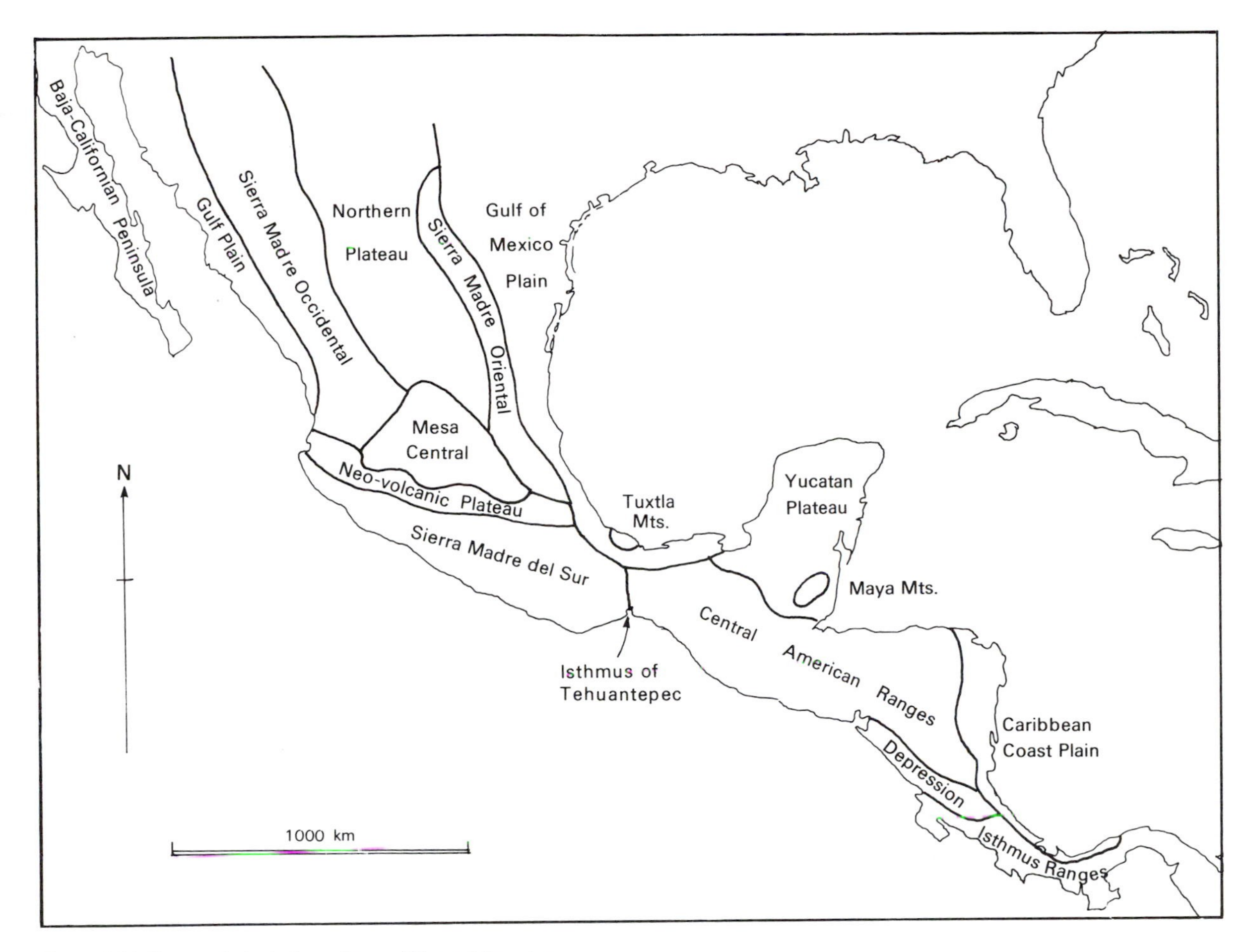


*Figure 4.26.* Geomorphological provinces of Central America.

most prominent mountains of this province are the High Sierras, formed of tightly folded limestones eroded into ridges and crossed by deep canyons where rivers flow through the range. Relief ranges from the coastal plain to about 2000 m with the highest peaks reaching 2700 m.

### *Sierre Madre Occidental*

The western margin of the northern plateau is formed by the Sierre Madre Occidental. These mountains have a maximum elevation of 3500 m in the centre of the range and towards the south, and enclose a rhyolite plateau with a height of over 2000 m.

Dissection of the rhyolite plateau is occurring from the west where rivers have cut spectacular gorges through the mountains and deep valleys across the pediments leading down to the Gulf of California.

A north-western sub-division of this province is the Buried (Sonoran) ranges, where large quantities of desert detritus have been moved westwards from the higher Sierra Madre Occidental and buried the lower mountains.

### *Northern plateau of Mexico*

This province of Central America is a continuation of the same structures observed in the basin and range province of the USA. It consists of series of faulted basins which have been partly infilled with desert detritus. Low ranges of hills separate the basins which have an average height above sea level of 1300 m. Arid pediplanation is the dominant geomorphological process taking place with extensive coarse fans or screes leading down to sands and clays in the centres of salty playas. The largest of these is the Bolson de Mapimi.

### *Mesa Central*

With Mexico City located in its southern part, this upland basin is really an extension of the northern plateau of Mexico, but an elevation of over 2000 m above sea level. However, it has an external drainage system and formerly there were shallow lakes on the surface of the plain. A number of smaller basins occur at higher elevations, such as the basin of Toluca. Recent lacustrine deposits, surrounded by pediments partly mantled by loess and volcanic ash are extensive landscape elements. Basaltic flows become dominant in the south-east of the basin.

### *Neovolcanic plateau*

Quaternary vulcanism, associated with a line of crustal weakness extends across southern Mexico. The lavas form a 2800 m plateau above which rise several large volcanic cones, reaching 2500 m above the plateau surface. The consolidated lavas, tuffs and scoria are covered with loose ash in which considerable erosion has occurred. The cones of Popacatapetl (5440 m) and Ixlaccihuatl (5290 m) are extinct and are undergoing erosion. The Tuxtla mountains on the Gulf coast are a small detached portion of the volcanic plateau; they range from Miocene to Recent in age and are composed of andesite and basalt flows.

### *Sierre Madre del Sur*

Where the north–south folding of North America meets the east–west folding of the Caribbean, the geological structure is inevitably complex and a rugged landscape is developed. Metamorphic rocks of Palaeozoic age crop out along the coast and the north-east section of this province is formed by Cretaceous limestone. The eastern boundary of the Sierre Madre del Sur coincides with the graben which forms the Isthmus of Tehuantepec.

### *Central American ranges*

East of the graben of the Isthmus of Tehuantepec lie two mountain ranges. The Sierre Madre are near the Pacific and the Sierre de Norte de Chiapas lie across the base of the Yucatan peninsula. Both ranges extend from southern Mexico across Guatemala and into Honduras. The mountain landforms result from a combination of block-faulting and vulcanicity, the volcanic cones surmounting the upthrust horsts. The Sierre de las Cachumatones is a horst of limestone with a summit plain of more than 3800 m. Karstic features occur and its margins are deeply incised by canyons. The north-eastern part of this province is a plateau with an elevation of less than 1500 m, sloping gently eastwards, composed of ignimbrites and tuffs with some sedimentary strata. The Maya mountains of Belize are a block-faulted upland area reaching 850 m with steep northern and eastern slopes.

### *The Nicaraguan Depression*

Across the southern part of Nicaragua, from the Gulf of Fonesca on the Pacific coast to the Caribbean lies a lowland, much of which is occupied by Lakes Nicaragua and Managua. Around the lakes is a moderately dissected plain, in places only 100 m above sea level, but there have been volcanic eruptions which have thrown up cones such as Momotombo (1260 m) on the shore of Lake Managua. Three volcanoes protrude from Lake Nicaragua. Andesitic and basaltic ashes have been re-sorted by fluvial action on the lowland where they have been strongly weathered to give clayey soils.

### *Isthmian ranges*

The Isthmian ranges extend along the southern shore of Lake Nicaragua and reach their greatest height in Costa Rica where the highest peaks occur, Irazu (3432 m), Turrialba (3328 m). The rocks of this province are mainly andesites of Mesozoic age which have been overlain by Tertiary and Recent volcanic deposits, as well as some carbonates of Upper Eocene age. The height of the mountains decreases in Panama to a lowland area where the Panama Canal crosses the 60 km Isthmus utilising the Gatun Lake, 28 m above sea level and formed by damming the Chagres river. The Isthmus ranges may be interpreted as a volcanic arc; offshore along the Pacific coast is the mid-American trench, 6662 m deep, where the Cocos plate is being subducted. On either coast of the Isthmus lie narrow coastal plains with rugged hills forming the Azuero and Nicoya peninsulas.

*The Gulf coast plain*

The Mexican section of the Gulf coast plain is a continuation of the Gulf coast plain of Texas. It extends 1000 km southwards from the mouth of the Rio Grande to the Yucatan peninsula. The province includes the 200–700 m limestone foothills of the Sierre Madre Oriental which tends to cause subsequent streams to develop in a north–south direction whereas the trunk streams drain directly to the Gulf of Mexico. The littoral is similar to the Texas coast with sand bars and lagoons.

*The Yucatan peninsula*

The foundations of this lowland peninsula are thought to be the stable basement platform underlying Florida and parts of the Caribbean. Uplift has raised the Yucatan section with its limestones and coral debris above sea level. Emergence has taken place since late Cretaceous and continued until Recent times, the oldest rocks are exposed at the base of the peninsula, the youngest at its northern tip. An indurated limestone crust developed before emergence and volcanic ash showers have provided material for the development of red soils over the limestone. On older parts of the landscape the crust has been dissected and a more undulating landscape developed.

*The Gulf of California coastal plain*

At its northern extremity, the Gulf coast plain is formed by the delta of the Colorado river, which has been built out into the head of the Gulf of California. Along the western Mexican coast the coastal plain of the Sonoran desert is between 10 and 80 km wide and is crossed by several rivers flowing from the Sierre Madre Occidental.

*Lower California*

The peninsula of Lower California (Baja-California) is formed of a linear granite batholith intruded into Cretaceous rocks and partly covered by Upper Cretaceous sediments. It has been block-faulted and tilted westwards and so has a steep, east-facing scarp culminating in the San Jacinto Peak (3600 m) and Cerro de la Encantada (3078 m). Irregular uplift during the Pleistocene has resulted in coastal terraces on the western side of the peninsula and on San Clemente Island. Up to 13 terraces are recognised ranging up to 400 m above present sea level.

Although the Gulf of California side of the peninsula is cliffed, the Pacific side has a narrow coastal plain with lagoons and sand bars. Desert geomorphology including steep walled canyons and detrital fans are common throughout the peninsula and mid-Tertiary rhyolitic lavas occur. The Sierra Vizcaino peninsula is composed of metamorphic rocks.

*The Caribbean arc*

The Caribbean arc extends from Grenada, northwards through the Lesser Antilles archipelago, curving around through Puerto Rica, Dominica and Jamaica. The Puerto Rico trench (−9218 m) lies to the north of that island and the Virgin Islands. Near Dominica and Haiti the trench is less deep, but between Jamaica and the Caymen Islands the sea floor again plunges to more than 5000 m deep. This is a complex area with the relative movements of the North American plate and the Caribbean micro-plate being opposed to each other. Subduction is occurring and results in the 17 active volcanoes of the Caribbean region. Monte Pelée in Martinique erupted violently in 1902 with a *nuée ardente* which killed 30 000 people in St Pierre. During the eruption a spine of solidified lava was pushed up out of the crater. La Soufrière (1171 m) on St Vincent has also been active recently. In 1971, the attractive lake in the summit crater was replaced with a domed mass of lava. An eruption occurred in 1979 and covered the upper part of the volcano with a fresh layer of ash. In contrast to the volcanic islands of the Lesser Antilles, Barbados is a plateau-like island of coral limestones. It rises to about 400 m above sea level and is surrounded by coral reefs. Ash falls from the neighbouring volcanic islands have been recorded, adding to the fertility of Barbadian soils.

## SOUTH AMERICA

The continent of South America has an area of 17.8 million km² and it extends from 12°N in the Caribbean to 55°S at Cape Horn and from 35°W at Reçife in Brazil to 83°W in the westernmost tip of Peru. The highest peak in the Andes is Mount Ojos de Salado (7084 m) on the Chile–Argentina border and Salinas Grandes in

Argentina is the lowest at 40 m below sea level. The continent has a mean elevation of 550 m and, according to Butzer, the largest area is covered by the ancient shields of Brazil and Guyana, 30%; erosional plains and plateaux on sedimentary rocks amount to 25% of the land surface. Depositional plains account for 21% but volcanic plains and plateaux only amount to 2% despite their importance in the Andean zone. Young mountains amount to 21% of the land surface area and old mountain belts only amount to 1%.

## Geological history

The existence of South America as a separate continental mass or lithospheric plate begins with its rifting away from Gondwana in the mid-Cretaceous about 100 million years ago. Since then, the South American plate has moved steadily north-westwards as the Atlantic has become wider through sea-floor spreading. In its movement the South American plate has encountered the Andean trench which it has displaced westwards without overriding it. Both to north and south of the South American plate, the Caribbean and Falkland trenches may represent the original position of the eastern Pacific trench system before it was displaced westwards by the continental mass. It has been suggested that the track of the north-westward-moving continent is marked by the undersea ridge which extends from the vicinity of Rio de Janeiro south-eastwards to a point near Tristan da Cunha. (Africa has similarly moved north-eastwards as shown by the Walvis ridge.) Both these undersea ridges are lines of basaltic outpourings from a deep-seated 'hot spot'. This hot spot poured out tholeitic basalts onto both the African and South American sides of the rift zone before they moved apart, and these have travelled with the continental rafts to their present positions. The continental plates moved but the hot spot remained in the same position, producing lavas which mark the track of the moving continental plates.

Interaction between the former Gondwana fragment and the Pacific plate resulted in the Andean folding, so the second major component of the South American plate is the Andean fold mountain chain which occupies the full length of the west coast of South America. Between the basement complex of the Gondwana remnants and these new fold mountains lies the sedimentary basins of the Gran Chaco and the Amazon. Thus the geomorphology of South America may be described in three major physiological divisions (Figure 4.27):

The ancient shield areas of Brazil, Guyana and Patagonia
The sedimentary basins of Gran Chaco and Amazonia
The Andean fold mountains

Landscape development in each of these three major physiographic divisions is quite distinct, leading to different landforms which in turn are reflected in the soils and vegetation. As South America is regarded as a part of the developing world, the facts of the physical environment should be taken into consideration when future developments are planned. At present, unfortunately, circumstances do not seem favourable for the rational development of many of the natural resources. Destruction of the natural environment in the tropical rainforest is causing international concern and doubtless the mistakes made elsewhere in the world will be made again in South America.

## The shield areas

Three main regions and one minor area of South America are underlain by ancient rocks originally cratonised on the former continent of Gondwana. The shield areas appear to have migrated together, even though there are wide sedimentary zones between them. The shield areas of South America are:

The Brazilian shield
The Guyana shield
The Patagonia shield
The Santa Marta massif

### *The Brazilian shield*

The Archean rocks of the Pre-Cambrian basement of the Brazilian shield crop out in two main areas: west of the Parana and Magdalene basins, and along the Atlantic coast from Uruguay to Cap San Roque (Figure 4.28). All these basement rocks have been metamorphosed, but the intensity varies from place to place. These oldest rocks consist of a series of banded gneisses and the younger, Protozeroic, rocks include quartzites, slates, dolomites and conglomerates. There are numerous intrusive rocks in the basement complex of the Brazilian shield, mostly granites, but extensive basaltic and andesitic rocks occur in the southern part of the shield in Uruguay.

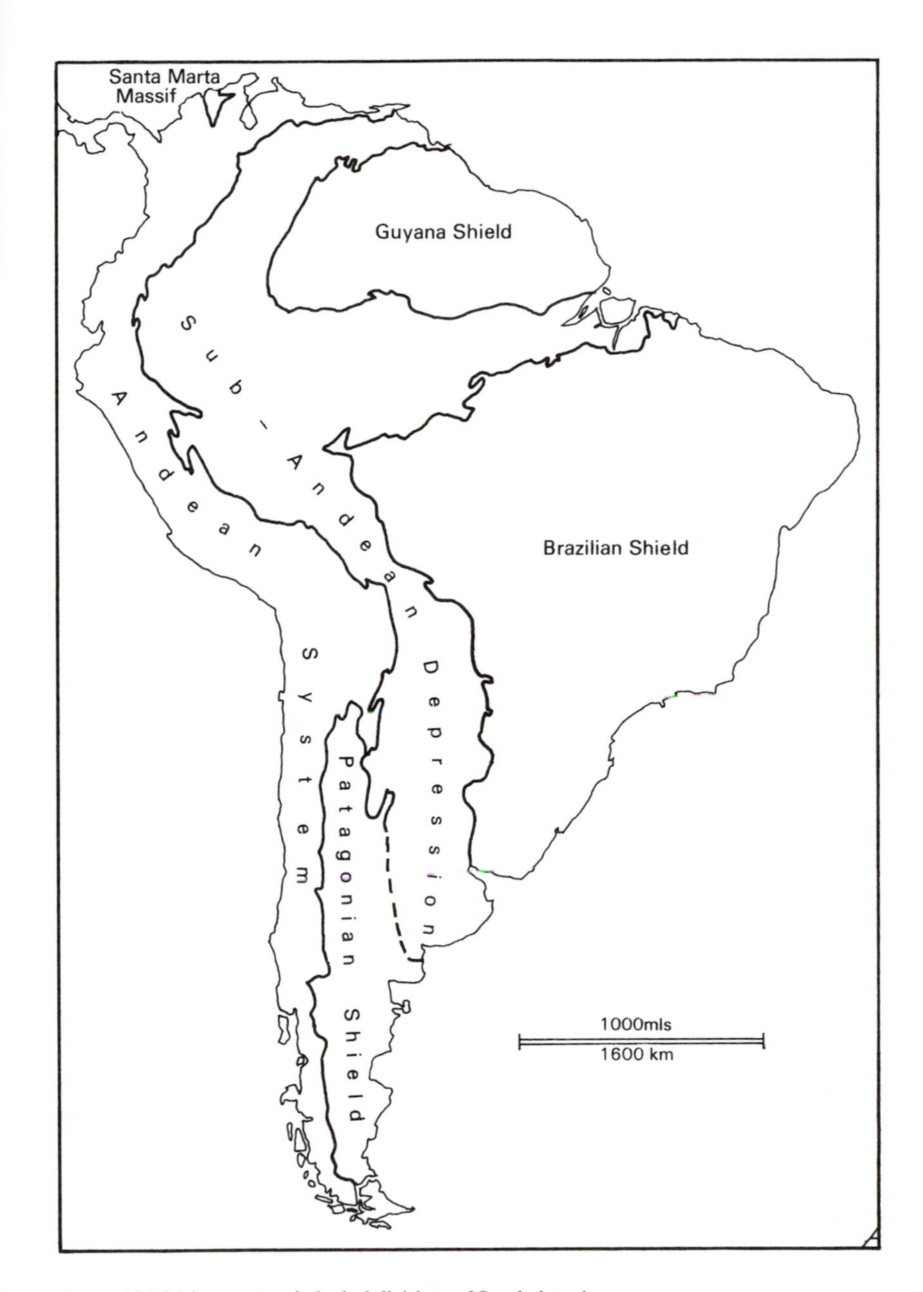


*Figure 4.27.* Major geomorphological divisions of South America.

The Brazilian shield is characterised by three large river basins, drained by the Paraná, Maranhão–São Francisco and Sergipe rivers. These basins resulted from the faulting of the original basement rocks of the shield and contain rocks from the Lower Palaeozoic to the present time. Other depressions occur where the basement complex has sagged downwards; these depressions contain mainly tertiary sediments. In both basins and depressions, alluvial deposits are widespread.

Along the east coast of Brazil, Cretaceous and Tertiary rocks have been laid down in marginal basins on the edge of the shield and south of Florianopolis, Quarternary and Holocene sediments occur in features such as the Lagoa Mirim and the Lagoa dos Platos in the extreme south of Brazil. Between the valley of the

*Figure 4.28.* The Brazilian shield.

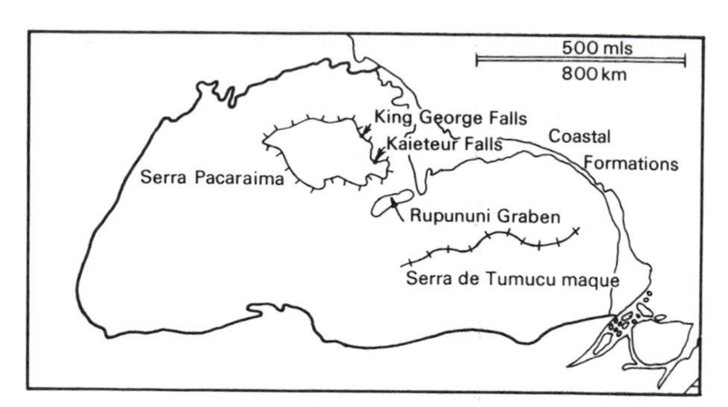


*Figure 4.29.* The Guyana shield.

Paraná and the south Atlantic Ocean, the Cretaceous Paraná basalts occupy 750 000 km², giving extensive plateau landscapes. The edge of the exposed shield in Bolivia is composed of Palaeozoic rocks of Cambrian and Devonian age where they form the faulted blocks of the Sierre de San José and Sierra de Santiago.

The old, hard rocks of the shield form upstanding areas upon which erosional landforms occur. King and Dresch have identified accordant surfaces which have been assigned to four cycles of erosion. The interfluve crests of residual relief with inselbergs, is thought to include areas of the Gondwana surface, developed before the rifting of the former continent began in the Jurassic. The most extensive surface, and the one from which most of the relief of Brazil has been developed is the early Tertiary 'Sul American surface'. Broad valleys excavated into this Sul American surface are referred to as the Velhas cycle of erosion. Remnants of these two surfaces occur over much of the Brazilian shield, but a fourth erosional episode which post-dates a Plio-Pleistocene uplift in eastern Brazil, is referred to as the Paraquaçu cycle of erosion. This last erosion cycle only affects the streams of the eastern coastal districts. In central Brazil the widely developed plains formed on the early Tertiary erosion surfaces support the cerradão vegetation and are associated with Ferralsols and Arenosols. The Panatal, at the headwaters of the Paraguay river, is a depression infilled with sandy alluvial materials, moulded into fossil dunes in response to north-east winds. Formerly a desert, the area is now under a humid climate.

*The Guyana shield*

The Guyana shield is a smaller, northward extension of the larger Brazilian shield, and is separated from it by the Amazon sedimentary basin (Figure 4.29). The Guyana shield is formed of Pre-Cambrian gneisses and migmatites, but there are no large down faulted basins as occur in the Brazilian shield. The Guyana shield lies between the Amazon and Orinoco river basins in the States of Guyana, French Guyana, Surinam, together with southern Venezuela and northern Brazil. The southern part of the shield is drained by tributaries of the Amazon and the western part by the Orinoco and the Essequibo, Corantijn and Maroni drain the northern slopes directly to the Atlantic. The Kaieteur Falls are the highest in the world, the Essequibo river dropping 270 m.

The landforms of the Guyana shield have been developed from an uplifted Plio-Pleistocene surface, the Rupununi surface, which is extensive north of the block-faulted Pacaraima mountains. Originating in the Cretaceous, these mountains are bounded by fault scarps. An older Kwitaro surface, between 300 m and 360 m above sea level, extends into Brazil. The Sierra Acarai are formed from the remnants of this surface. The oldest surface present is the Kopinang surface at 630–690 m in the Pacaraima mountains where it cuts across sandstones and intrusive rocks. Many of these old erosion surfaces are lateritised. Westwards the older rocks pass beneath the younger sediments of the Orinoco valley but islolated outcrops do occur on the Llanos of Venezuela.

The coastal region of the Guyanas is covered by

unconsolidated sediments, amongst which the Berbice formation is an irregular series of brown and white sand with lenses of lignite. The coastal plain is composed of Holocene blue and grey clays with sandy ridges rich in shells, the Demerera formation.

### *The Patagonian shield*

Two stable shield areas occur in Patagonia, lying east of the Andes, and south of Bahia Blanca. The Patagonian massif lies north of the Chubut river and the Desado massif to the south of the river. The rocks of these shield areas are Pre-Cambrian gneisses and migmatites which have been extensively covered by volcanic deposits of Jurassic age, including tuffs, ashes and both andesitic and rhyolitic lavas. In the north of this province, the basin of the Rio Negro contains Cretaceous sedimentary rocks and in the eastern part unconsolidated Quaternary deposits are widespread.

The main landform of this region is a tableland lying at 1500–1700 m in the west and at 450–500 m in the east. Rivers flowing from the Andes to the south Atlantic Ocean are incised into the tableland surface in narrow valleys. The tableland is semi-arid, so some areas of it have endoreic drainage, and surface drainage is absent over areas of porous lavas and sandstones. The western limit of this tableland is the sub-Andean depression marked by lower elevation and the presence of several lakes.

### *The Santa Marta massif*

A small portion of the Pre-Cambrian basement complex appears to have been caught up with the Andean folding and faulting. The Santa Marta massif is composed of Pre-Cambrian gneisses and Palaeozoic schists which are in part covered by Triassic and Jurassic sedimentary and volcanic deposits. The highest point of the massif is the Pico Cristobal Colom (577 m) and the general surface of the block is tilted towards the south-east. On all sides the Santa Marta massif is bounded by faults.

## The sedimentary basins

Between the ancient fragments of the former Gondwana continent and the newer fold mountains of the Andes are a number of interconnected sedimentary basins which are an important part of the physical geography of South America. Their landforms are usually plains or alluvial lowlands and they are characterised by sedimentary deposition. There are three subdivisions of this major landform region of South America:

The Orinoco basin
The Amazon basin
The Chaco–Pampa basin

### *The Orinoco basin*

The Orinoco, which flows into the Atlantic south of Trinidad, has its headwaters in the Eastern Colombia cordillera and the Venezuelan Andes. North of the river lie the extensive plains of the Llanos. These plains rise gently from the river but nowhere exceed much over 250 m above sea level before the land rises rapidly into the north coast cordillera.

The Orinoco lies in an asymmetric tectonic trough with a steep northern limb and a gently inclined southern limb. Erosion products from the Andes and the Caribbean coastal ranges have been brought down by the rivers since the late Cretaceous to give a considerable thickness of continental alluvial deposits. Quaternary sediments, including gravels, sands and clays have been laid down in alluvial fans associated with the rivers. In the Llanos these fans have coalesced to form an extensive pediment. South-west of the Rio Meta, in Colombia, the alluvial deposits have been partly covered by aeolian materials in the form of loess and dunes.

King states that the interfluve between the Orinoco and the Amazon has been part of the Sul American surface upon which many river captures have taken place. The most interesting of these is a link between the two great rivers at the head of the Orinoco and the Rio Negro in the southern province of Venezuela.

### *The Amazon basin*

The Amazon river, which flows into the Atlantic apoproximately where the Equator crosses the South American coast, occupies a large natural downfold in the Earth's crust. Its headwaters rise in the great arc of mountains from central Bolivia in the south to central Colombia in the north and its mean annual discharge is 200 000 m$^3$/second. Many tributaries, such as the Marañón, rise only 100 km from the Pacific coast and then flow for several hundred kilometres parallel to the cordilleras before managing to break through into the Amazon lowlands. Although the last major uplift of the Andes took place in the early Pleistocene, rivers have already carved deep valleys to carry their load of sediment onto the lowlands.

Sedimentation began in the Amazon basin in the Cambrian. The older rocks may be found as narrow outcrops around the northern and southern limbs of the downwarped basin. However, the majority of the Amazon basin is floored with Tertiary deposits derived from the erosion of the Andes since the Oligocene. These sediments comprise coarse clastic materials at the foot of the mountains in the west where there is considerable intermixing of alluvial and pyroclastic material. Further east the Tertiary deposits are made up of friable sandstones and mottled clays which are referred to as the Barrieras or Alter de Chao series. These are overlain by a clay, the Balterra clay which resulted from sedimentation of 10 to 120 m of kaolinitic material in a shallow inland sea. This has resulted in large areas of the Amazon lowlands having a uniform, level surface between 150 m and 250 m above sea level, known as the Amazon planalto. Near to the Andes this planalto surface decreases in elevation, probably caused by continued subsidence since it was deposited. This planalto has been dissected during Pleistocene and Holocene time, so that terraces are common along major streams. Pleistocene deposits are thin and are mostly confined to reworked Tertiary materials except for the Ilha de Marajó where they amount to 250 m thick. Holocene alluvium is found only along the courses of the rivers and in the estuary. Usually they are non-calcereous silts and clays, but along the Rio Solimoes alluvium is calcareous as a result of erosion of limestone in the Peruvian Andes. Altogether over 1 billion metric tonnes of sediments is transported to the sea every year by the Amazon.

### *The Chaco–Pampa basin*

Between the Andes east of Santa Cruz in eastern Bolivia and the Brazilian shield southwards to the south Atlantic Ocean lies the Chaco–Pampa basin. It is occupied by the Paraná, Paraguay and other rivers which flow into the estuary of the Rio de la Plata. The watershed between the Amazon and the Parana–Paraguay river system is only about 300 m above sea level. In a similar manner to that seen in the Amazon basin, sediments have been received from Palaeozoic times onwards. However, the older sediments are rarely seen at the surface as there is a covering of several hundreds of metres of late Tertiary and Quaternary conglomerates, sands and muds eroded from the deeply incised canyons in the Andes. West of the Paraguay river, the Gran Chaco is mainly a sandy lowland of extensively pedimented alluvial materials including some lacustrine deposits. The drier areas are characterised by the accumulation of soluble salts in the soils. Further south in the Argentinian Pampa there is considerable loessial element derived from volcanic ash which becomes finer in texture from west to east and which also provides the parent material for some of the most fertile soils of South America.

The plains west of the river system have no visible drainage courses and the landform consists of a featureless lowland of less than 200 m above sea level.

## The Andean fold mountains

The Andean mountain system extends from Trinidad, Venezuela and Colombia in the north to Tierra del Fuego at the extreme tip of South America. These mountains represent the results of a succession of orogenic episodes which have taken place along the western edge of the South American plate as it has moved westwards. As disruption of the former continent of Gondwana did not begin until the Cretaceous, the Andean mountain system is mainly a feature of the Alpine orogeny. However, the eastern cordillera comprise Palaeozoic sedimentary rocks, folded in the Hercynian orogeny.

A relatively simple model can be used to represent the development of the Andes, but in detail the structure and timing of many of the episodes of mountain building are complex. Essentially, the Andes consists of two parallel mountain ranges with a series of intermontane basins and plateaux between them. The mountain belts run parallel to the Pacific coast and vary in width from about 100 km in Ecuador to over 600 km in Bolivia. The highest peaks, usually volcanic, occur in the widest part in Bolivia and in general terms average elevation of the mountains declines both north and south of Bolivia.

In the south of Chile, the central basin is flooded by the sea to give the often quoted example of a 'parallel'

*Figure 4.30.* Cotopaxi. (By courtesy of R.D.F. Bromley.)

type of coastline where the sea has partly submerged the western ranges, the Cordillera Costanera. The central plateaux of the Andes in northern Argentina and Bolivia are high altitude basins with thick alluvial infills and lakes, such as Titicaca and Poopo at heights of over 4000 m above sea level. In Peru, the central Andean basins are smaller than those of Bolivia and are overshadowed by the volcanic peaks of Chimborazo (6310 m) and Cotopaxi (5344 m) and other volcanoes which have been active in recent times (Figure 4.30). The ejecta from these volcanoes have added to the alluvial filling of the basins between the mountain ranges. After being somewhat constricted in width in Peru and Ecuador, the Andes splay out in Colombia and Venezuela to enclose low level basins which contain the Magdalena river, Lake Maracaibo and the Gulf of Venezuela as well as the horst of the Santa Marta massif.

A section across the Andes from the Pacific Ocean to the Amazon basin would typically include:

1. The stumps of older Pacific coast ranges, probably of Hercynian age.
2. The western cordilleras composed of Mesozoic sediments and volcanic peaks.
3. The intermontane plateaux and basins.
4. The eastern cordillera and Palaeozoic sediments.
5. The sub-Andean depression with Tertiary and Quaternary alluvial infill.

Although the Andes are the result of earth movements mainly during the Alpine orogeny, there is evidence for earlier mountain building along similar lines. An earlier Palaeozoic phase of uplift, equated with the Taconic (Hercynian) of North America, at the end of the Ordovician is thought to have affected the region of the Andes from Venezuela to approximately 30°S, where it divided, one branch turned eastwards towards the Atlantic between Buenos Aires and Bahia Blanca and the other continued southwards in the coastal ranges of

Chile. Further earth movements took place along this alignment in the late Palaeozoic and again in the closing stages of the Triassic when the folding of the Sierra del Tandil was reactivated. (These ranges of mountains are continued in the Cape ranges of South Africa.)

The evolution of the present Andes ranges began in the late Cretaceous when folding of Mesozoic rocks took place, accompanied by the eruption of lavas and tuffs and the intrusion of acid igneous batholiths. The irregular landscape which resulted was eroded and the hollows infilled with flysch-like materials. The eastern cordilleras were uplifted in the Oligocene accompanied by volcanic activity in the western cordilleras which has persisted to the present day.

A further upheaval occurred in the Miocene when volcanic eruptions reached a peak of activity. In the succeeding Pliocene, erosion occurred reducing parts of the central Andes to an accordance of level, know as the Puna surface. Mammalian fossils found in the Pliocene deposits of the Altiplano and Puna de Atacama basins indicate a low-altitude fauna. So it is assumed that these montane plateaux were not then elevated to their present positions. The fourth and final phase of building the Andes mountains took place during the Pleistocene, when violent tectonic uplift occurred in what King described as the arching of the whole Andes region. Faulting occurred which emphasised the previous structural arrangement of parallel mountains with basins between. In the case of the Altiplano and the Puna de Atacama the basins were thrust up to a mean elevation of 3750–4000 m. The original landsurface, of which these plains were part, has now been deeply dissected by rivers, particularly by the headwaters of the Amazon which have cut back into the mountains and destroyed the initial surface of the anticlinal flexure. Finally, during the Pleistocene an extensive ice cap covered the Andes and glaciation still affects many of the highest peaks today.

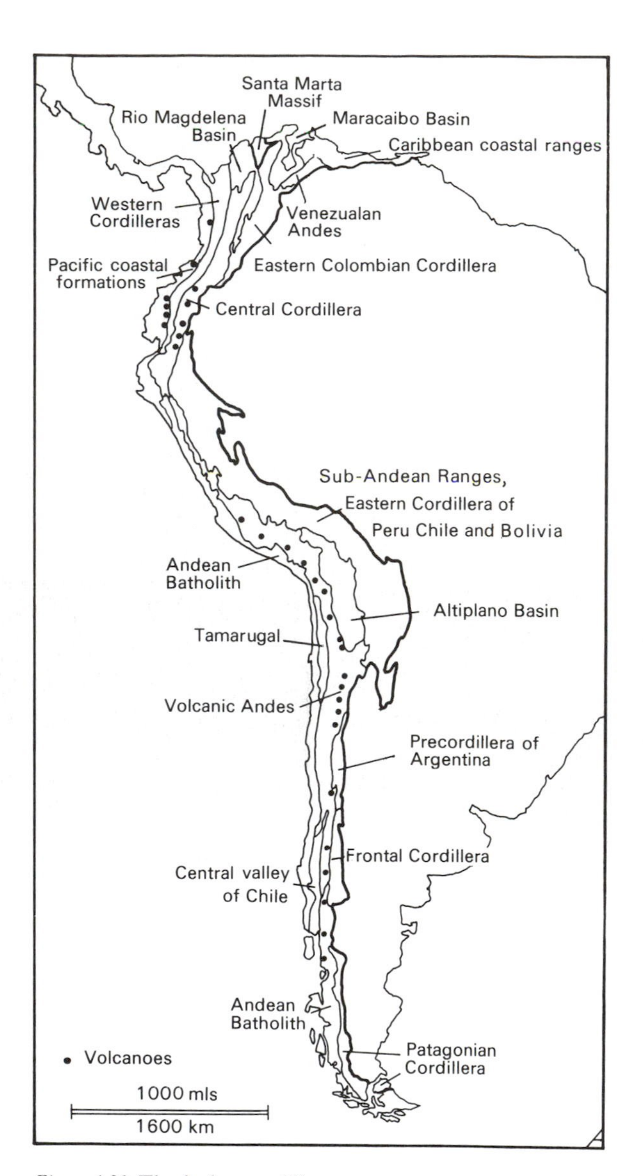


*Figure 4.31.* The Andean cordilleras.

The geomorphology of the Andean physiographic region is presented in the following provinces (Figure 4.31).

The Caribbean coastal ranges
The Venezuelan Andes
The Maracaibo basin
The eastern Colombian cordillera
The Rio Magdalena basin
The central cordilleras
The western cordilleras
The eastern cordilleras of Peru and Bolivia
The volcanic Andes
The Altiplano basin
The Pampa de Tamarugal
The precordillera of Argentina
The central valley of Chile
The Andean batholith
The frontal cordilleras and Patagonian cordilleras

*Figure 4.32.* The North Coast ranges of Trinidad.

## *The Caribbean coastal ranges*

Commencing on the northern coast of Trinidad, the Caribbean coastal ranges are interrupted by the sea in the Golfo de Paria, but resume in the Paria peninsula of Venezuela (Figure 4.32). West of the town of Barcelona, the coastal ranges are broken by the Unare lowland, but continue south of Caracas, eventually swinging southwards in the Cordillera del Merida. These mountains are formed from metamorphic Mesozoic rocks which appear from beneath the alluvia of the Orinoco lowland to the south. Folding took place in the late Cretaceous.

The Caribbean coastal ranges are nowhere more than 2500 m in elevation and consist of two parallel ranges with intervening basins. There is a steep descent to the coast from the northern range, but slopes are more gentle on the southern side and the southern range has many gaps providing easy access to the Llanos sedimentary plains to the south. Drainage of these southern slopes is to the Orinoco, and the intermontane basins are drained in an easterly direction or to Lake Valencia.

## *The Venezuelan Andes*

South-west of the town of Barquisimeto, the trend of the coastal mountains turns south-west in a range known as the Sierra de Merida. According to King, these mountains show positive evidence of late Palaeozoic mountain building and associated granitic intrusion in their structure. The present relief, though, is formed from various Eocene marine or brackish water sediments, Pliocene lacustrine materials and similar Quaternary deposits.

North of the Cordillera de Mérida and to the east of Lake Maracaibo in the provinces of Falcón and Lara, the trend of the mountain ridges is similar to that of the Cordillera de Mérida, consisting of sharply folded anticlines and synclines. The rocks are mostly of Tertiary age with Eocene and Miocene marine deposits with Pliocene rocks consisting of part marine and part continental facies.

*The Maracaibo basin*

The basin occupied by Lake Maracaibo lies between the Venezuelan Andes and the eastern Colombian cordilleras. It is reported as having a considerable thickness of pre-Quarternary sediments below recent alluvium. The Lake is approximately 160 km from north to south and 120 from east to west and it is linked to the Gulf of Maracaibo by a narrow channel 32 km long and 8 to 16 km wide. Oil was found on the eastern shores of the Lake and production has taken place over a prolonged period.

*The eastern Colombian cordillera*

These mountains lying between the western Llanos of Colombia and the Magdalena valley form the eastern limb of the Andes. The trend of the range is similar to that of the Venezuelan Andes, but at the north-eastern end the ranges swing northwards to form a natural prolongation of the Sierra de Perija, finally ending in the Peninsula de Guajira on the northern side of Lake Maracaibo. The Cordillera Occidental rise to their highest point in the Alto Ritacurvo (5900 m) and are composed mainly of Cretaceous sediments. This range has not been affected by volcanic activity, but the Sierra de Perija has a core of Palaeozoic and Pre-Cambrian crystalline rocks.

*The Rio Magdalena basin*

The Magdalena is a strike valley between the eastern and western cordilleras of Colombia. The valley originated during the late Tertiary and it contains marine sediments of late Tertiary and non-marine facies further inland. West of the Magdalena river and south of the town Barranquilla, the strata become strongly folded as the western cordilleras are approached.

*The central cordilleras*

The central cordilleras increase in height from north to south. They rise from the alluvium of the Magdalena river and comprise mainly metamorphic and igneous rocks, ranging in age from Palaeozoic to Mesozoic. Volcanic activity took place during the Miocene and still persists today to a minor degree. On the western side of the cordilleras a deep strike valley is occupied by the Rio Cauca, a tributary of the Magdalena. Further south in Ecuador, the central cordilleras became the eastern cordilleras as the mountains of Peru merge with the central range south of Pasto. The line of the Cauca valley southwards is marked by the faulted basin or graben between the two main ranges. This basin is infilled with abundant Tertiary and Recent volcanic material. The folding of the mountains has incorporated a Pre-Cambrian core of metamorphic and igneous rocks which eastwards become buried beneath the later Palaeozoic, Mesozoic and Tertiary rocks.

*The western cordilleras*

The range of the Andes west of the inter-cordilleran depression in Ecuador and Colombia includes peaks of between 4000 m and 5000 m. In northern Colombia this range splits in two, one branch continues northwards and ends against the coast of the Caribbean and the other swings north-west into the Isthmus of Panama to join the Darien mountains. Most of the rocks at the surface are volcanic, but these overlie Jurassic and Cretaceous sedimentary strata. Plutonic intrusion of granodiorites occurred when folding took place in the late Cretaceous.

*The eastern cordilleras of Peru and Bolivia*

The Front ranges of the Andes have been compared with the Front ranges of the Rockies in North America. Like the Dakota sandstone in the USA, these South American equivalents are formed by escarpments which rise to heights of 1600 m or even 2000 m and are steeply incised by gorges which carry antecedent streams from the higher mountains to the plains of the sub-Andean depression. A little to the south of Santa Cruz in Bolivia (at about 19°S) is the continental divide between streams draining to the Amazon and those to the Rio de la Plata in the south. The eastern cordilleras of Peru and Bolivia lie on the eastern side of the Altiplano and reach heights of over 5000 m. It is on the tops of these mountains that geomorphologists have attempted to trace a warped erosion surface from the plains of the Altiplano basin, the late Pliocene Puna surface. Throughout this large area the rocks of the mountains are mainly sedimentary from all formations

*Figure 4.33.* The Altiplano. (By courtesy of R.D.F. Bromley.)

ranging back to the Pre-Cambrian; some of the older sediments are locally metamorphosed, but there is a lack of volcanic activity compared with the ranges to the west.

### *The volcanic Andes*

From Ecuador to Central Chile the western Cordilleras of the Andes are characterised by a covering of late Tertiary to Recent ejecta. In northern Peru the cover is discontinuous, but in the south of that country it reaches its maximum development and width. The amount of lava and other material thrown out by the volcanos amounts to between 2000 m and 2700 m thick, and in southern Peru, 175 km wide. The crest line of the western Cordilleras is dominated by the great volcanic peaks of Coropuna (6425 m), Chachani (6075 m) and El Misti (5822 m). Older lavas, the Tacaza group, can be distinguished from more recent eruptive rocks, the Sillaopaca group, because the latter have been effected by earth movements. The most abundant lavas are andesite, trachyandesite and trachyte, but in Chile and Argentina Quaternary eruptions produced olivine basalts.

### *The Altiplano basin*

This elevated intermontane basin extends for approximately 1000 km from southern Peru, across Bolivia and into Argentina and Chile (Figure 4.33). The floor of the basin is at an altitude of 3750–4000 m and is some 200 km wide. The basin originated during the Miocene orogeny and had been infilled before it was finally uplifted during the early Pleistocene. The lower deposits of the basin are mainly sedimentary in origin and have been folded, but more recent accumulations were of volcanic material. King described the floor of the Altiplano as being ringed with raised beaches and floored with the sediments of shrunken or extinct lakes. Two large lakes, Titicaca and Poopo and several large salt flats (salars) occur, including the Salar de Uyani and the Salar de Coinasa (Figure 4.34).

*Figure 4.34.* Lake Titicaca. (By courtesy of R.D.F. Bromley.)

*The Pampa del Tamarugal*

This region is a tectonic depression between the coast ranges and the western cordilleras in northern Chile. It runs at an angle to the coast, intersecting it south of Arica. Tertiary and Quaternary volcanic deposits and alluvial materials have partially infilled the depression which also has extensive deposits of salts including chlorides, nitrates and sulphates.

*The precordillera of Argentina*

Emerging from beneath the volcanic Andes in the extreme west of Argentina, south of 27°S in the province of La Roja and extending through San Juan and Mendoza provinces is a series of upfaulted ridges and downfaulted valleys. These ridges and valleys of the Argentinian precordilleras are formed from a thick sequence of marine and continental rocks. They were first faulted in the Pliocene and then uplifted in the early Pleistocene.

*The central valley of Chile*

South of Santiago in Chile there develops a long depression between the coastal ranges and the High Andes with an elevation of less than 200 m. The central valley came into being with the Plio-Pleistocene uplift of the Andes. With a block-faulted graben structure the central valley has been partly infilled with alluvium derived from volcanic deposits in the High Andes. In the southern part of the depression glacial deposits occur, but further north a hummocky landscape was caused by debris and mud flows. Some eroded tertiary volcanoes lie in the central valley between 37° and 48°S. South of Puerto Montt the central valley continues as an inlet of the sea, the coastal ranges forming the Archipelago de los Chonos.

*The Andean batholith*

Extending the length of Peru and Chile is a batholith mainly composed of granodiorite, granite and diorite.

This continuous belt of plutonic rocks, including some Palaeozoic remnants, forms the western coastal range of the Andes. Marine terraces have been observed on the coasts of Peru and Ecuador at three different levels. These decline in elevation southwards but differential uplift means they show little constancy of height. In the extreme south of Patagonia, the Andean batholith forms the main mountain divide, rising to 2500 m near Ushuaia.

*The frontal cordilleras and Patagonian cordilleras*

Continuing southwards from the precordilleras through Mendoza and Neuquen provinces of Argentina are the frontal cordilleras. These mountains are formed from Mesozoic rock from Triassic to Cretaceous in age together with volcanic rocks. They form the eastern ranges of the Andean system and lie between the Pre-Cambrian basement of the Patagonian massif and the western Chilean mountains. After a short break in the west of Chebut province these mountains continue as the Patagonian cordilleras the crest of which forms the political boundary between Argentina and Chile.

## Oceanic landforms of the South American plate

The South American plate extends from the mid-Atlantic ridge westwards to the west coast of South America where it overrides the Pacific plate. The southern margin of the South American plate lies north of the Scotia plate and it continues northwards to the Caribbean plate. Its eastern margin is constructive, the western margin is destructive but the northern and southern boundaries are conservative. As the continental material of the South American plate has overridden the oceanic Pacific plate, probably forcing the trench and subduction zone to migrate westwards, all the oceanic terrains associated with the South American plate lie beneath the south Atlantic Ocean.

*The Guyana basin*

There is no clear demarcation between the Guyana basin and the north western Atlantic basin. After trending south-westwards from the Azores the mid-Atlantic ridge turns south-eastwards at latitude 10°N in a series of transform faults towards St Paul's Rocks. This part of the mid-Atlantic ridge forms the eastern margin of the Guyana basin.

On the western margin, the continental shelf at the mouth of the Amazon is 400 km wide, but its width decreases both north and south of the estuary. This continental shelf and slope is crossed by a major submarine canyon related to the northern distributary of the Amazon and which extends down on to the abyssal plain. Fresh water from the Amazon can be traced over 150 km from its mouth.

On the floor of the basin, a number of seamounts rise from the abyssal plain; some are scattered on the slopes of the mid-Atlantic ridge but others cluster near the Brazilian coast north-east of Cape San Roque. Two of these, Rocas and Fernando de Noronha, emerge as islands. The southern limit of the Guyana basin is also indistinct, marked only by a slight rise in the ocean floor between Cape San Roque and the mid-Atlantic ridge.

*The Brazil basin*

At the equator the mid-Atlantic ridge is strongly offset to the east along the Romanche fracture zone; this forms the northern margin of the Brazil basin. The eastern boundary is the mid-Atlantic ridge and the southern boundary the Bromley plateau (Rio Grande rise), at one point only 638 m below the sea surface. Much of the basin is taken by the abyssal plain, over 6500 m deep, but of the seamounts which occur, only Trinidad forms an island. The Bromley plateau is comparable to the Walvis ridge below the eastern south Atlantic formed by volcanic outpourings which mark the tract of the westward-migrating South American plate.

*The Argentine basin*

This basin contains an extensive area of abyssal plain at about 6500 m depth extending between the Bromley plateau and the Scotia arc, south of the Falkland (Malvinas) Islands. The mid-Atlantic ridge forms the eastern margin and along the Brazil–Uruguay coast there is an extensive continental shelf. North of Rio de Janeiro it is relatively narrow but south of Bahia Blanca it attains a width of over 400 km. In the extreme south the Falkland Islands, formed of continental rocks, lie on the edge of the shelf.

A submarine canyon which crosses the continental shelf is associated with the River Plate. Seamounts are not so common in this basin, although one near the Uruguay coast has only 11 m of water over its summit. The southern margin of the Argentine basin lies north of the islands of the Scotia Arc which includes South Georgia.

### *The Scotia Arc*

Between Tierra del Fuego and the Antarctic peninsula the Andean structures make a large loop, 2000 km eastwards and 800 km from north to south, through South Georgia, and the South Sandwich Isles. These Islands mark the edge of an oceanic micro-plate which terminates at the Meteor Deep (−8264 m), an oceanic trench where the Atlantic sea floor is being consumed. Over the micro-plate, the sea is 3500–4000 m deep and it appears that the South American plate has pushed westwards past this part of the ocean floor, leaving the trench and volcanic arc behind. The trench may originally have been part of the Pacific subduction zone, most of which has been displaced westwards by the movement of the South American plate.

5

# Antarctica

The continent of Antarctica is almost completely ice-covered and so presents a very different appearance from the other land masses of the world. It is 13.3 million km$^2$ in area, centred almost on the South Pole and from 90°W to 90°E it is approximately 4400 km.

Extension of sea-floor spreading has resulted in Africa, India and Australia all being forced away from Antarctica, as the Southern Ocean widened. The extent of the Antarctica lithospheric plate is shown on Figure 5.1. The triple-junctions are the points where the Atlantic, Indian and Pacific mid-oceanic spreading centres join the margin of the Antarctic plate. Unlike the Pacific Ocean, there are no trenches or subduction zones associated with the Antarctic plate.

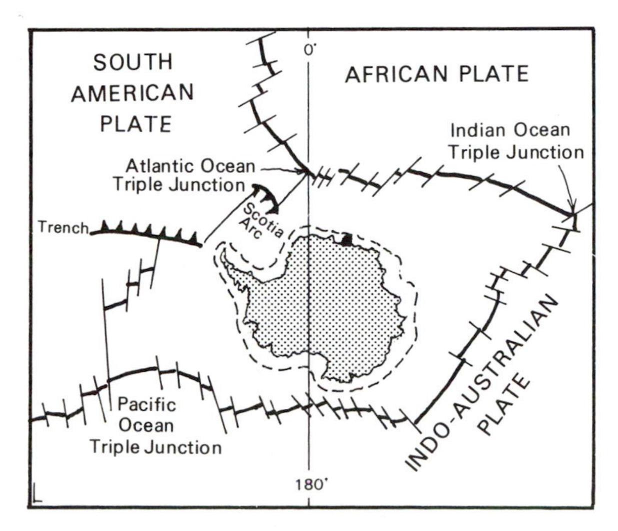


*Figure 5.1.* The Antarctic plate.

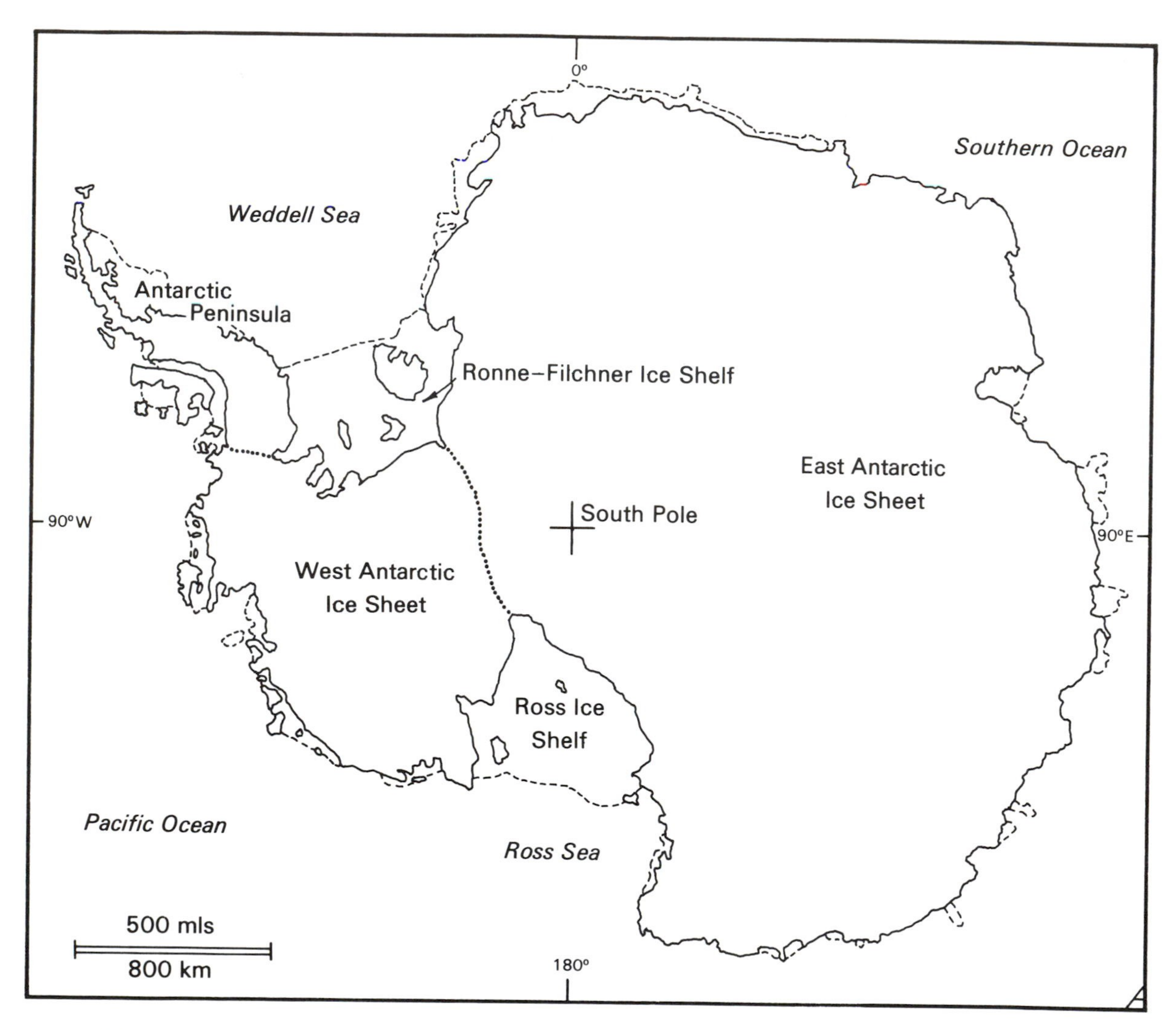


*Figure 5.2.* Major divisions of Antarctica.

The continent has a mean elevation of between 2000 and 2500 m above sea level which means that its plateau-like surface is the most consistently high elevated surface on the Earth. The almost circular shape of the continent is broken by the large embayments called the Ross Sea and the Weddell Sea which effectively separate the smaller West Antarctica from the larger, compact, East Antarctica. The only areas free of ice lie around the periphery or on the Antarctica peninsula.

The relief of Antarctica at present is virtually the same as the relief of the ice sheet which covers the continent. East Antarctica has a large single dome of ice, 10.35 million km$^2$, which reaches 4000 m in its highest part, but West Antarctica has three smaller separate ice caps, altogether only 1.97 million km$^2$, and the underlying bedrock is much more evident, including the highest point on the continent, the Vinson massif, at 5140 m (Figure 5.2). Beneath the ice the bedrock contours, mapped by seismic or microwave profiling of the ice sheet, show that Antarctica has an upstanding rim with a hollowed-out centre. At the South Pole there is 2700 m of ice overlying a rock surface at 100 m below sea level. Under the highest parts of the east Antarctic plateau there may be up to 4000 m of ice as hollows occur in the bedrock which go to 600 m below sea level. Without the ice cover and if no isostatic compensation or rise in sea level took place, the maximum relief throughout most of east Antarctica would be 2000–3000

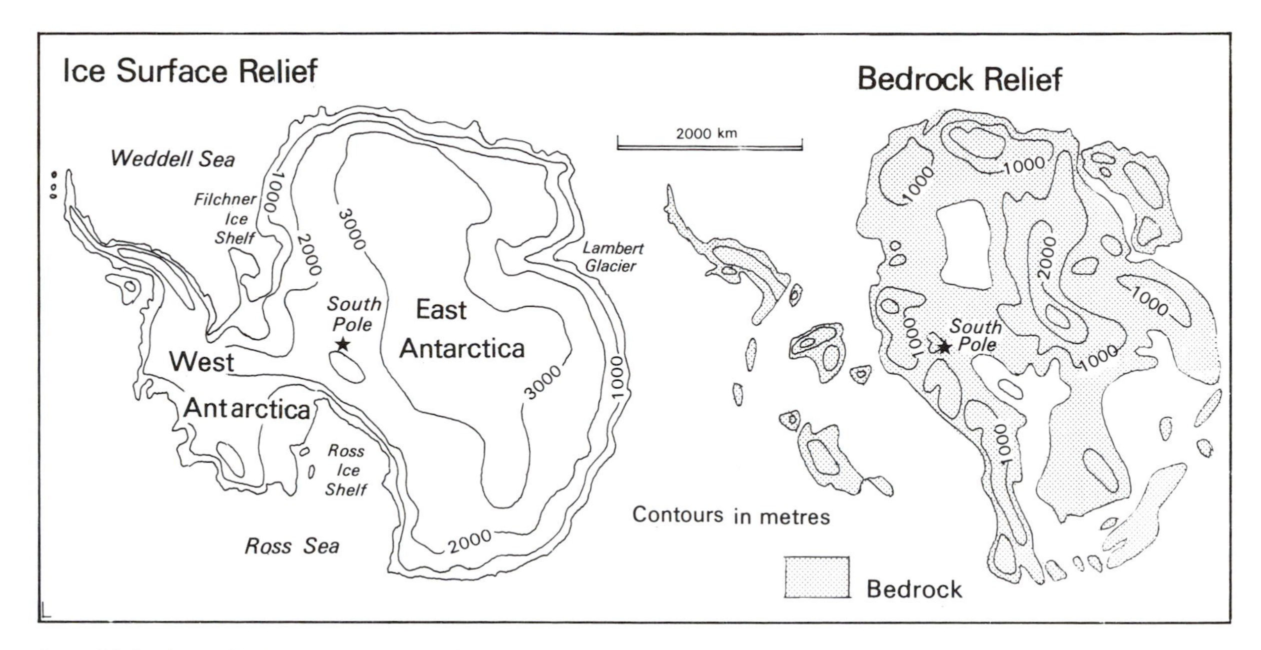


*Figure 5.3.* Surface and bedrock relief of Antarctica.

m and west Antarctica would become an island archipelago (Figure 5.3).

The geology of Antarctica is mainly known from exposures in the mountains which emerge as nunatacks around the margin of the ice sheet. East Antarctica is underlain by schists and gneisses of the basement complex, and was formerly part of Gondwana. Cambrian limestones lie unconformably on the basement complex, and in west Antarctica Devonian sandstones and shales occur, including the characteristic *Glossopteris* plant fossils. The range of fold mountains which constitutes the west Antarctica peninsula is a continuation of the same trend of folding which loops around the South Georgia micro-plate to form the South Sandwich island arc. These mountains continue across west Antarctica in the Ellsworth mountains to Marie Byrd Land on the coast of the Ross Sea. Jurassic igneous intrusive rocks occur along the line of a former early Palaeozoic orogenesis which links with similar rocks near Adelaide in Australia, the Cape folding of South Africa and the fold mountains of Uruguay. Cretaceous basaltic lavas occur on the coast of Coats Land which match those of the South Africa–Mozambique border district. More recent volcanic activity on Antarctica occurred 4 million years ago when volcanic cones erupted in the glacially scoured valleys of Victoria Land, and Mount Erebus is still active at the present day.

The ice on the Antarctic continent gradually flows seawards under the influence of gravity in an almsot centrifugal pattern. Where the underlying relief facilitates it, the ice makes use of valleys such as the Lambert glacier which is over 400 km long and occupies a 50 km wide valley leading down to the Amery ice shelf. Other glaciers such as the Beardmore glacier have been used by explorers to gain access to the high continental plateau surface through the nunatak mountains of the continental rim. For most of Antarctica, the ice sheet completely covers the land surface and ends in sheer ice cliffs at the coast.

In the Weddell and Ross Seas the ice forms floating shelves, which are 200 m thick at the sea cliff and 1000 m thick at the landward margin. As atlas maps of Antarctica are usually on a small scale, it is worth commenting that the 520 000 km$^2$ Ross ice shelf has the same area as France; the Ronne–Filchner ice shelf in the Weddell Sea is only slightly smaller with an area of 400 000 km$^2$. During winter the pack ice around the Antarctica may extend as far north as 55°S, doubling the size of the continent.

Research points to the first major accumulation of ice taking place on east Antarctica between 11 and 14 million years ago, but there is evidence in deep sea cores of the presence of glaciers as early as 22.5 million years ago. The ice accumulation of West Antarctica is younger, probably dating from between 4–5 million years ago. At its greatest extent during the Pleistocene, the Antarctic ice sheet may have contained up to 80% more ice than at present. It is the best example of a continental ice sheet available for study. The east Antarctic sheet is land based, frozen to the surface, and relatively stable. The west Antarctic ice caps are resting on bedrock below sea level and could be more readily influenced by a rising sea level or loss of ice through climatic warming, either of which could trigger the final stages of deglaciation.

Studies of the ice-free areas, especially in Victoria Land, have shown evidence of a former erosion and the presence of boulder clays and outwash materials indicating the former greater extent of the ice. Deposits of marine till are also present on the submerged continental shelf. On land, wind action has played an important part in landform development; fine materials have been removed to leave a desert pavement, sand has been blown into dunes and ventifacts produced through the effects of sand-blasting. Although the rate of evaporation is not high, it is sufficient to gradually concentrate salts in the small lakes present. They are floored with evaporites and have a distinctive biology.

# 6

# Asia

Asia is the largest landmass, amounting to 44 million km$^2$. It extends across 120° of longitude from the Urals to the Bering Straits, and from 80°N in Siberia it extends to 10°S of the Equator in Indonesia. Bounded by the Arctic Sea to the north, the Pacific Ocean to the east and the Indian Ocean to the south, it is only in the west where there is any doubt about Asia's boundaries. Traditionally, geographers have included Asia Minor, the Middle East and the Indian sub-continent in Asia. The Ural mountains have likewise been the traditional boundary between Asia and Europe. The continent of Asia includes the highest mountain, Everest (8848 m) and the lowest place on earth, the Dead Sea rift valley (−396 m).

In view of its large size and complexity, it is proposed to discuss the geomorphology of Asia in two parts. The first part will consider the plains, plateaux and mountains of northern Asia in Siberia, China and south-east Asia. The second part will be devoted to the plateaux of India and Arabia and the fold mountains of Turkey, Iran, Afghanistan, Pakistan and India. The island arcs along the east coast of Asia are described in Chapter 9.

The presence of the high plateaux of central Asia raises the mean elevation of the landsurface of the continent to 915 m above sea level. According to Butzer, young mountains only amount to 19% of the land surface but old mountain belts cover 33% and crystalline shields 10%. Volcanic plateaux and plains only cover 4% but depositional plains occupy 22% and erosional plains and plateaux in sedimentary rocks 12%.

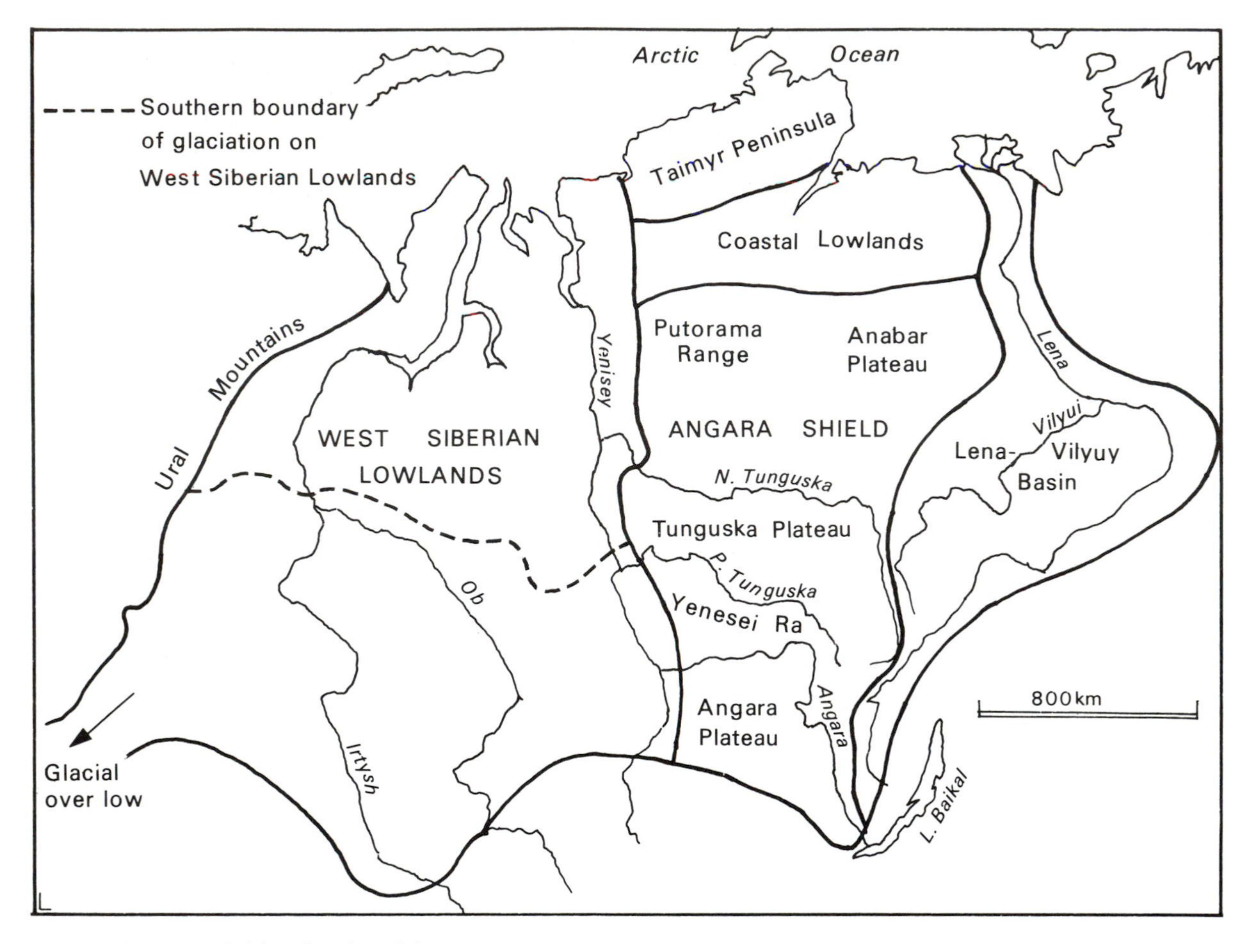


*Figure 6.1.* Plateaux and plains of northern Asia.

## NORTHERN ASIA

The boundaries of the northern part of the Asian plate may be traced beneath the Arctic Ocean where new sea floor has been created in the Siberian basin. To compensate for the Arctic sea floor extension, Asia appears to have pivoted about a point in the New Siberian Islands and compression occurred in the region of the Verkhoyansk mountains which were uplifted during the Mesozoic along the eastern margin of the Angaran shield. A southern boundary may be traced along the northern margin of the Alpine folds of Iran, Afghanistan, India, Nepal and Bhutan. East of the Brahmaputra, the boundary turns south at the eastern end of the Himalayas to the Bay of Bengal along the line of the Naga hills and Arakan Yoma. It continues around Indonesia and follows the edge of the continental shelf on the eastern seaboard of China. The boundary between the Eurasian and North American lithospheric plates occurs across the 'neck' of Alaska continuing the line of the Aleutian trench rather than at the Bering Straits.

### Geological history

Northern Asia has developed from fragments of the ancient landmass known as Laurasia. Break-up of this former continent gave the core areas of the present continents in the northern hemisphere. The rocks comprising Laurasia were mainly pre-Cambrian crystalline rocks, gneisses and schists. In Asia these are the Angara shield, the Inner Mongolian–Korean shield, the Ordes shield and the south-east Asia shield. Since the break-up of Laurasia, the segments which have recombined to form the present continent have been subjected to orogenesis around their margins to give the complex of mountain ranges and plateaux of different elevations which characterise northern Asia today. Outcrops of these ancient rocks are confined to upfolded or upfaulted sections of the shield but their presence has been confirmed below Mesozoic and later sediments where these overlap the older rocks.

Fragments of the former continent of Gondwana have also contributed to the building of northern Asia. The northwards movement of the Arabian and Indian shields closed the Tethys Sea and the collision resulted in the production of the high fold mountains and elevated plateaux of central Asia.

Mountain building has affected northern Asia many times but for convenience it can be attributed to three main periods of geological history. Around the ancient shields, the older fold mountains are attributed to the Caledonian and Hercynian orogenies of the middle and late Palaeozoic. These only affected Asia north of the line of the Himalayas. The more recent earth movements of the Alpine orogeny resulted in extensive folding and faulting of Mesozoic and early Tertiary sediments from the Tethys geosyncline. Major east–west lithospheric faults delimit the Tibetan and Mongolian plateaux as well as the structural basins of Tarim, Qaidam and Junggar. These major faults probably result from the stresses caused by the impaction of the Indian plate into the Laurasian landmass. Erosion of the high mountains and plateaux produced by the Alpine orogeny has resulted in large quantities of sediment which the rivers have transported to produce extensive alluvial plains in India, China and Cambodia. Great thicknesses of sediment have been deposited in areas of internal drainage such as the Tarim and Dzungarian basins.

Pleistocene glaciations of northern Asia, although extensive, play a less significant role in the recent geological history of the continent compared with North America or Europe. The Scandinavian ice sheet continued east of the Urals, covering the northern two-thirds of the Ob basin and extending onto the elevated parts of the Angara shield between the Yenesei and Lena rivers. Elsewhere, mountain glaciation left a legacy of glacial features on the east Siberian mountains, the mountains of Kamchatka, the Altai, Tien Shan, and other smaller areas of high mountains. Ice caps remain today in the islands of Severnaya Zemlya and Novaya Zemlya, and individual glaciers occur on many central Asian mountains. Permafrost is widely distributed across Siberia varying from 30 to 600 m in depth and covering an area of 9.6 million km$^2$.

Volcanic activity characterises several of the mountainous regions of northern Asia. The Koryat range and the Kamchatka peninsula bordering the Bering Sea contain active volcanoes and the Anadyr plateau is formed from igneous rocks. On the Mongolian plateau there is an area of basaltic lavas and volcanic cones, probably associated with one of the large lithospheric faults which occasionally are the site of major earthquakes. Except for Indonesia, the major volcanic province associated with the island arcs of the western Pacific is discussed in Chapter 9.

Discussion of the geomorphology of northern Asia will be in seven major geomorphological divisions, subdivided into many provinces, based on structure and landforms:

Plateau and plains of northern Asia
Mountains and plateaux of north-east Asia
Mountains and plateaux of east-central Asia
Mountains and plateaux of central Asia
Uplands and plains of eastern Asia
Mountains, basins and plateaux of south China
Mountains and plains of south-east Asia

Although the deposits of the mountain ranges of Asia are well known and the regional structure understood, the detailed geomorphology of this large land area is imperfectly known. Thus it is not always possible to provide much more than the broad structural background upon which the geomorphology is developed.

## Plateaux and plains of northern Asia

Northern Asia is built around the Angara shield which occupies the land between the Yenisey and Lena rivers (Figure 6.1). The shield also underlies the lowlands of the Ob river to the west but folded and faulted mountains of Lower Palaeozoic rocks occur to the south and east in the central Asian mountains and the East Siberian mountains. There is a limited area of coastal plain west of the Lena basin and the shield rocks crop out again in the Taimyr peninsula.

The plateaux and plains of northern Asia include:

The Angara shield and Taimyr peninsula
The west Siberian plain
The Arctic coastal lowlands

*The Angara shield and Taimyr peninsula*

The Angara shield occurs between the Yenisey and Lena rivers where it forms a dissected plateau on Archean and Protoerozoic rocks of Pre-Cambrian age. These rocks of the basement complex are schists and gneisses but they are extensively covered by Cambrian sandstone and conglomerates. Despite resting upon the shield, some of these rocks have been affected by folding of Caledonian age, and Tertiary lavas lie unconformably upon them.

The relief of this province is mainly between 300 and 800 m above sea level but the higher mountains rise to 1500 m. The divides between the major rivers have an accordance of level at 700–800 m and lower erosion surfaces have been identified at 300–400 m and 150–200 m above sea level. The present relief was developed following the Alpine orogeny when this province underwent a general uplift, followed by Pleistocene glaciation.

Several major rivers drain this province including the Angara, the lower Tunguska, the upper Tunguska, tributaries of the Yenisey, and the headwaters of the Lena and its tributary the Vilyui. The Kotuy, Anabar and Olenek rivers drain directly to the Arctic Sea.

Within this central Siberian plateau province several sections may be identified. These include, from north to south, the Taimyr peninsula, the Anabar plateau and the Putorama range which rises to 1500 m above it; the Lena–Vilyuy basin; the Yenisey ranges which rise from the Tunguska plateau and the Angara plateau itself which lies between the eastern Sayan mountains and the upper Tunguska river.

*The west Siberian plain*

The west Siberian plain lies between the Urals and the plateau of the Angara shield. It is widest in the south (2000 km), narrowing in the north to 800 km, and in a north–south direction it extends 2200 km. The plain has a very low relief, even in the southern part, 1800 km from the Arctic Sea, the land is only 130 m above sea level.

Most of the plain is developed upon Mesozoic sedimentary rocks with overlying superficial deposits. Jurassic and Cretaceous rocks crop out at the edges of the plain and in the Yenisey valley where they cause rapids in the lower course of the river. Occasional outcrops of Devonian or Carboniferous rocks form low, gently rounded hills above the drift-covered plain immediately east of the Ob river, but the basment complex is deeply buried. During the marine transgression of the Mesozoic, rivers flowing northwards from central Asia formed deltas at the southern edge of the plain. These deltaic deposits remain as sandy, lignite-bearing rocks forming a pediment at the foot of the mountains. North of these sandy pediments lies a zone of Quaternary alluvial deposits, but further north Pleistocene glacial deposits cover the northern two-thirds of the plain. The low relief and the nature of the surface deposits mean that the west Siberian plain is poorly drained with many swamps. The occurrence of swamps is encouraged by the presence of permafrost throughout much of the plain, and also by the headwaters of the Siberian rivers thawing before their lower courses. In southern districts, the swamps are smaller, occurring in shallow depressions and it has been suggested that these may be remnants of more extensive wetlands which have become infilled by wind-blown sediments. In the extreme south-west of the Ob basin, there is an extensive area of flat lacustrine sediments. A proglacial lake is thought to have inundated this area when the mouths of the northward-flowing rivers were blocked by Pleistocene ice. The lake overflowed southwards into the Aral–Caspian depression across the Turgai plateau.

From north to south the following sections of the Siberian plain may be identified. At the mouths of the Ob and Yenisey rivers a recent coastal marine alluvial plain lies adjacent to the Arctic Ocean. A plain of glacial deposition extends south to the latitude of 52°N, south of which outwash plains extend to meet the pediments from the Kazakh uplands described previously.

*The Arctic coastal lowlands*

During the Cretaceous, a marine transgression affected the Arctic coastal lowlands of northern Siberia and the valleys of rivers such as the Lena, Yenisey, and Anadyr. Marine sediments were deposited and subsequently covered with Tertiary and Pleistocene materials. These areas now form extensive plains characterised by swampy conditions which are encouraged by the presence of permafrost. The Lena basin covers 2.3 million $km^2$, extending from its delta 1200 km inland to beyond Yakutsk.

Between the Lena and Yenisey rivers and the

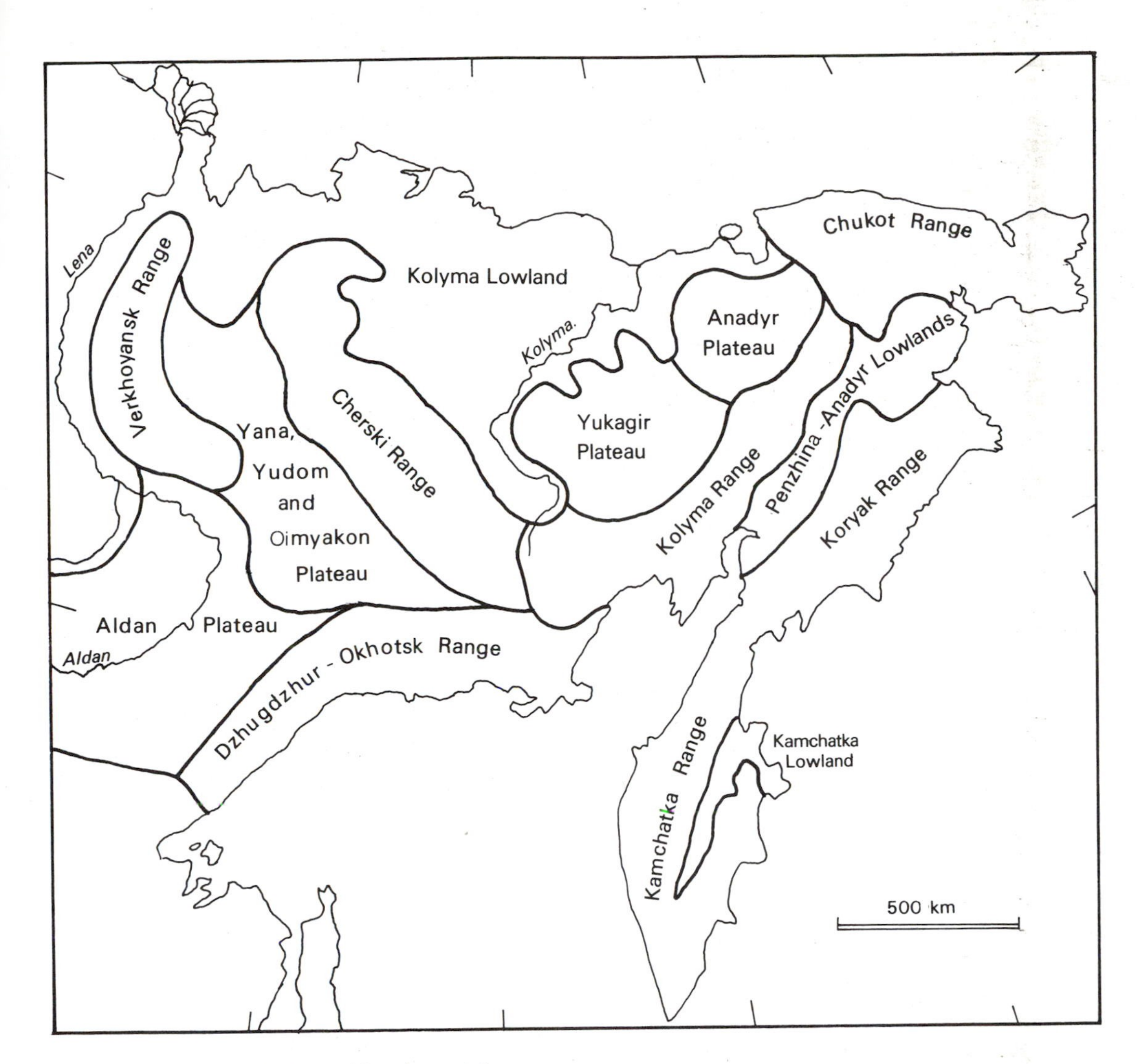


*Figure 6.2.* Mountains and plateaux of north-east Asia.

upstanding areas of the Angara shield and the Taimyr peninsula, the coastal lowland extends westwards to the west Siberian plain along the Khatanga depression.

## Mountains and plateaux of north-east Asia

The fold mountains of east Siberia comprise the sediments of a former shelf sea on the eastern flank of the Angara shield which were deposited and folded at various intervals during the Palaeozoic and Mesozoic eras. Except for the Verkhoyansk and Cherski ranges which have a north-west–south-east alignment, these fold mountains trend south-west to north-east continuing the alignment of the south Siberian mountains east of Lake Baikal (Figure 6.2). Provinces within this major geomorphological division include:

The Verkhoyansk and Cherski ranges
The Yana, Yudom and Oimyakon plateaux
The Yukagir and Alazeya plateaux
The Kolyma lowland
The Kolyma range
The Dzhugdzhur–Okhotsk mountains
The Anadyr plateau
The Penzhina–Anadyr lowland
The Koryak range

The Kamchatka peninsula
The Chukot range

*The Verkhoyansk and Cherski ranges*

The Verhoyansk and Cherski ranges, together with the plateaux between them may be considered as a single structural and morphological unit. The Verkhoyansk range is formed from Upper Palaeozoic rocks, folded during the Lower Cretaceous. This S-shaped range of mountains follows the line of the Lena river from which the land rises very steeply to a general height of 2000 m and peaks of 2700 m. The general summit level is a smooth, undulating plateau with relict peneplain surfaces, but the higher peaks support corrie glaciers. The Cherski range is the eastern margin of this fold zone with Palaeozoic rocks forming the eastern part and Mesozoic rocks the western part of the range. Granite intrusion has also occurred in this range which rises to 3300 m. As in the Verkhoyansk mountains, glaciation has been experienced, but active glacial features are restricted to corries on the higher peaks. The Suntar–Cherski mountains are a smaller range to the south of the Verkhoyansk–Cherski ranges with peaks at 3000 m, but these mountains retain a considerable ice cover with over 100 glaciers and firn fields. The glaciers are reported to be 200–250 m thick and to descend to 2000 m.

*The Yana, Yudom and Oimyakon plateaux*

Between the ranges of the Cherski and Verkhoyansk mountains, these intermontane plateaux lie at between 700 and 1000 m above sea level. They are formed from Triassic and Jurassic rocks with some igneous intrusions and are covered by Quaternary glacial deposits. The Oimyakon plateau in the centre is drained by the Indigarka river which has caused a more dissected relief. The valleys of this region are said to experience the coldest temperatures on Earth during winter. The Yudom plateau in the south forms the continental divide between rivers flowing to the Arctic and those flowing to the Sea of Othotsk.

*The Yukagir and Alazeya plateaux*

These plateaux in the basin of the Lena river rise to about 1000 m above sea level, but mostly have an elevation of between 400 m and 700 m. They are formed from Palaeozoic rocks covered by horizontal Mesozoic shaley sandstones, but a granite intrusion has also taken place below the Alazeya plateau. The Quarternary terraces of the Kolyma and Omolu rivers merge upstream into the surface level of the Yukagir plateau.

*The Kolyma lowland*

This lowland, less than 50 m above sea level has a cover of unconsolidated Quaternary sediments with permafrost. There are many swamps and lakes, some of which are the result of thermokarstic features in the permafrost. The Kolyma river has an estuary into the Arctic Ocean. The Indigarka river has an extensive delta.

*The Kolyma range*

The structural trend from north-east to south-west is re-established in the Kolyma range which lies to the south-east of the Yukagirsk plateau, and at its northern end gradually merges into the Anadyr plateau. The Kolyma range is formed from Permian, Triassic and Jurassic rocks which have been well dissected with the higher parts having an alpine relief.

*The Dzhugdzhur–Okhotsk mountains*

There is only a narrow coastal lowland on the shore of the Sea of Okhotsk and Shelekov Bay as the Okhotsk mountains, developed from Mesozoic sedimentary and acid igneous rocks lie close to the sea. Although called mountains, this area is a faulted and dissected group of sedimentary rocks intruded by granite which forms a plateau-like area, 2000 m high, which descends in a series of fault scarps to the sea.

*The Anadyr plateau*

At the northern end of the Kolyma range, and between the Anadyr lowland and the Kolyma lowlands, the 800–1000 m Anadyr plateau is formed from Quaternary extrusive rocks. Lake Elgythyn is in a dormant caldera, 11.5 km in diameter.

*The Penzhina–Anadyr lowland*

The Penzhina-Anadyr lowland lies between the folded Anadyr–Chukot ranges and the Koryat mountains of the east coast. This is a lowland of less than 200 m above sea level which is also an area of subsidence infilled with volcanic materials and alluvial sediments.

### *The Koryak range*

The Koryak mountains, north of the Kamchatka peninsula, are formed from Mesozoic shales and sandstones with intrusive igneous rocks. Many of the peaks of the Koryak mountains are volcanic, some of which are still active. These mountains are subject to glaciation and many corries on the alpine highlands feed valley glaciers up to 5 km in length.

### *Kamchatka Peninsula*

The 1000 km long peninsula of Kamchatka extends southwards from the Koryak mountains to link with the Kurile island arc north of Japan. It has parallel ranges of mountains, enclosing the valley of the Kamchatka river. The western mountain range forms an unbroken ridge of 1000–1300 m and includes the extinct volcano of Ichinskaya (3550 m). Its western slopes are relatively gentle but the eastern side slopes steeply into the Kamchatka valley. This western ridge is formed from crystalline gneisses and schists overlain by Mesozoic sandstones and capped by extinct volcanic peaks that have been glacially eroded. The easternmost of the ranges has steep asymmetrical relief, the steeper slopes also facing the Kamchatka valley. Descriptions refer to three or four river terraces alongside major rivers.

Out of the 74 volcanoes present in the eastern range of Kamchatka, 13 are still active. Altogether, igneous rocks make up about 40% of the peninsula. Hot springs and geysers are present. The west coast is composed of Tertiary material, either horizontally disposed or weakly folded. The coast itself is formed of gravel and sands and is without indentation. In contrast, the east coast has rocky headlands and bays.

### *The Chukot range*

The trend of the Chukot (Chuckchi) mountains is from north-west to south-east, extending from Cape Shelagsk to the eastern most point of Asia, Cape Dezhneva. The mountain range is formed from Palaeozoic and Mesozoic strata which after folding in the Mesozoic, has acted as a massif without further disturbance. The eastern part of the mountain range has rounded summits of 800–1000 m elevation with U-shape glacial valleys and fjord-like valleys on the Cape Chukotski peninsula.

## Mountains and plateaux of east-central Asia

West of Lake Baikal on the borders of Mongolia and the Soviet Union lies a mountainous region known as the Altai-Sayan (Figure 6.3). On the crests of the mountains throughout much of southern Siberia, the remnants of a plateau-like surface may be seen extending from the Kazakhstan uplands towards Lake Baikal and in the country beyond, known as Transbaikalia. This surface averages about 1350 m, although locally there are more elevated areas such as the Vitim plateau (1650–2000 m), and summits rise above it in Munku Sardyk (3491 m) in the Sayan range and Mount Belukha (4506 m) in the Altai range. It is dissected by many rivers including the upper Ob and the Angara. The provinces include:

The Sayan mountains
Lake Baikal
The Baikal and Transbaikalian mountains
The Stanovoy range and Aldan plateau

### *The Sayan mountains*

The Sayan mountains may be sub-divided into the western and eastern branches. The more southerly western Sayan mountains are strongly dissected highlands formed from Palaeozoic rocks folded during the Caledonian orogeny and intruded with granite. The area was eroded to a peneplain during the Tertiary, uplifted in the Alpine orogeny and again subjected to erosion and glaciation to give the higher peaks an alpine scenery. The eastern Sayan mountains have a north-west–south-east trend, and a 450 m scarp separates them from the lowlands to the north. Different erosional levels are apparent on the Sayan range: 1000–1300 m, 1600–2000 m, 2000–2500 m and a summit surface of more than 2200 m. The eastern Sayan mountains are formed from Pre-Cambrian slates, limestones and basalts. Glaciation has been more intense than in the western Sayan.

The Sayan ranges form the northern margin of the Tuva basin and the southern boundary is the Tannu Ola range, part of which also forms the border of Siberia and Outer Mongolia. The basin is at 500–800 m above sea level and it is drained by the upper Yenisey.

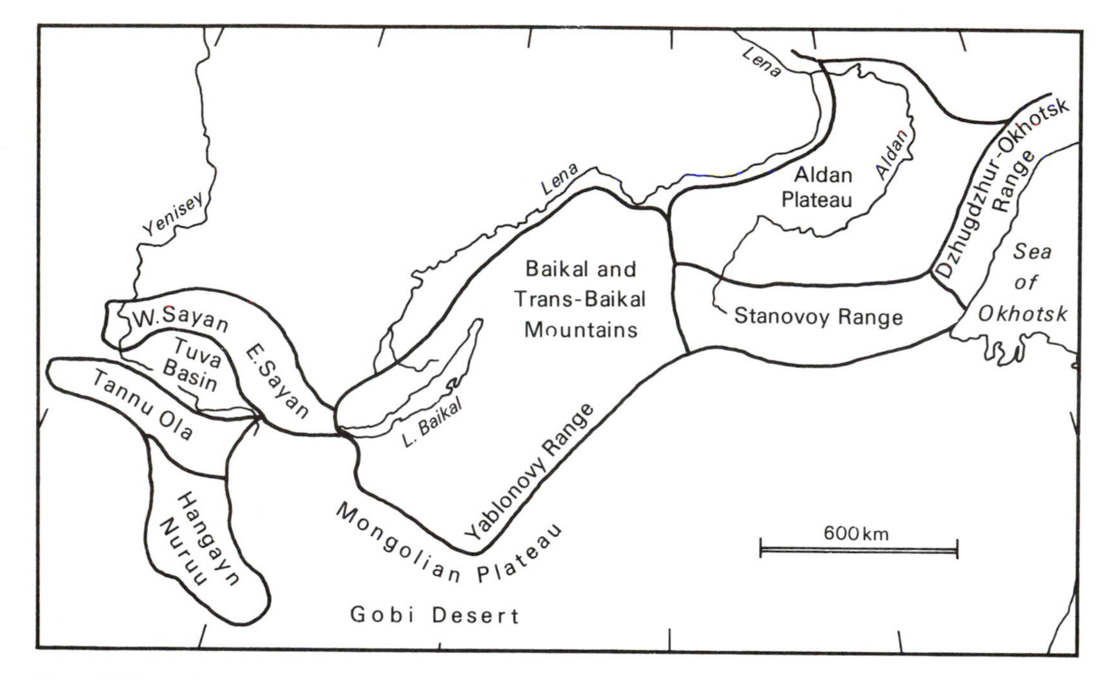


*Figure 6.3.* Mountains and plateaux of east-central Asia.

*Lake Baikal*

Lake Baikal has an area of 31 613 km² and extends for 630 km in length and its greatest width is 80 km. Most of the lake is more than 600 m deep and the greatest depth, approximately in the centre is 1700 m. It is a tectonic depression with fault scarp margins which was formed in the mid-Tertiary. The western shore is strongly uplifted and the rivers flowing from it rejuvenated. Terrace features occur at three levels around the lake: 2–9 m; 20–27 m and 55–90 m above the present lake level. Lake Baikal freezes over in January to a depth of 1 m and is free of ice for about 110 days during the year. The lake is one of the oldest lakes in the world and it has an interesting fauna; 73% of all faunal species and 44% of the fish species are endemic. The fauna includes relict fresh water species which evolved in Lake Baikal and species which have migrated up rivers from the Arctic. There is a unique Baikal seal and endemic species of goby live in the depths of the lake.

*The Baikal and Transbaikalian mountains*

East of Lake Baikal the remnants of a summit plateau were disrupted by mid- and late Tertiary block-faulting into a series of north-east–south-west trending horst and graben features. These include the graben of Lake Baikal itself, the lake surface of which is at 455 m, the downfaulted 'amphitheatre' of Irkutsk and the complex faulted structures of the Yablonovy range. In this area there is no extensive summit plain but an accordance of summit levels exists, the height and width of which decreases towards the north-east. East of the Yablonovy range the Vitim plateau is an exception to this generalisation. This plateau lies at 1650 m and has upon its surface a number of volcanic cones and lava flows. Geothermal activity continues to the present day.

*The Stanovoy range and Aldan plateau*

Between the Transbaikalian mountains and the Dzhugdzhur–Okhotsk range along the coast of the sea of Othotsk lies the Stanovoy range. Its highest point is at 2500 m above sea level and the range forms the continental divide between streams draining to the Arctic and those to the Pacific. The Stanovoy range is formed from rocks of Pre-Cambrian age, extensively intruded by granite. Many of the crests of the range are rounded, but the higher peaks have been eroded by corries to give an alpine relief. To the north of the Stanovoy range lies the Aldan plateau. At 1000 m above sea level it has been described as a warped erosional

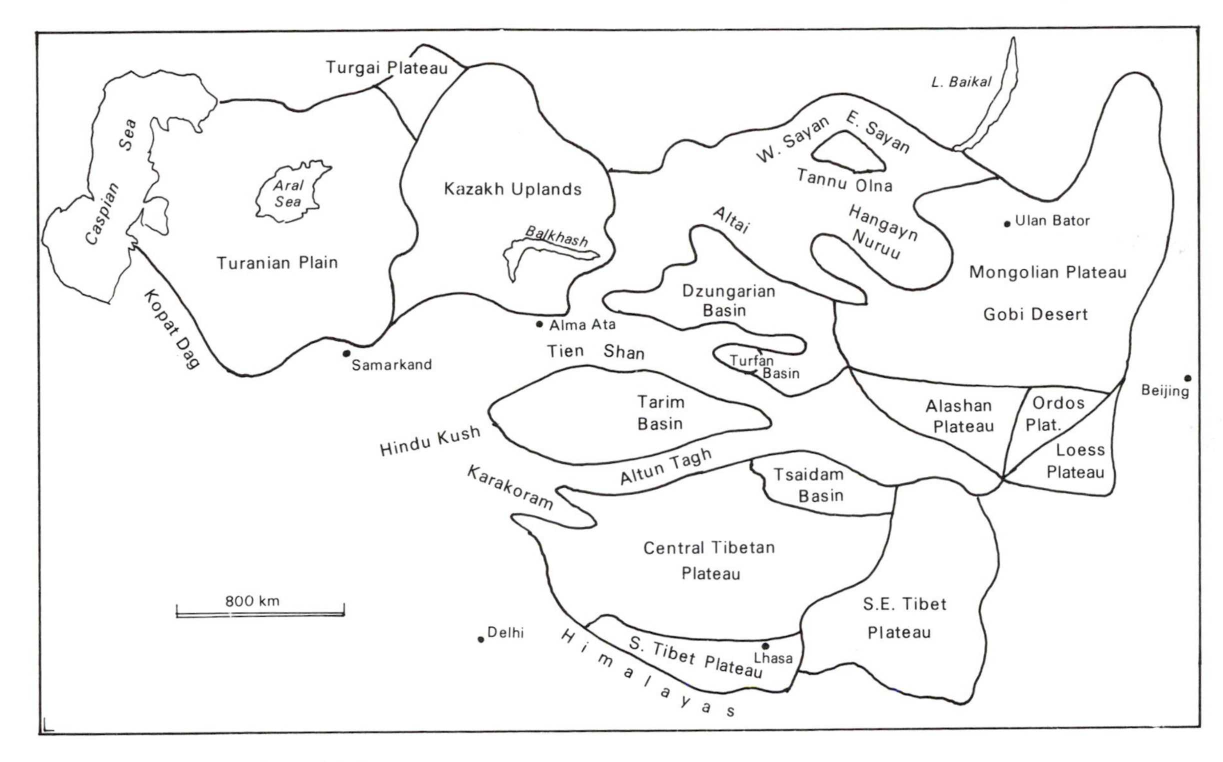


*Figure 6.4.* The plateaux of central Asia.

plain and is separated from the plains to the north by a fault scarp.

## Mountains and plateaux of central Asia

The elevated plateaux and mountain ranges of central Asia result from the interaction of the Asian and Indian lithospheric plates in a continent to continent collision. The thickness of crustal rock suggests that there is a double layer where the Indian plate has thrust beneath the Asian plate. The subsequent uplift of this part of the Earth's crust to form the high plateaux of central Asia is the highest, and folding on its southern margin has resulted in the Himalayan mountain ranges (Figure 6.4).

A secondary effect of the impact of the Indian plate upon Asia has been to force south China eastwards, the movement taking place between a fault system associated with the Altyn Tagh mountains and another along the line of the Red river.

These plateaux are described from west to east as they occur in increasing elevation from the 200 m Turgai plateau in the west to the Tibetan plateau at heights of 4500 to 5000 m above sea level. The provinces are:

- The Turanian plain
- The Kazakh uplands
- The Turgai plateau
- The Dzungarian basin
- The Tien Shan mountains
- The Tarim basin
- The Mongolian plateau (Gobi desert)
- The Alashan plateau
- The Ordos plateau
- The Loess plateau
- The Tibetan plateau
- The south-east Tibetan plateau

### *The Turanian plain*

Surrounding the Aral Sea is an area of inland drainage of 5.2 million km² known as the Turanian plain. Between the Aral and Caspian seas the plain is known as the Ust Urt plateau. Most of the Turanian plain lies at an elevation of less than 200 m, although piedmont areas

*Figure 6.5.* Loess hills, Tadjikistan.

adjacent to the surrounding highlands rise to greater heights. The form of this plain can be attributed to erosion during the Cretaceous which cut across outcrops of Palaeozoic marine sediments and Jurassic continental deposits. Eocene sediments were laid on this surface during an early Tertiary marine transgression; some of these remain, but most have been removed and replaced by Quaternary and Recent deposits. During the Quaternary, ice blocked the north-flowing rivers of Siberia which overflowed into the Aral–Caspian basins, resulting in higher water levels. Since the Pleistocene, water levels have fallen and surface erosion has occurred by flash floods from the surrounding mountains. The only rivers with sufficient water to cross the desert to the Aral Sea are the Syr Darya and Amu Darya. The middle courses of both are terraced and as they cross the piedmont zone they have cut into mud-avalanche deposits and loess. The central desert flats lie at an elevation of less than 125 m and the main areas of sand desert lie to the south and east in the Kara Kum and Kyzyl Kum where shifting barchan dunes have developed. Loess has been deposited in the upper Syr Darya valley but the northern part of the plain has a gravelly surface.

### *The Kazakh uplands*

This is an area which lies between the west Siberian lowland and the Turanian plain, centred on the town of Karaganda. In the west it is separated from the southern Urals by the less elevated Turgai plateau and the Kara Tau, Talassky Ala Tau and Dzungarsky Ala Tau ranges of the Tien Shan lie to the south. The Irtysh river forms the north-east boundary and the Syr Darya the south-west, but most drainage of the area is endoreic, flowing either to Lake Tengiz or to Lake Balkash. The underlying rocks are of Palaeozoic age; they have been block-faulted in the Hercynian orogeny and Tertiary marine sediments laid in depressions. Further faulting took place during the Alpine orogeny and subsequent ero-

*Figure 6.6.* Yazob Gorge, Tadjikistan.

sion has given a hilly upland area, the highest parts of which just reach 1000 m above sea level. The Tengiz depression in the north-west and the Chu depression in the south are both less than 500 m above sea level. The surface of Lake Balkash is at 342 m and this large lake is situated in a synclinal structure north of the Tien Shan ranges. Dune landscapes lie to the west and south in the Munyun Kum desert at 375 m above sea level and loessial landscapes characterise lowlands between mountain ranges in Tadjikistan (Figure 6.5). Despite tectonic instability, the deep valleys of the Pamir mountains have been dammed for power generation (Figure 6.6).

### *The Turgai plateau*

The Turgai plateau lies between the Urals and the Kazakh uplands; it is mostly below 450 m in altitude and slopes towards the Aral Sea. Its surface geology is formed by Oligocene–Miocene marine and continental deposits with horizontal bedding which are eroded into flat-topped mesas with gullied sides. A tectonic trough with salt lakes runs from north to south through this section and took drainage waters from the west Siberian lowlands to an enlarged Aral Sea during the Pleistocene.

### *The Dzungarian basin*

The Altai, Tien Shan and Tarbagatai ranges surround the Dzungarian basin on the north, south and west respectively. Westwards it opens out through a 16 km-wide pass to the Lake Balkhash depression. The Dzungarian basin contains four separate drainage systems, two internal and two, the Irtysh and the Ili, which flow out of the region. The internal drainage systems are those of the Barku and Turfan basins. The floor of the Barku basin is at 1500 m and extends for 80–100 km east and west of the lake; it is up to 30 km wide. The sides of this basin are steep, rising abruptly 2500 m to the Tarbagatai range. The floor of the Turfan depression lies at −155 m and extends 65 km east and west of a central intermittent salt lake. It is flanked by precipitous slopes which rise abruptly to the adjacent peaks, but pediments up to 25 km wide occur around the margin. Sand dunes of 200 m in height occur in the south-east part of the Turfan basin, but elsewhere the desert surface is characterised by gravelly hamada.

### *The Tien Shan mountains*

The Tien Shan mountains lie between the Dzungarian and Tarim basins. They consist of several ranges of mountains rising to over 7000 m above sea level and extend for 1700 km in an east–west direction. Glaciers occupy the higher parts of the mountains above 3800–4200 m covering 9500 km$^2$ and glacial tongues extend as low as 2500 m. Glaciers were formerly present over a large area and fluvial erosion has caused five terraces to be developed along the mountain rivers.

### *The Tarim basin*

This province has as its foundation a portion of the basement complex which became involved in the strong uplift associated with the Alpine orogeny and the development of the Himalayas. The shield rocks are covered by a deep accumulation of rock debris eroded from the surrounding mountains and carried into the

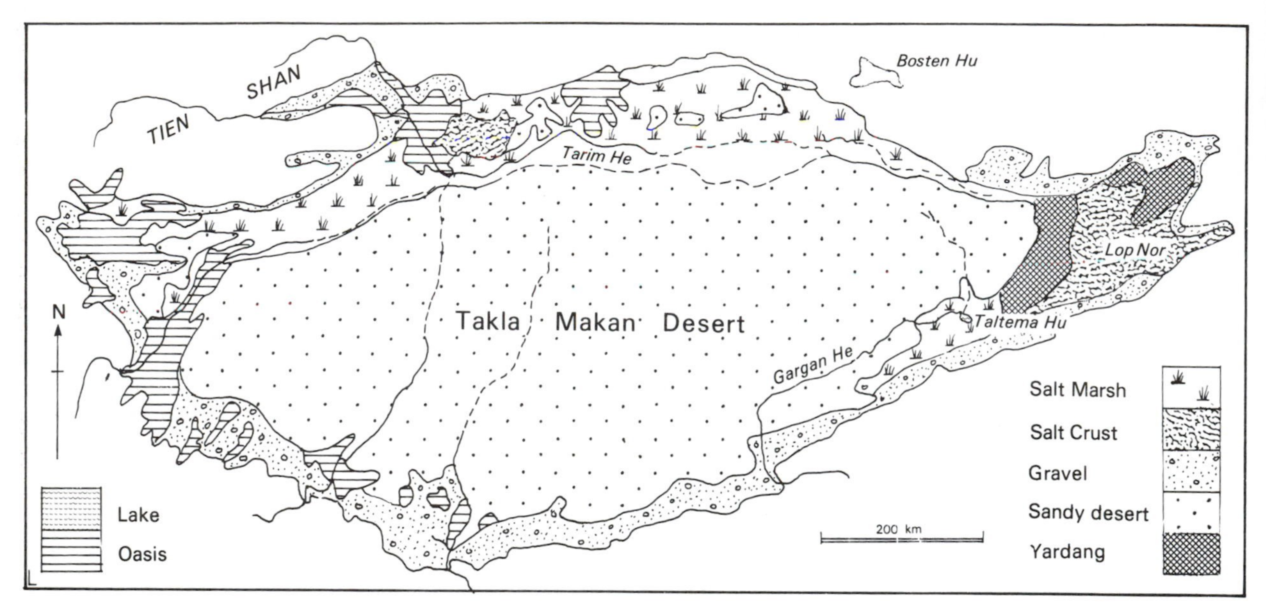


*Figure 6.7.* The Tarim basin.

basin by fluvial action. Around the margin of the Tarim basin are alluvial fans and a pediplane which slopes gently towards the centre and pass beneath the sands of the Takla Makan desert (Figure 6.7). Aeolian action funnels material through a 200 km gap between the Pei Shan and Nan Shan mountains near Lop Nor into the eastern part of the basin where dunes of 100 m high occur.

In the western part of the basin the dunes are largely replaced by a blanket of silty loess which is also being deposited upon the adjacent Tien Shan mountains. Where dunes occur in the western part of the basin they have a confused pattern because winds are deflected by the surrounding mountains. Alluvial deposits have been laid down by the Tarim river, its permanent tributaries and intermittently flowing wadis. The Tarim flows for 500 km across the floor of the northern part of the basin, but over this distance it only descends 200 m. Because of its low gradient the river has meandered and frequently changed its course to leave numerous abandoned channels. The low gradient also facilitates irrigation on land adjacent to the river.

The Tarim basin is one of internal drainage to the lake Lop Nor, but between the fourth century and 1921, the Tarim river flowed into two shallow 'wandering' lakes, Karakoshun and Taitema. In 1921 these lakes dried up because the river abandoned them for Lop Nor. Around the periphery of the basin, only the larger mountain streams manage to cross the pediment before they sink into the desert sands.

### *The Mongolian plateau (Gobi desert)*

The Mongolian plateau coincides with the region known as the Gobi desert which lies mainly within the state of Outer Mongolia. It has a plateau surface at about 1500 m but is surrounded on all sides by a higher mountain rim which makes an elevated saucer-shaped feature. The lowest part of the plain occurs towards the east, at about 800 m, where the eastern margin is unwarped in the Great Khinghan range.

The desert surface has a series of major depressions containing sedimentary infill, called talas (Dalai Tala in the east and Iren Tala in the centre), separated by ridges trending north-west to south-east. These ridges are flanked by pediments and frequently show examples of badland topography. The floors of the depressions are characterised by playa lakes and have been referred to as the Pang Kiang erosion surface. Some drainage in the east and north is integrated, either to Lake Baikal or to lakes Hulin or Buir Nor and so to the Amur in the extreme east of the province. The Hulin lakes are situated west of the Da Khinghan range in an area which lies between 550 and 700 m above sea level. Sand dunes are restricted to the western part of the plateau north of the Altai range; elsewhere the desert floor is covered with a hamada of gravel and small stones. An area of volcanic hills and basalt flows known as the Xilin Goh is situated along the Sino-Mongolian border. In the extreme south of the province the Ulanqab plateau slopes northwards from the Yin Shan fault scarp overlooking the great bend of the Huang Ho.

The western margin of the Gobi desert and the Mongolian plateau is marked by the Altai mountains which can be traced from a knot of mountainous country at the western extremity of Outer Mongolia eastwards for almost 1000 km. The core of these mountains was folded in the Caledonian orogeny and sharply uplifted in the Miocene with the Himalayas. There is evidence of multicyclic erosion features at elevations of 1800–2000 m, 2600–2700 m and 2900–3000 m on the Altai mountains and above 3200 m glaciers occupy 270 km$^2$.

### *The Alashan plateau*

This plateau lies west of the great bend in the Huang He river and north of the Qilian mountains, forming a southwards extension of the Mongolian plateau. It has been peneplained at a height of between 1800 and 2500 m above sea level and its surface is mostly a type of stony hamada. The western part of the Alashan plateau is also known as the Black Gobi because of the presence of desert varnish on the stones. However the marginal basins of Badain Jarau, Tengger and Ulan Buh are sandy; the sand dunes of the 40 000 km$^2$ Badain Jarau reaching 420 m in height, the largest in the world. The alignment of the mountain ridges is from north-east to south-west and they are surrounded by pedimented slopes which pass down to playas with salt pans or takyr surfaces.

### *The Ordos plateau*

The Ordos plateau lies completely within the great bend of the Huang Ho on the northern side of the Great Wall of China. It is a stable fragment of Laurasia which was submerged in the Mesozoic and covered with sandstones, conglomerates, shales and Tertiary red sandstones. Subsequent uplift and denudation has produced an undulating surface between 1000 and 1500 m with aeolian landforms. In the north of the Ordos plateau is the rocky Hobq desert, 16 000 km$^2$, and to the south is the Mu Us desert of about 25 000 km$^2$, a third of which is shifting dunes and deflation hollows. The western and northern limbs of the Huang Ho bend are grabens infilled with up to 1000 m of sediment. These alluvial lands lie on the outside of the bend and are capable of being irrigated.

### *The Loess plateau*

In its middle reaches the Huang Ho crosses one of the most famous geomorphological districts of China. Over an area of 30 000 km$^2$, the loess plateau occurs in the provinces of Shanxi, Shensi, Gansu and Ningxia. The loess plateau has an elevation of between 1200 and 1600 m above sea level and the depth of loess can be as much as 200 m, although it is normally between 30–60 m thick. The configuration of the loess mantle follows the bedrock surface and extends to an elevation of 2400 m on the western slopes of the Luipan mountains. The Wei He river occupies a fault graben extending east–west across the base of the great bend in the Huang Ho river. It has an extensive flood plain and a flight of three river terraces.

There are three main layers within the loess. The Wuchang loess is the oldest, a reddish loess 17.5 m thick. This is covered by the Lishi loess, 80–100m thick. The upper layer of the loess is pale yellow in colour, 20–40 m thick and is known as the Malan loess. The particle size of the loess becomes finer towards the south-east (Figure 6.8), reflecting the sorting influence of the wind.

Chinese geomorphologists recognise three major landforms on the loess. These are: liang, loessial flat ridges; yuan, loessic high plains; and mao, loessic gentle slopes. Unfortunately the loess is easily eroded and soil losses are severe, sediment yields from small and medium sized basins frequently exceed 5000 tonnes/km$^2$ and may range up to a maximum of 35 000 tonnes/km$^2$ (equivalent to a surface lowering of 25 mm per year). Despite erosion control measures, gulleying is the chief source of sediment supply.

### *The Tibetan plateaux*

The Tibetan or Qinghai-Xizang plateaux have an area of 2.5 million km$^2$ and comprise the most extensive areas of the Earth's surface at high elevation. Their elevation ranges from 300 m in the south-east rising to 4500–5000 m above sea level in the north-west, surrounded by even higher mountain ranges (7000–8000 m). In central Tibet, these mountain ranges trend east–west but in eastern Tibet, their alignment changes to a

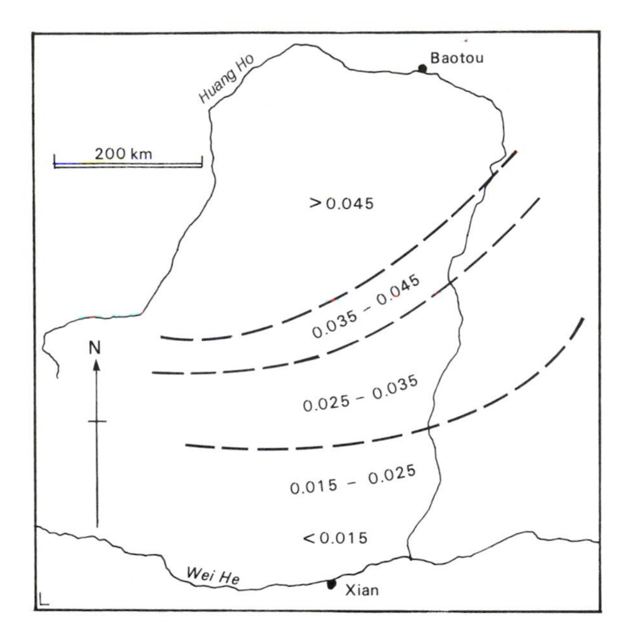


*Figure 6.8.* The particle size of loess decreases in a south-easterly direction. (After Z. Songqiao, 1986.)

north-south direction. Southern Tibet is drained by the Brahamaputra (Tsangpo) and the eastern part is drained by the headwaters of the Salween, Mekong and Yangtse. Elsewhere, the high plateaux of Tibet are characterised by areas of internal drainage. The southern boundary of this elevated region is the north-facing slope of the Himalayas and its northern boundary is the Kunlun–Qilian mountain range. In the west the high plateaux are pinched out between the Great Himalayas, the Ladakh range and the Karakoram range in Kashmir; to the east the high plateau drops to the lower plateaux of Yunan and Guizhou in China.

The central Tibetan plateau lies between the Kunlun mountains to the north and the Nysingentanglha mountains in the south. It is between 4500 and 4800 m above sea level and the endoreic drainage of the area is to lakes such as Nam and Siling. Extensive pediments occur around faulted lake basins. The lake basins tend to be enriched with carbonate and sulphate salts in the north and chloride salts in the south. North-east of this region the Altun and Qilian mountains bifurcate from the Kunlun mountains to enclose the Qaidam (Tsaidam) basin. The floor of this upland basin is at about 2000 m and is characterised by sand dunes in the west and swamps in the east. The Qinghai (Tsing Hai) lake at 3200 m above sea level has an area of 4400 km$^2$ and is another area of internal drainage although the Huang Ho passes only about 100 km to the east in a terraced valley upstream from Lanchow.

South Tibet is characterised by high mountain ridges of 5000–6000 m with major rivers flowing in gorges at 2000–4000 m above sea level. There is a residual surface at 3500–4000 m above sea level where the major rivers have broad valley floors above knick points which are working upstream. The Tsangpo in particular, has a broad open valley above 4500 m with an alluvial plain, braided channels and some barchan dunes. Periglacial activity occurs on the slopes, and footslopes are pediments. On many mountain ranges glaciers are present.

### *The south-east Tibetan plateau*

On the eastern borders of Tibet and western China, the east Himalayan plateau is the name given to the eastern part of the high plateaux of Tibet. It is separated from the low plateaux of eastern China by the Yungling fault zone, but the boundary with the high plateau of central Tibet is less well defined. The mountain ridges of Amne Machen Shan and Bayan Karsa Shan at 5000–6000 m elevation with valley floors at 3500–4000 m, are drained by the headwaters of the Huang Ho, Yalung, Yangtse, Mekong and Salween. These last three great rivers 'converge without uniting, flow parallel a short distance apart for hundreds of miles and then separate to enter three different seas' (Barbour, 1936). In this province the Himalayan trend swings from east–west to north–south and the mountain ridges gradually decrease in elevation, becoming the Hengduan mountains of western Yunan. The parallel gorges of the rivers widen out above knickpoints where a residual surface occupies the broad marshy floors of valleys. Glaciers occur on the mountain ranges, the Karchin glacier being the largest, 32 km in length. Periglacial features are common.

## Uplands and plains of eastern Asia

In eastern China and Manchuria, the ancient shield rocks occur as several isolated 'nuclei': Inner Mongolia–Korea, the Shangtung peninsula and south-east China. The basement complex in these areas consists of metamorphic and plutonic rocks upon which younger sedimentary rocks are laid unconformably. Between these areas of old rocks extensive riverine plains have been laid down by the Huang Ho and Yangtse Kiang (Figure

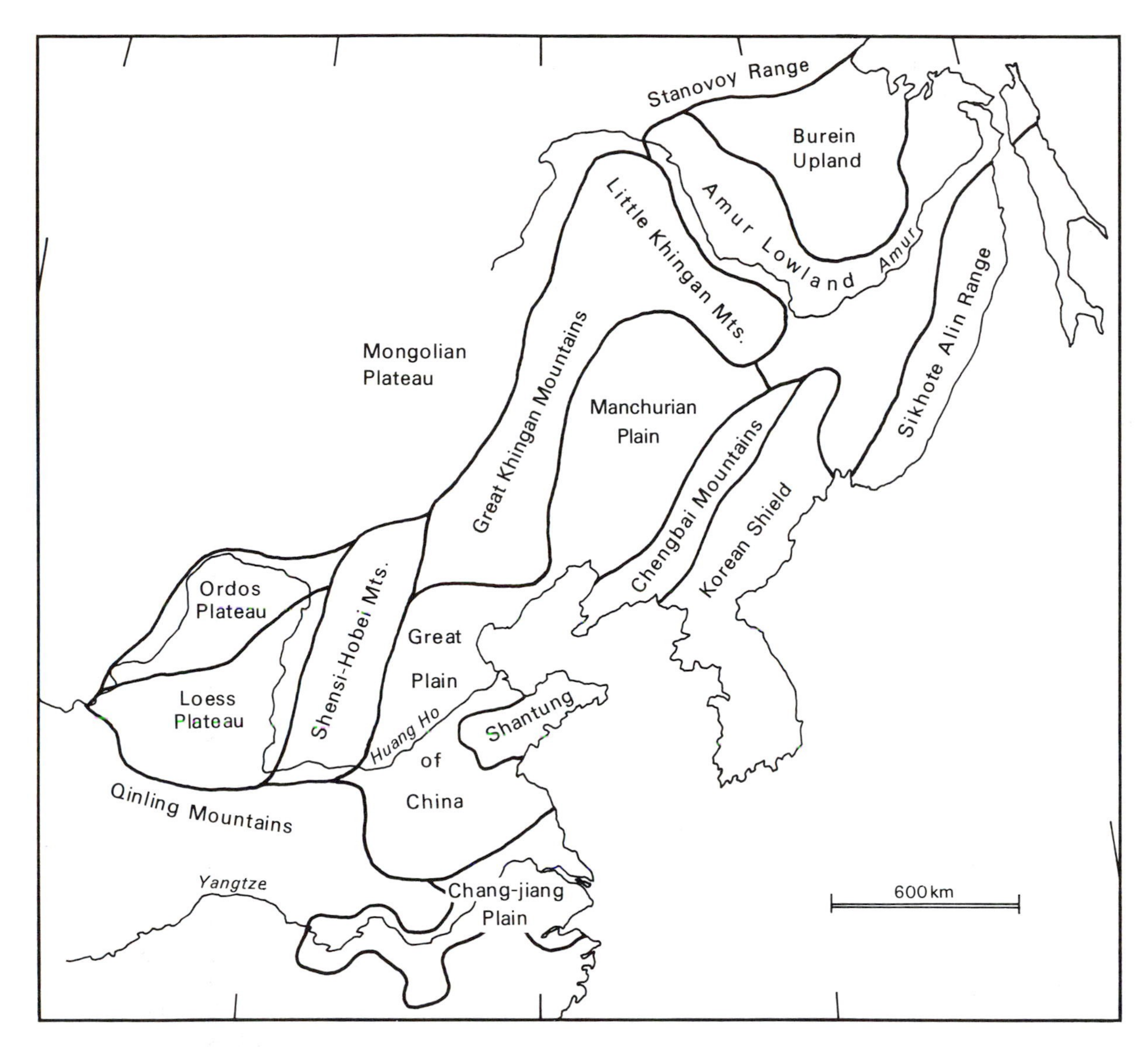


*Figure 6.9.* Uplands and plains of eastern Asia.

6.9). The area will be considered under the following provinces:

- The Korean shield
- The Chengbai mountains
- The Manchurian plain
- The Amur lowlands and Burein uplands
- The Sikhote Alin range
- The Khingan mountains
- The Shantung peninsula
- The Great Plain of China
- The Shensi–Hobei uplands

*The Korean shield*

This continental nucleus is partly overlain by Upper Palaeozoic and Mesozoic strata which lie unconformably on the schists and granites of the basement complex. In North Korea, Ordovician sediments cover part of the shield but the ancient rocks form a rolling upland with deeply incised rivers. In South Korea gneiss forms the highest ground with peaks of 3000 m and granite forms lower, rounded hills between 650 and 1300 m. A volcano with a crater lake is situated on the Korean–

Chinese border, Paektu-san (3000 m) and deep dissection by rivers reveals sharp, faulted escarpments which bound mountain blocks such as the Taebaek range of the east coast.

### *The Chengbai mountains*

A major fault marks the limits of the Shield rocks. The down-thrown north-western side has been subsiding throughout the Mesozoic and now forms a large synclinorium centred on the Manchurian plain. Parallel to this major fault the Chengbai mountains consist of Lower Palaeozoic rocks folded and uplifted in the Hercynian orogeny and peneplained in the early Tertiary. Volcanic activity, including basaltic flows, and further uplift occurred in the Pleistocene. These mountains form the south-eastern part of Manchuria and extend into the Liaodung peninsula, decreasing in elevation to Port Arthur. The coastline of the Liaodung peninsula is rocky.

### *The Manchurian plain*

The rectangular Manchurian plain lies between the Chengbai mountains and the edge of the Inner Mongolian plateau (the Great Khinghan mountains) to the west, the Little Khinghan mountains to the north and the gulf of the Yellow Sea known as the Bohai Sea to the south. It is drained by two rivers, the Sungari, a tributary of the Amur and the Liao He which flows directly into the Gulf of Bohai Sea. The plain has a synclinal structure which has periodically sunk to allow sediments to accumulate throughout the Mesozoic. There are between 30–50 m of Quaternary deposits as well. Extensive pedimented surfaces surround the lower-lying land which is poorly drained and swampy. As a result the plain has an undulating surface ranging from 300 m in the Sungari valley rising to some 600 m on the divide between it and the valley of the southwards flowing Liao He. The Sungari and its tributaries have incised into the surface with steep-sided valleys, the rivers flowing up to 150 m below the surface of the plain.

### *The Amur lowlands and Burein uplands*

The Amur valley forms a northwards extension of the plain of Manchuria through a relatively narrow gap where the Sungari river passes between the Little Khingan and Chengbai mountains. During the Tertiary, the lower Amur valley was downwarped and marine sediments were laid down. It is an area affected by permafrost and so swampy conditions occur on the valley floor. The Amur valley broadens where the Bureya and Zena, its major tributaries, join the trunk stream. On the northern side the valley is enclosed by the steep slopes leading up to the Burein uplands and the summits of the Stanovoi range but the eastern slopes to the Sikhote Alin are less precipitous. The Burein uplands have an erosion surface at 1200–1300 m. As the lower course of the Amur flows in a north-easterly direction, it crosses the structural trend of the coastal mountains and there are many rapids along its course.

### *The Sikhote Alin range*

Folding occurred in the late Palaeozoic on the edge of the Laurasian landmass to give the core of the Sikhote Alin. Further uplift occurred in the early Jurassic and basalts were erupted onto a peneplained Tertiary surface, now of 1200–1300 m above sea level. Summits of the range mostly lie between 1000–2000 m. The southern part of the island of Sakhalin has had a similar geological history.

### *The Khingan mountains*

The Great Khingan mountains extend northwards from Beijing in the western part of Manchuria. The mountains are a dissected fault scarp which forms the eastern rim of the Mongolian plateau. The rocks were folded and intruded by granite in the Hercynian earth movements, faulted during the Mesozoic and basaltic eruptions occurred in the Tertiary. Post-Miocene erosion surfaces have been uplifted to give plateau features at 1000 m, 600 m and 500 m. The Little Khingan mountains form the northern rim of the Manchurian plain, lying south of the Amur lowlands. North-east of Beijing the ancient fold ranges of the Yashan mountains have been referred to as the Beijing Grid; these ranges of low mountains are blanketed with loess.

### *The Shantung peninsula*

The Shantung peninsula was formerly a rocky island in the Yellow Sea. It has been linked to the mainland by deposition of sediments of the Huang Ho in the Yellow Sea. This upland area has summits of 1600 m formed by rocks of the basement complex.

### *The Great Plain of China*

The Great Plain is one of the larger alluvial lowlands of the world, resulting from the infilling of the shallow Yellow Sea with sediments. Relief everywhere does not exceed 100 m above sea level and the plain extends from the Pei He north-east of Beijing southwards to the Chang Jiang, a distance of 1200 km. It continues inland along the major river valleys to include the lake basins south of the Chang Jiang.

The Great Plain comprises the deltas of the Pei He, Hwai He, Huang Ho and Yangtse Kiang. These rivers have meandered and frequently changed their courses during floods, often with great loss of life. The deltas of the Huang Ho form the largest part of the Great Plain of China. Before 602 BC the river flowed north of Shantung but between 602 BC and 70 AD it flowed south of the peninsula until it again established its present course north of Shantung in 1851. The construction of artificial levees along the lower course has reduced the extent of flooding but has increased sedimentation in the delta. The sediment load of the Huang Ho is the largest of any river in the world: 1600 million tonnes per annum, of which 800 million tonnes are deposited in the delta and 400 million in the channel and flood plain. The mean concentration of suspended sediment (37.6 $kg/m^3$) is the highest found in any major river in the world and maximum values recorded reach 666 $kg/m^3$. Of the sediment of the Huang Ho, 90% is derived from erosion of loess in the middle part of the river's catchment in Shansi. The Great Plain has been extended seawards in the delta areas by the great quantity of sediment carried by the river. However, deposition in the channel has raised the bed of the river until it is often between 3 and 5 m above the surrounding plain and in some places as much as 10 m which means great efforts must be made to maintain the effectiveness of the flood protection banks.

### *The Shensi–Hobei uplands*

The Shensi–Hobei uplands lie west of Beijing where the basement complex re-emerges from beneath the plains. It is block-faulted to give a series of parallel ridges of between 1000 and 2000 m elevation. Between the ridges, the valleys are downfaulted graben with a north-east–south-west alignment. The ridges are breached by the Yongding river which drains the northern part of these uplands. Many of the deep narrow valleys near Beijing are used for the provision of water supplies for the city.

The southern part of the Shensi–Hobei uplands is made up of the Shansi plateau, 800–1200 m above sea level. The eastern boundary of the plateau is formed by the fault scarp of the Taiching mountains and the western limit is the Liliang mountains, east of the great bend of the Huang Ho. The area is drained by the Fen He river, a tributary of the Huang Ho.

## Mountains, basins and plateaux of south China

The underlying structure of this region is not fully understood but it is thought fragments of the Pre-Cambrian basement complex underlie Palaeozoic and Mesozoic strata subjected to orogenesis in the Hercynian and Alpine orogenies. The contrast between creamy-white limestone and red Mesozoic sandstones is a feature of south Chinese landscapes. The dramatic landforms of the Chinese karstic areas of Nanling province along the Zhujiang river in Guizou plateau province are famous throughout the world (Figure 6.10).

The provinces used to describe the geomorphology of this area are:

The Quinling Shan mountains
The Red or Szechuan basin
The Yunnan plateau
The Guizhou plateau
Changsha region
The Nanling mountains
Fukien uplands
Coastal lowlands

### *The Quinling Shan mountains*

South of a line made by the Huang Ho and Wei He the Quinling mountains extend across central China in an east–west direction between the valley of the Wei, a tributary of the Huang Ho and the Yangtze. The northern slopes are faulted and steep rising to a general height of 2000–3000 m above sea level; Mount Taibai

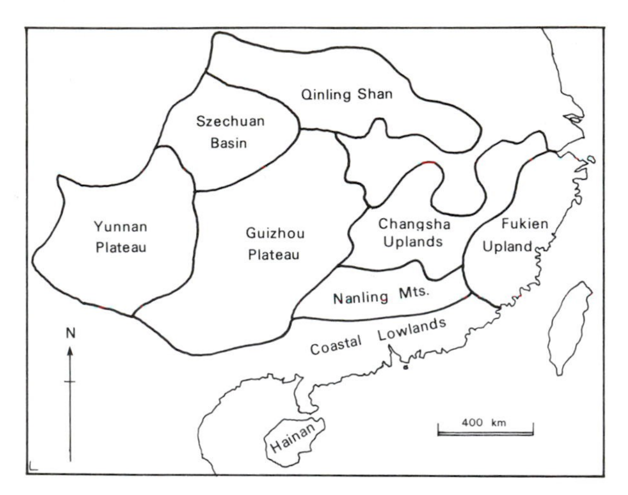


*Figure 6.10.* Mountains and plateaux of south China.

(3767 m) is the highest peak. Included within this mountain region is the Nanyan–Xiangfan basin which includes red Mesozoic sandstones. These mountains are sometimes referred to as the Szechuan 'Alps'.

*The Red or Szechuan basin*

The Red basin lies at the foot of the Szechuan Alps; its floor is at an altitude of 250–500 m and it is drained by the Yangtse river. It takes its name from the red-coloured Mesozoic sandstones which contrast strongly with the surrounding limestone hills which rise steeply from the basin floor to mountains of 3000–4000 m. The western part of the basin, the Chengdu plain is formed by Quaternary sands laid down in alluvial fans between 450 and 750 m above sea level. In the centre of the basin the reddish sandstones are horizontally bedded and erosion has produced mesas, but in the eastern part of the basin the red sandstones form north-east–south-west trending ridges.

The southern edge of the Red basin is marked by a section of the middle Yangtze valley, and the Kialing Kiang tributary flows across the central part of the basin to join the mainstream at Chungking. Another tributary of the Yangtze forms a small alluvial plain near Chengtu.

*The Yunnan plateau*

South of the Yangtze, situated in the extreme south-west of China and continuous with the karstic plateau of Guizhou, is the Yunnan plateau. It is formed from sandstone and shale and is less dissected than the karstic plateau. Erosion surfaces have been identified at 4000–4100 m, 3600–3700 m, 2400–2500 m and 1800–2100 m. The western boundary is the Salween river and the plateau is crossed by the Mekong and Red rivers. There are several faulted basins containing lakes.

*The Guizhou plateau*

South of the Red basin Devonian and Permian limestone crop out to form the Guizhou plateau, a dissected Tertiary erosion surface 1000 to 2000 m above sea level. The limestones are 3000–5000 m thick and extend over 300 000 km$^2$. The northern part of the Guizhou plateau has many 'sugar loaf' hills (round-topped but vertical-sided), typical of tower karst. In the south, the classic Chinese karstic landscapes are developed where there is a relative relief of 300–700 m in the valley of the Zhujiang river. Generally, the Guizhou plateau surface is tilted southwards, descending from the mountains to the coastal plain adjacent to Hainan Island.

*Changsha region*

South of the Yangtze and extending as far south as the Nanling mountains is the Changsha region of uplands and basins, lying east of the Guizhou plateau. Two large lowlands are linked with the valley of the Yangtze; these contain the Poyang and Tung Ling lakes. Around Hengyang is the largest of several basins containing Jurassic–Tertiary red sandstones similar to those of the Red basin of Szechuan. The uplands between the lakes are the Hunan–Jiangxi hills which have a north-north-east–south-south-west structural trend and reach elevations of 1000–1500 m. Erosion surfaces have been identified at 1000 m and 1500 m.

*The Nanling mountains*

These mountains consist of five mountain ranges which form the watershed between the streams flowing north to the Yangtze and those flowing to the coast or to the Si Jiang river. These ranges reach 2000 m elevation and include spectacular karstic scenery between Guilin and Yangshao.

*Fukien uplands*

In the province of Fukien, the eastern coastal uplands have a north-east to south-west alignment and enclose many small intermontane basins, some of which contain red sandstone. The age of the rocks in the uplands of south-eastern China is the subject of controversy, but the cores of the mountains include fragments of the basement complex.

Some authorities attribute the structure to the Alpine orogeny, thus creating an eastward extension of the folding of the Himalayan ranges; others suggest they may be of late Carboniferous age. These uplands reach elevations of 500 to 1000 m, and except for numerous small delta plains, slopes descend abruptly into the sea with flooded valleys or rias. The hills are aligned in parallel ridges near the coast, but inland the mountainous topography of the Wuyi Shan gives an intricate network of ridges rising to 1000 m above sea level and long valleys between them. The coast is highly indented with many rocky islands of which Hong Kong is the most famous.

### *Coastal lowlands*

The coastal plain where it exists is narrow, except around Canton where the Si Kiang, Peh Kiang and Tung Kiang form a deltaic plain of about 10 000 km$^2$.

The Leizhou peninsula and Hainan Island have extensive Quaternary basaltic flows which form a terrace, 150 m above sea level and along the coast marine terraces occur at 10–15 m, 25–35 m, 45–55 m, 60–80 m, and 100–150 m. Southern Hainan is composed of granite with erosion surfaces at 300 m, 500 m and 800 m. Coral reefs occur along the coast.

## Mountains and plains of south-east Asia

The south-east extremity of Asia comprises the states of Thailand, Vietnam, Cambodia, Indonesia and the Malay peninsula. These countries lie adjacent to the South China Sea, the southern part of which is shallow, less than 150 m deep. Although a diverse area, it forms a major geomorphological unit because it is underlain by a stable crystalline Pre-Cambrian nucleus. Like other areas of basement complex it is mostly covered by later sedimentary rocks but in this case it is also partly submerged beneath the sea. The area is known as the Sunda shelf and its upland areas form the peninsulas and islands of this region (Figure 6.11).

The boundary of the lithospheric plate lies along the line of the Andaman Islands of the Indian Ocean, Sumatra and Java and the smaller islands which surround the Banda Sea. It continues east of Sulawesi and the Philippines and Taiwan. The peripheral island arcs are discussed in the chapter on the Pacific Ocean, but the Indonesian arc, backed by continental crust, is considered here. The provinces of this part of Asia include:

Shan plateau
The Irrawaddy lowlands
The peninsular ranges
The Annam range
The Red river
Chao Phraya lowland
The Cambodian lowland
The Korat plateau
The Sunda shelf
The Indonesian arc

### *Shan plateau*

The stability of the area is reflected in an old peneplained and now uplifted surface of the Shan plateau 1000–1300 m above sea level. Archean gneiss crops out in the western part of this province and granite in the east. These are succeeded by slates, quartzites and limestones ranging from Pre-Cambrian to Jurassic in age. Mid-Tertiary clays and former lake deposits continue the geological sequence into the Pleistocene. The structure of the area is said to be simple with hills and valleys formed by block-faulting. The major river draining the province is the Salween which has cut a deep valley, 300 m below the plateau surface. The Shan plateau is renowned for its lapis lazuli, rubies, jade, silver, lead and zinc.

### *The Irrawaddy lowlands*

This basin is drained by the Irrawaddy river and its major tributary the Chindwin, which become confluent south-west of Mandalay. The basin lies between the range of the Arakan Yoma to the west and the edge of the Shan plateau to the east. Separated from the main lowland area is the smaller basin of the Sitang river plains. The Irrawaddy basin of central Burma is a geosynclinal downwarp which has been infilled with sediments during the Tertiary and it contains some 6000–10 000 m of Eocene sandstones and shales followed by a further 3000 m of sandstones with coal seams of Oligocene age. The geological succession is

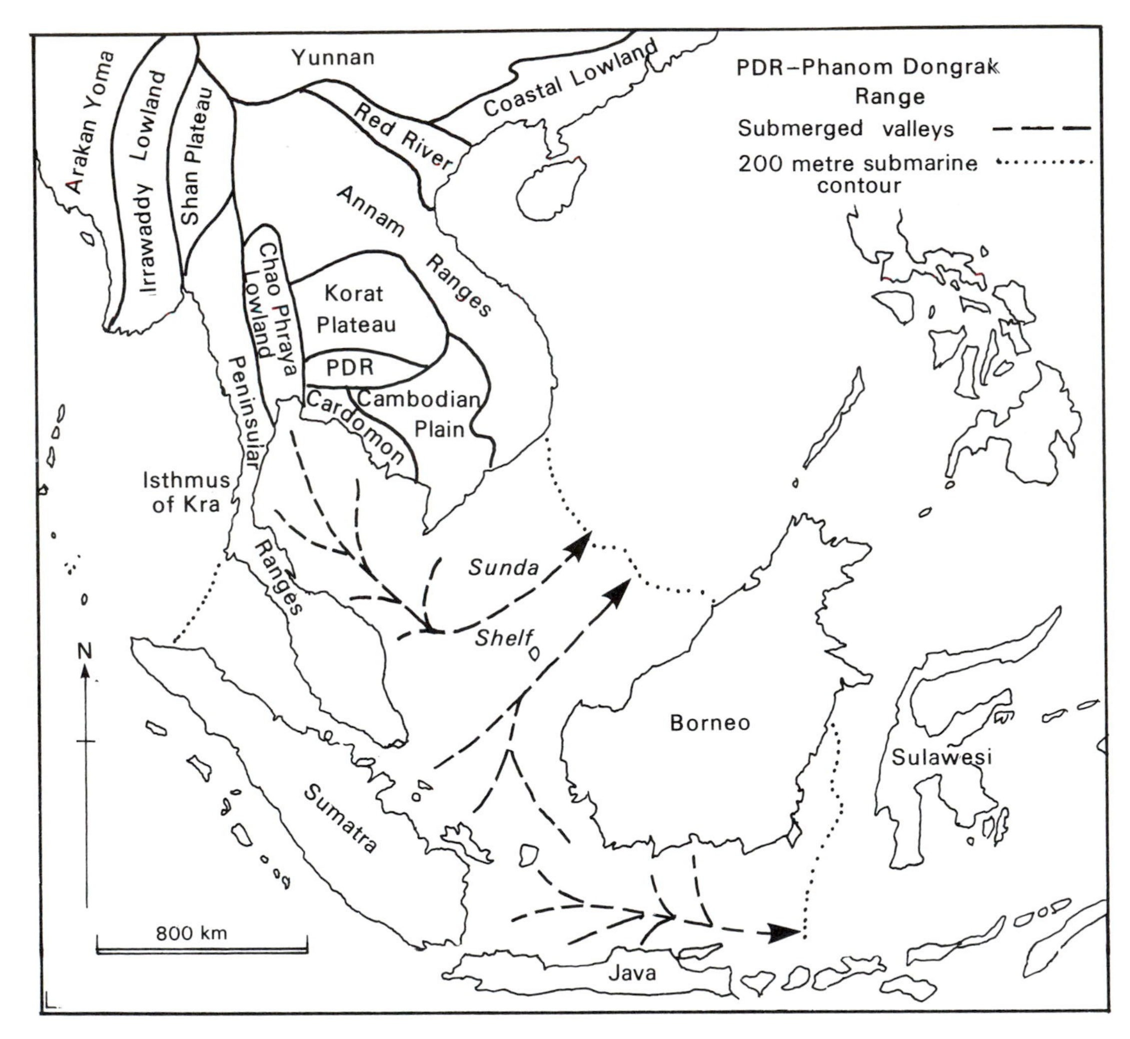


*Figure 6.11.* The Sunda shelf.

completed by Pliocene fluviatile deposits and laterites. Intrusion of peridotite took place in the Oligocene, and a number of extinct volcanoes can be seen, such as Mount Popa along the line of the Pegu Yoma mountains. Oil occurs in a series of small anticlinal structures aligned north–south parallel to the Irrawaddy river. The deltas of the Irrawaddy and Sitang rivers are used extensively for rice cultivation.

### *The peninsular ranges*

The north–south trend of the eastern Himalayas is continued in the Malay peninsula. The line of mountains continues southwards along the Burma–Thailand border in the Dawna range which rises to a maximum height of 2000 m although most of the crests are usually below this level. Rivers, such as the Tenasserim, follow the same trend, flowing parallel to the coast before entering the sea. Offshore, the islands of the Mergui archipelago are the peaks of a drowned range of mountains with the same structural trend.

The peninsula is at its narrowest at the Isthmus of Kra, but then widens again in Malaysia. Malaysia is surrounded by a coastal plain of varying width, fringed with mangrove swamps on the south-west coast and sandy beaches on the north-east coast. Inland the western range of mountains includes the Cameron highlands and the central, eastern mountainous area is formed by the Trengganu hills, both rising to just over 2000 m. These ranges are composed of sandstones, shales and limestones, folded during the alpine orogeny and intruded with granites. Where limestones crop out, karstic features are developed. Associated with the granitisation are lodes of tin and the alluvial gravels on the western slopes of the Cameron highlands are rich in cassiterite, a tin ore. Some gold, iron and tungsten are mined as well as bauxite.

### *The Annam ranges*

The Annam mountains lie between the South China Sea and the Mekong river. South of the Red river and its delta, the spurs extending from the eastern Himalayas attain less than 3000 m so that the highest peaks are Phou Lai Leng (2710 m). Pou Atouat (2499 m) and in the south B'nam M'hai (1647 m). Sediments of the Tethys geosyncline were folded in the Miocene and during the early Pliocene an erosion surface was cut and subsequently uplifted to 1000–1500 m in North Vietnam. This surface was subsequently covered with conglomeratic gravels west of the mountain Fan Si Pan. A second, lower erosion surface, developed in the late Pliocene, occurs in South Vietnam and was uplifted to 600–1000 m. In North Vietnam deposition of loess took place (contemporaneously with the Malan loess of China) during the early Pleistocene on the higher erosion surface.

### *The Red river*

The major river of this coast is the Red river which rises in the high mountains of the Himalayas and flows in a trough-like valley to the Gulf of Tongking where it has produced a delta of 14 000 km$^2$. The upper reaches of the Red river appear to have been captured by the Yangtze which, after capture, has entrenched itself 600 m. The capture took place at Shihku where a broad col and a valley led southwards past Tali and the lake Erk Hai. The river is only navigable a short distance upstream. Along the coast of the Gulf of Tongking, coastal terracing indicates uplift continued into the late Pleistocene.

### *Chao Phraya lowland*

Western Thailand is a lowland area lying east of the main peninsular range of mountains (Dawna and Bilanktaung ranges) drained by the Chao Phraya river. Alluvial lowlands extend nearly 500 km inland from the Gulf of Thailand.

### *The Cambodian lowland*

The Cambodian lowland is dominated by the delta of the Mekong river, occupying 38 700 km$^2$. The Mekong, which rises 2700 km away in China, floods during the monsoon season and water backs up and flows into Lake Tonle Sap which extends from 2580 km$^2$ in the dry season to 20 000 km$^2$. The sediment brought by the Mekong is advancing the front of the delta an average of 60 m annually. The land surface has been lateritised; laterite being used to construct the Khmer temples at Ankar. On the western margin of the lowland are the Cardomon mountains bordering the Gulf of Thailand with peaks of 1500–1600 m.

### *The Korat plateau*

The eastern plain of Thailand is a plateau of horizontal Triassic sandstones covering 154 800 km$^2$. It is separated from the Cambodian lowland by the east–west oriented Phanom Dongrack range. The Korat plateau is drained to the Mekong and is a slightly dissected plateau at about 200 m above sea level.

### *The Sunda shelf*

The shallow sea enclosed by Cambodia, Malaysia and Borneo is everywhere less than 200 m deep, and the southern part less than 150 m deep. The crystalline basement is gently folded into several basins and swells. The surface is thought to be a late Cretaceous peneplain and in the basins up to 800 m of Tertiary sediment has accumulated. An apron of deltaic sediment has been deposited from the Mekong river, 160 m thick. On the sea floor the course of former rivers may be traced, draining to the South China and Flores seas. The main emergent part of the Sunda shelf is the island of Borneo.

Little is known of the geomorphology of the Borneo highlands other than what has resulted from expeditions. The backbone of the island is a ridge of mountains trending from north-east to south-west. Mount Kinabalu (4101 m) in northern Sabah is the highest peak, but other mountains in Kalimantan reach 2000 m height. In the Gunung Mulu National Park of northern Sarawak sandstone and shale of early Tertiary age are succeeded by limestone of upper Eocene to Miocene age. The sandstone and shales form Guning Mulu (2376 m) and the limestone forms a ridge on the north-western flank

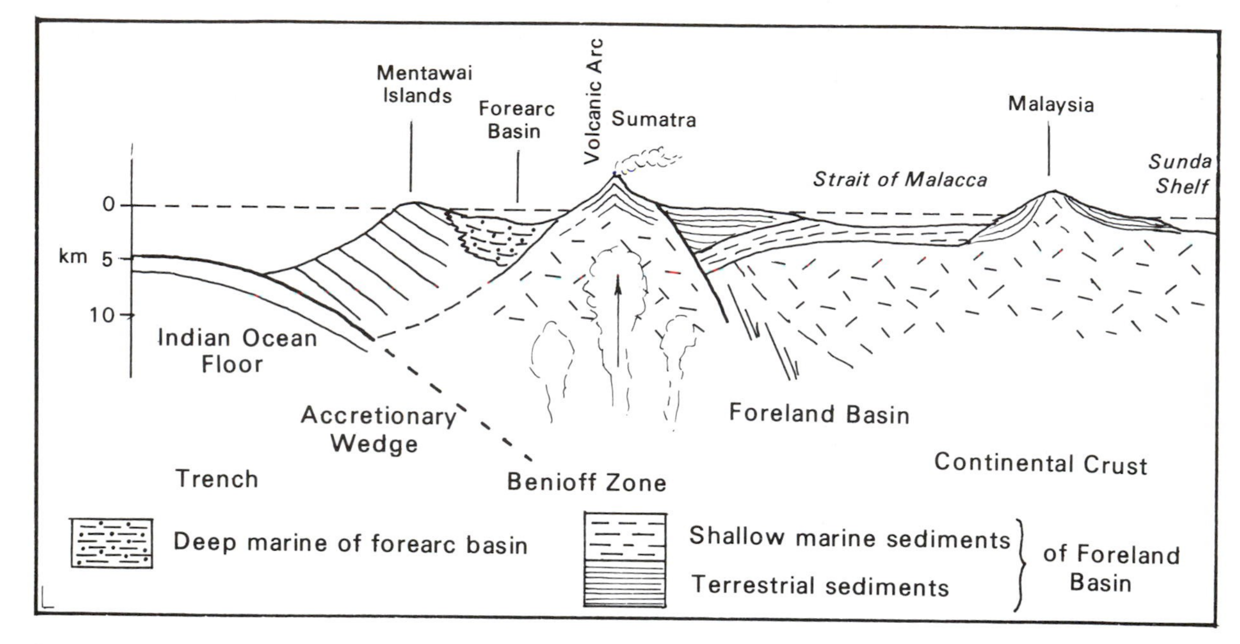


*Figure 6.12.* Section through the Indonesian volcanic arc.

of the mountain which drops down steeply to alluvial lowlands on the Sarawak–North Borneo border. The area is deeply dissected by the Melinau river which crosses the limestone in a gorge, and on the outcrop limestone pinnacles (karren) up to 50 m in height have developed. On the adjacent lowland river terraces indicate recent uplift.

Coastal plains extending well upstream on the major rivers occur on the coast of Sarawak and the southern coast of Kalimantan. These coasts are low-lying with a natural vegetation of mangrove swamp. Along the east coast of Kalimantan, thick Tertiary deposits accumulated which are now the site of oil reserves.

Sulawesi (Celebes) is composed of strongly disturbed younger sedimentary rocks, folded in the Alpine orogeny associated with vulcanicity, especially along the northern peninsula of this island of peninsulas. Despite its narrow configuration, Sulawesi has two large lakes, Poso and Towuti, situated towards the centre of the island. The mountain ranges reach just over 3000 m above sea level, and evidence of both high and low pressure metamorphism has been identified indicating the site of a subduction zone. This is further supported by the presence of a trench in the Mollucca Sea which curves into the Gulf of Tomini parallel to the volcanoes of the northern peninsula. The strange shape of Sulawesi may reflect the presence of a detached fragment of the Australasian plate on which the islands of Obi and Soela rest and which connects Sulawesi with West Irian (New Guinea) beneath the Ceram Sea.

### *The Indonesian arc*

The Indonesian Arc comprises the islands of Sumatra, Java and the smaller islands to the east of Java. The Sunda trench lies about 200 km from the southern coasts of Java and Sumatra, reaching a known depth of 7450 m. The Mentawai islands off the south coast of Sumatra are part of the forearc ridge and a forearc basin can be traced closer to the coast. The main arc is formed of the volcanic peaks along the length of Java and Sumatra where 70 active volcanoes occur. The highest peaks include Leuser (3381 m) and Kerintji (3805 m) on Sumatra and Slamet (3428 m) on Java.

The Indonesian arc occurs where oceanic material of the Indo-Australian plate is passing below continental material of the Sunda shelf (Figure 6.12). The deposits associated with the trench include a wedge of siallic material in a forearc ridge. Vulcanicity can be traced back as far as Mesozoic times and the lavas tend to be of the calc-alkaline type. The most spectacular eruption of historic times took place in 1883 on the island of Krakatoa, between Java and Sumatra, when the whole cone of the volcano was destroyed. Later a smaller new ash cone rose out of the sea, Anak Krakatoa. There is no marginal basin behind the Indonesian arc comparable to those of the western Pacific as it rests on the edge of continental material of the Sunda shelf.

## SOUTHERN ASIA

The second part of this chapter considers that part of Asia which lies south of the Himalayas. From east to west it includes Turkey, the Levantine countries, the Arabian peninsula, Iraq, Iran, Afghanistan, Pakistan and India. Although not directly connected with each other these areas have become joined to Asia, and unlike the areas to the north of the Himalayas had their origin in the sediments of the Tethys or as fragments of the former continent of Gondwana.

### Geological history

The early history of southern Asia then, is the history of Gondwana, which began to break up about 200 million years ago when rifts first appeared between Antarctica and Africa, and between India, Africa and Antarctica. Both Africa and India migrated northwards as the Southern and Indian oceans gradually widened. About 5 million years ago, in the Pliocene, the Red Sea rifts became operative and the smaller Arabian plate began to split away from the major African plate. As they moved northwards, Africa and India gradually closed the Tethys Sea, a large geosynclinal gulf between Gondwana and Laurasia. Into this sea were poured the sediments derived from the erosion of the continental masses on either side and a great thickness of sedimentary strata accumulated. The Tethyan zone of the Himalayan geological sequence is a 10 m thick accumulation of sedimentary rocks from Cambrian to mid-Eocene age. The lower Himalayas consist of a 20 km thick thrust pile of Proterozoic gneisses, Palaeozoic and Mesozoic sediments. It has been estimated that the Indian lithospheric plate moved northwards at an average rate of 15 cm per year from 70 to 40 million years ago, and then slowed to its present rate of 5 cm per year. There is dispute about the time when the Indian plate first began to collide with Asia, some authorities say 110 million years ago, others 55 million years ago. Which of these two dates is correct is an academic matter, the practical outcome is that there has been considerable crustal shortening as the two landmasses collided. As much as 2000 km has been suggested, achieved by under-thrusting of the Indian plate beneath the Asian plate to give a double thickness of crust, and by folding of the former Tethys sediments.

As the Indian plate impacted and pushed beneath the Asian plate a deep trough, the Indo-Gangetic depression, developed at the surface which has been infilled with sediment from the rising Himalayas. Uplift of the mountains is continuing at the present day at the rate of 1 cm per year. This is approximately balanced by erosion which has provided 7 km thickness of molasse type sediments, the oldest of which have been folded in the Himalayan foothills as the Siwalik hills. The younger sediments are in the alluvial terraces, flood-plains and deltas of the Indus and Ganges valleys.

From the end of the Cretaceous and during the Eocene, basaltic lava welled up on to the surface of the Indian shield. This occurred in the Deccan, the north-western part of the shield. where 500 000 km$^2$ was covered at its maximum extent. It has subsequently been reduced in area by erosion. Mountain glaciation affected the Himalayas during the Pleistocene and glacial action continues to influence landform development on the high peaks at the present day.

The core of the Arabian peninsula is a fragment of shield rocks which has split away from Africa since the Pliocene. Part of a former remnant of Gondwanaland, it is formed of Archean rocks. Brought northwards with Africa, and now moving independently, it has come into contact with the Asian landmass as the sediments of the Tethyian geosyncline were transformed into the mountains of Iran during the Alpine orogeny. As Afro-Arabia has pushed northwards, it has interacted with the Turkish micro-plate which is being forced westwards. Sedimentation continues in the Mesopotamian plain, an alluvial depression lying at the foot of the Zagros mountains.

### The Middle East

The Arabian peninsula, the Levantine countries of Lebanon, Syria, Israel and Jordan, Iraq, Turkey, Iran and Afghanistan comprise the area discussed under the heading of the Middle East. It is not a discrete geomorphological entity such as Africa or Australia, but the major geomorphological divisions are interrelated sufficiently well to justify description together. Economically, it is an area which has provided large quantities of fuel oil and still has the largest known reserves. Unfortunately, it is an area plagued with political instability and international tension. Despite these problems, the geomorphology of the Middle East has much to offer, containing many examples of geomorphological processes and resultant landforms in a relatively small area. The major geomorphological divisions of the Middle East are:

The Arabian shield
The Mesopotamian plain
The mountain ranges and plateaux of Iran
The Turkish peninsula

## The Arabian shield

In broad terms, the Arabian and Indian shields have many similarities. Both are fragments of Gondwanaland, from which they were separated by rifting and then both moved north as new sea floor was created. Both areas also have an elevated south-western margin and have a general easterly or north-easterly sloping surface which passes beneath Mesozoic and later rocks. Flows of basalt forming extensive plateaux have taken place in both areas and large areas have in common an arid climate.

The Arabian shield has a basis of Archean rocks comprising schists, gneisses, phyllites, cherts, marbles and slates into which granites have been intruded. Basic rocks have erupted during the Tertiary and Quaternary to cover large areas of the shield, particularly in the western part of the Njad plateau (Figure 6.13).

The rifting which affected East Africa continues northwards through the Red Sea, creating the abrupt western scarp of the Arabian shield where it is known as the Hijaz and Asir mountains. Block-faulting has affected the eastern side of the shield in the Jebel Tuwaiq where the overlying sedimentary strata have been uplifted to form an arcuate escarpment east of Riyadh. In the north of Saudi Arabia the ancient rocks are covered by the sands of the Nafud desert, beyond which an increasingly thick sequence of Mesozoic limestones, dolomites, marls and shales overlie the basement rocks. The south-eastern coast of the Arabian peninsula is also deeply covered by Mesozoic and later rocks which form an extensive plain in the Al Summan district. The provinces identified on the Arabian peninsula are:

The Hijaz and Asir mountains
The Njad plateau
The Al Nafud desert
The Rub al Khali desert
The Hadhramaut plateau
Yemen high plateau
Yemen plains
Jebel Tuwaiq
The Al Summan region
The Syrian desert
The Levantine uplands and the rift valley

### *The Hijaz and Asir mountains*

The northern part of the western fault scarp of the Arabian shield rises abruptly from the Red Sea to elevations of over 1000 m in Jebel Shefa. The scarp slope has been dissected by many wadis, some of which have basaltic lava flows confined within them. Eocene lavas occur at several places in this northern part of the shield but it is not covered by Mesozoic and later strata.

Further south in the Asir mountains, the fault scarp rises to 2700 m and in Yemen to over 4000 m in the vicinity of San'a. Beyond the crest the relief slopes gently eastwards in the regions of Njad and the Yemen plateaux.

Along the Red Sea littoral at the foot of the scarp there is a narrow coastal plain. It is formed by gravel- and silt-covered pediments overlying Tertiary and Pleistocene sediments. This coastal plain is best developed south of Jiddah where it is known as the Al Tihamah plain.

### *The Njad plateau*

This north–central part of the Arabian peninsula lies between 600 and 1000 m above sea level. The Pre-Cambrian basement complex crops out over wide areas with inselbergs. The area is drained by a system of wadis which converge on the Wadi Rumma which breaks through the Jebel Tuwaiq to form a large detrital fan in the northern part of the Al Summan province. In the southern part of the Njad plateau the network of wadis converges on the Wadi Dawsir which discharges around the southern end of Jebel Tuwaiq into the north-west part of the Rub al Khali desert. Volcanic plateaux occur in the west of this region and many small volcanic cones representing the last phase of eruption are present.

### *The Al Nafud desert*

The northern margin of the Njad plateau merges imperceptibly into the Al Nafud desert. This is an area of mobile barchan dunes between 10 and 30 m high.

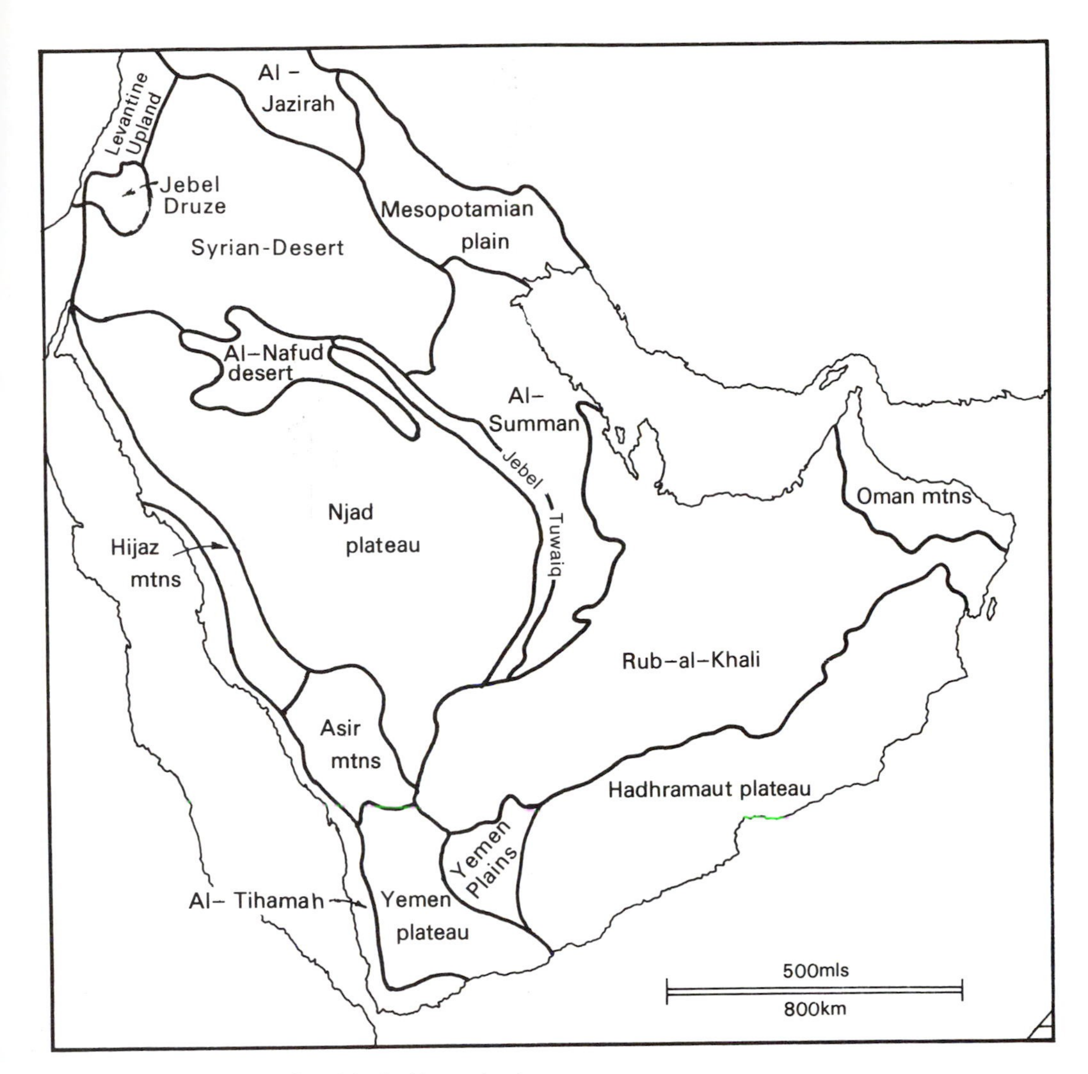


*Figure 6.13.* Plateaux and plains of the Arabian peninsula.

The Al Nafud desert extends in a long narrow strip, up to 70 km wide which passes east of Riyadh to link with the sand seas of the Rub al Khali.

### *The Rub al Khali desert*

The 'empty quarter' of Saudi Arabia is a large desert area extending 500 km from north to south and 1000 km from east to west. It is characterised by gravel plains and sand deserts where extreme aridity, strong winds and high summer temperatures are experienced. The winds have moulded the sands into sief dunes, 30–150 m high which can be up to 40 km long with a north-east–south-west trend. These long ridges are partly stabilised by sparse vegetation on their flanks but the crests remain mobile. One report describes 'sand mountains' 50 to 300 m in height where sands had been heaped up into stellate dunes by the wind. Barchans occur, particularly in the eastern and southern parts of the desert, lying on a desert surface of gravel plain or saline flats. Three hundred kilometres south of the Gulf of Bahrain is the site of a meteorite impact crater.

### *The Hadhramaut plateau*

Adjacent to the south coast of Arabia and north of the scarp which marks the edge of the plateau, Cretaceous and Tertiary rocks have been dissected to give a mesa and cuesta landscape on the surface of the Hadhramaut plateau. The plateau is about 1000 m in elevation and slopes gently eastwards, eventually reaching sea level. Drainage of the area is channelled in to the Hadhramaut valley which narrows eastwards before reaching the sea. In the eastern part of southern Arabia, the south coast of Oman is formed by the Jiddat el Harasis plain characterised by low hills of Eocene age. Gravel-covered Cretaceous strata crop out in the eastern part of this plain, some of which is covered by the Wahiba sands.

### *Yemen high plateau*

In this province the Pre-Cambrian basement complex is covered with horizontal Mesozoic and Tertiary rocks accompanied by massive basaltic flows, andesites and rhyolites. This high plateau has been matched with a similar elevated surface in Ethiopia on the other side of the Red Sea rift valley.

### *Yemen plains*

The Yemen plains lie east of the high plateau and are a lower continuation of it. In the western part, Archean rocks crop out with numerous inselbergs, but further east gravel plains with low hills and mesas of Cretaceous rocks occur. The Yemen plains gradually narrow eastwards and drainage from them is funnelled into the upper reaches of the Wadi Hadhramaut.

### *Jebel Tuwaiq*

The eastern edge of the Njad plateau is marked by the escarpments of sedimentary strata which have been uplifted to form the Jebel Tuwaiq. These escarpments are of silicified sandstone and limestone. South of Riyadh, the large Wadi Rumna, breaks through the escarpments, crosses the 70 km wide belt of the El Nafud sand desert to form a large detrital fan in the Al Summan province south of Hofuf. The eastern margin of the Jebel Tuwaiq province is delimited by the belt of dunes of the Al Nafud.

### *The Al Summan region*

The crystalline basement rocks are deeply buried beneath rocks of Palaeozoic, Mesozoic and Tertiary ages. Slight relief on the plain is attributed to buried block-faulting at depth which affects the overlying strata. The landsurface is extensively covered with limestone gravel. This region is where oil is found in Jurassic strata at depths of between 1500 m and 3000 m below the surface.

### *The Syrian desert*

The Syrian desert lies between the rift valley of Jordan and Israel and the Mesopotamian plain. It comprises the Al Hijarah, Al Widyan and Al Hamad regions which are a continuation of the Al Summan region between the deserts of Al Nafud and the lowland of the Mesopotamian plain. The land surface is gravel-covered or with dunes overlying Eocene limestones and Cretaceous strata. Oil is produced from the Cretaceous rocks at depths of 1000 m. The surface of this region is deeply scoured by roughly parallel wadis leading to the Euphrates.

### *The Levantine uplands and the rift valley*

The Levantine uplands lie between the Mediterranean Sea and the Syrian desert. Lowland areas follow the line of the Jordan rift valley northwards through the Bekaa and Orontes valleys. Over 1000 m of Jurassic and Cretaceous rocks accumulated over the Archean basement in this region, although the Archean basement rocks are closer to the surface in Israel. Development of the Jordan rift valley began in the Cretaceous, but it was not until the Miocene when the eastern side was displaced 40 km northwards. Further movement has taken place in the Pliocene and the Quaternary. The rift valley can be traced as a continuation of the Red Sea along the Gulf of Aqaba, through Israel and into the Beqaa valley of Lebanon and the Orantes valley of Syria. In Lebanon the downthrow of the rift valley is 1500 m but further south 7000 m of alluvial infill has been recorded. Associated with the rifting process basaltic lavas erupted to form a plateau lying between Damascus and Amman, known as Jebel Druze.

The rift valley dominates the geomorphology of the eastern Mediterranean seaboard. It creates a north–south pattern of features with plateaux lying between the coast and the rift valley and further plateaux lying to the east of the rift valley. Jebel Ansaryie and Jebel Liban (Mount Qornet es Saouda, 3083 m) rise abruptly from the Mediterranean to plateau surfaces and then drop with equal abruptness into the rift valley. This pattern is continued southwards into Israel where Jerusalem is located on a similar plateau between the coast and the rift valley. Southwards from Haifa there is also a coastal

*Figure 6.14.* The Dead Sea, Jordan.

plain, the Plain of Sharon. Some volcanic activity occurred on the western side of the rift, Lake Galilee is impounded by basaltic flows but the greatest amount of igneous activity occurred on the east bank in the Jebel Druze. The presence of hot springs and solfataras indicates that not all activity in the area has ceased.

The Dead Sea (−395 m) is approximately 70 km long and 20 km wide and occupies the lowest part of the rift valley (Figure 6.14). It is highly saline and it has deposited salts of potash, bromine and magnesium which are being mined. The Dead Sea receives the water from the River Jordan and several other ephemeral streams but has no outlet to the sea. Its level has declined in recent years and suggestions have been made to channel water from the Mediterranean to make up for the losses by evaporation while at the same time to using the difference in elevation to generate power. During the Pleistocene the water level was higher; a former shoreline occurs at −180 m., the level when the Lissan marls were deposited.

## The Mesopotamian plain

The two major rivers of the Mesopotamian lowland rise in the Armenian mountains and follow tortuous courses in deep gorges before they cross the uplands of Kurdestan and emerge onto the plain of Al Jazira. After leaving the mountains, the Euphrates, length 2720 km, only receives the Kharbar tributary but the Tigris, length 1840 km, flowing nearer to the Zagros mountains, receives many left bank tributaries.

The rivers flood in spring time, influenced almost entirely by the melting snows of the Armenian mountains with the Tigris rising by as much as 7 m at Baghdad. Sedimentation by the two rivers has infilled the head of the Arabian Gulf so the coastline has moved more than 250 km south-eastwards in 5000 years. The two rivers join together 100 km inland and reach the open sea through the waterway known as the Shatt al Arab. Extensive irrigation works have operated on the plain for the past 6000 years, and archaeological evi-

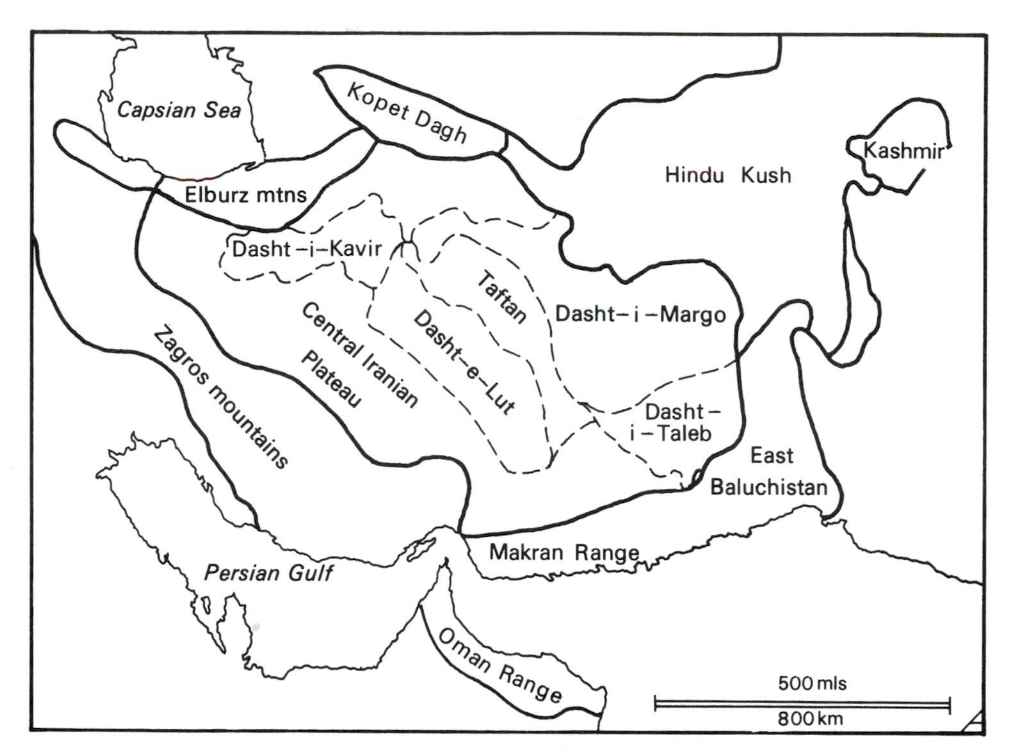


*Figure 6.15.* Mountains and plateaux of Iran.

dence suggests 10 m of silt have accumulated in some alluvial areas in the last 5000 years.

## The mountain ranges and plateaux of Iran

West of the Pamir knot of mountains, north of Pakistan, mountain ranges extend across Afghanistan to Iran where they enclose upland plateaux. These ranges include the Hindu Kush, the Kopet Dagh, Elburz, Zagros and Makran mountains (Figure 6.15). The approach and collision of the Arabian shield with Eurasia has resulted in tightly folded mountains in the Zagros ranges with a sharp bend around the head of the Mesopotamian plain. The Arabian plate has been forced beneath the Eurasian plate along the line of the Mespotamian valley and the vulcanicity of the Iranian plateau indicates subduction may be taking place. The subdivisions of the Middle East mountains are:

The Zagros–Oman mountains
The Makran range and east Baluchistan mountains
The Elburz mountains
The central Iranian plateaux

### *The Zagros–Oman mountains*

The Zagros mountains lie to the north of the Mesopotamian plain and extend from the Taurus mountains of Turkey through Iran. This province is also thought to include the mountains of Oman across the straits of Hormuz from Iran. The highest parts of the Zagros mountains reach 4000 m above sea level, but in Oman the peaks only attain 3000 m.

The Zagros mountains were formed from sediments which accumulated in the Tethys sea from Cretaceous to Miocene times. Throughout this period, slight folding occurred which has given anticlinal reservoirs in which oil has accumulated, particularly the Oligocene and Pliocene rocks. This gentle folding affected the southern and western parts of the Zagros mountains, but in the late Tertiary strong thrusting and faulting disturbed the northern and eastern parts of the mountains, leaving the gently folded autochthonous zone relatively unaffected. Within the succession, beds of salt and gypsum occur in the Eocene and Cambrian, which may have assisted the overthrusting by supplying a lubricant in the form of salt. Metamorphism has resulted in the production of schists and gneisses in the core of these nappes. Volcanic activity in a linear zone from Tabriz to Kerman has occurred as is demonstrated by the presence of basaltic flows, volcanic cones and fumeroles. The structurally similar Oman mountains also consist of Cretaceous strata and Pre-Cambrian core which have been uplifted. There is pronounced thrusting of the Cretaceous rocks and intrusive gabbros and other ultra-basic rocks are present.

Within the Zagros mountains there are many longitudinal valleys with internal drainage between the ranges. The rivers which reach the Mesopotamian plain either cross the grain of the country obliquely, zig-zag through the mountains, or cross them in a straight line. This has been used as evidence that the rivers were superimposed onto the underlying rocks and were antecedent to the formation of the mountains. Some doubts are now expressed about this suggestion as there is no evidence of a suitable cover material from which superimposition could occur. It is possible to explain the stream courses without resorting to superimposition by referring to the thickness of easily erodible beds which could have the same effect morphologically as superimposition.

### *The Makran range and east Baluchistan mountains*

This province lies between the Straits of Hormuz at the entrance to the Arabian Gulf and the town of Quetta. It is composed of simply folded mountains of Tertiary sandstones, shales and mudstones which reach an elevation of up to 2000 m or 3000 m above sea level.

The east Baluchistan mountains have a more north–south alignment which is pinched into a sharp bend near Quetta. This feature and others like it is referred to as syntaxis. The western parts of these mountains are formed from Mesozoic sediments, folded into long anticlinal ridges, unconformably resting on limestones of Permian age. The eastern ranges of these mountains are the Kirthir and Suliaman ranges which overlook the Indus valley. In these mountains folding has been more severe and sharp anticlinal features in Jurassic strata give cuestas and hogbacks rising to between 300 m and 1000 m above sea level. The configuration of the ranges and the frequency of earthquakes strongly reflect the stresses which affect this region.

### *The Elburz mountains*

The Elburz mountains attain a maximum height of 6000 m and lie in an arcuate line around the southern shore of the Caspian Sea, north of the city of Tehran. The structural trend of these mountains may be traced into north-western Iran from Turkey where they link with the Pontic mountains. Eastwards they link with the Kopet Dagh mountains of the Iranian–Soviet border.

The rocks of these mountains are of Jurassic and Cretaceous formations, deposited in the Tethys Sea and not folded until the late Miocene as some Tertiary sediments occur on the flanks of the Elburz. Metamorphosim affected the north-western part of the range and the highest peak, Mount Damavand (1850 m), is an extinct volcano. A narrow coastal plain lies adjacent to the Caspian Sea.

### *The central Iranian plateaux*

Between the Zagros and Elburz mountains, the interior of Iran is characterised by a number of plateaux and basins, areas which were caught up in the Alpine orogeny, faulted but not intensely deformed, and uplifted to various heights. Surrounding the town of Isfahan, the central Iranian plateau is composed of Oligocene and Miocene rocks which are separated from the Zagros mountains in the west by strong faults along which Eocene extrusive volcanic activity took place. The eastern margin of the central Iranian plateau is marked by the elevated area of the Kuh Rud.

Much of the Iranian plateau drainage is endoreic and many enclosed basins formerly contained Pleistocene lakes which are now dry, leaving a floor covered with silts, sands and in some cases, salts. During the Pleistocene there were glaciers on the 4000 m peaks of Shir-i-Kuh and Kuh-i-Jupar, south of Isfahan.

The largest of the interior basins of Iran, the Dasht-i-Kavir lies south and east of Tehran and is approximately 500 km by 300 km in extent. It is a desert basin into which detritus has been deposited after erosion from the surrounding mountains. The Dasht-i-Lut is a similar basin situated in the eastern part of Iran and there are other smaller basins enclosed within the mountain ranges. The border between Iran and Afghanistan is

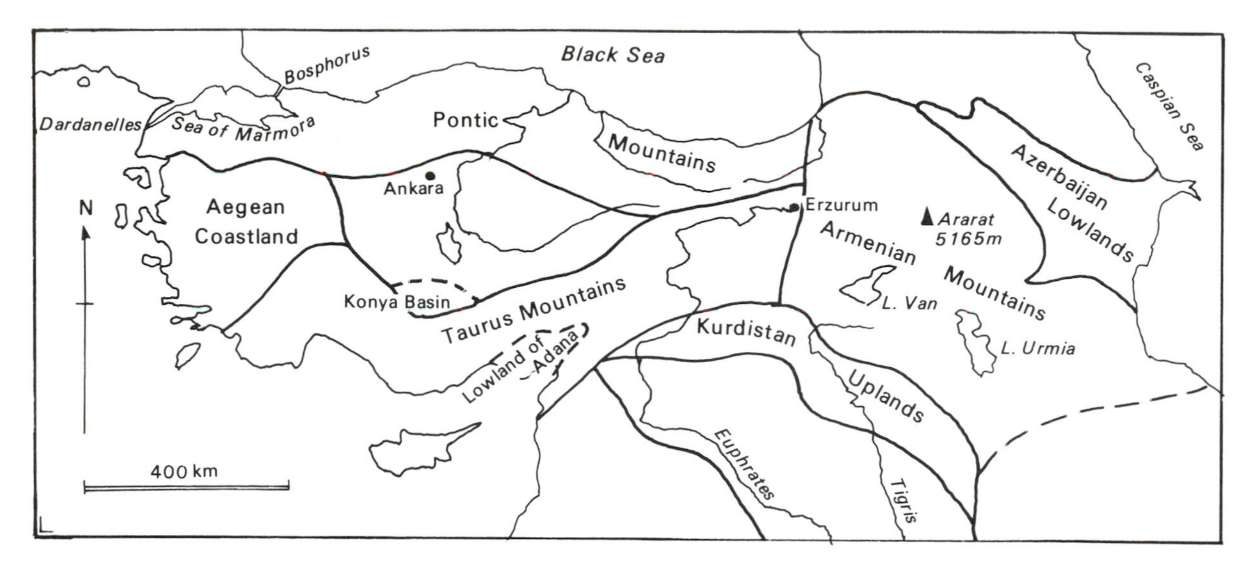


*Figure 6.16.* Mountains and plateaux of Turkey.

marked by the irregular highlands of the Taftan region. This is an area of volcanic rocks with some active cones in the Kuh-i-Taftan which reach an elevation of 4000 m above sea level.

Similar landforms may be seen in western Afghanistan and western Pakistan. The Dasht-i-Margo is a synclinal basin lying between the Hindu Kush to the north, the Taftan region to the west and the Makran ranges to the south. The floor of this depression is at 300 m above sea level and the lowest parts are occupied by dry plains which flood occasionally from the several streams which originate in the snows of the Hindu Kush mountains. The eastern part of the basin, covered in aeolian sand deposits, is known as the Registan.

The Dasht-i-Taleb is a smaller basin with similar geomorphology to the larger Dasht-i-Margo further north. Studies in the western part of Pakistan indicate that deposits which range from Pleistocene to Holocene have been partly dissected during the latter part of the Holocene, so that formerly extensive pediments are now only represented by terrace features. Many of these basin deposits are silty and loessial material is an important component of the sediments. Early Pleistocene deposits have been disturbed by tectonic activity.

## The Turkish peninsula

The structure of Turkey comprises two mountain ranges: in the north the Pontic and in the south the Taurus mountains. These mountain ranges are separated in the west but east of the town of Erzincan the two ranges come together to form the mountains of Armenia. South of the mountains, around the town of Urfa a less elevated area occurs in the middle Euphrates valley, the Kurdestan uplands. Small coastal lowlands at Adana and Antalya on the south coast and in the valleys leading to the west coast (Figure 6.16).

The geomorphology of the Turkish peninsula will be considered in the following provinces:

The Pontic mountains
The Aegean coastlands
The Taurus mountains
The central plateau
The Armenian mountains
The Kurdistan uplands

### *The Pontic mountains*

The line of the Pontic mountains extends along the north coast of Turkey from the knot of mountains in Armenia. At the Bosphorus the line of mountains is broken by the sea but the trend continues into Europe in the Stranca mountains. There are two ranges, the elevation of which is about 2000 m, and they are terminated abruptly on the northern side by a series of arcuate faults which mark the edge of the Turkish micro-plate, there being no coastal plain. Inland from Trabzon the mountains reach their maximum height of

4000 m in the Cakirgol and Tatus Dagh. Faulting has strongly affected the whole range which consists of a series of horsts. In the eastern part vulcanicity has occurred. A lower series of downfaulted graben lie between the two ranges, extending from the Sea of Marmora, following the line of the Kelhit and Coruh valleys. Cross-faulting has enabled these rivers to break through the coastal range to reach the Black Sea. The largest river to flow into the Black Sea is the Kizil which drains a large area of the upland plateau east of Ankara.

The western part of the Pontic mountains enclose the Sea of Marmora. There is more lowland in this area and downfaulting which has allowed the sea to flood the synclinal structure of the Sea of Marmora, and to connect the Black Sea with the Aegean Sea. Granite rocks crop out in the northern part of the Bosphorus which varies from 200 m to 1.5 km in width and is 25 km in length. At the south-western end of the Sea of Marmora, the Dardanelles is a strait 40 km in length and between 4 and 7.5 km wide with an abrupt fault scarp along the northern shore.

### *The Aegean coastlands*

The Turkish Aegean coast is very irregular with many islands and long inlets. Lowlands penetrate the valleys of rivers which flow in an east–west direction parallel to the mountain ranges. Faulting has affected the area in a north–south direction. The valley sides are steep, with the interfluves rising to 1000 m; the valley floors have deep infills of alluvial sediments over which the rivers meander. Penck reports a Tertiary summit plain near Bursa, east of the Sea of Marmora, with a subdued relief and wide valleys with terraces cut at lower levels on the mountain sides. Erosion in Holocene times has cut deeply into the landscape to give the steep valley sides.

### *The Taurus mountains*

Unlike the block-faulted mountains of northern Turkey, the Taurus mountains of the Mediterranean coast are fold mountains thrown up by the Alpine earth movements. The ranges continue the line of the Cretan arc through the islands of Rhodes and Karpathos, proceeding inland almost to the town of Afyonkarahisar where they make up a sharp loop southwards to enclose the Konya basin on the inland plateau before turning north-eastwards to join the mountains of Armenia. This part of the Taurus range is the highest, reaching over 3500 m and in marked contrast to the coastal plains of Antalya and the Adana. A number of streams have cut valleys through the main Taurus range from the interior. One of these provides access from the inland plateau to the Adana lowland and is referred to as the Cilician gates.

### *The central plateau*

The central plateau of Turkey is surrounded by mountains 1800–2000 m high and the plateau surface is between 1000 and 1500 m. The surface consists of an undulating plain in which numerous shallow depressions occur, some of which are salt marshes, others are water-filled and their extent is dependent upon the rainfall. The largest of these lakes is Lake Tuz at 980 m above sea level. The plateau surface has been faulted to give small elevated areas (horsts) and volcanic cones indicate past volcanic activity. Towards the east earthquake activity and vulcanicity increase in intensity, some of the cones reaching over 3000 m.

The northern part of the plateau is drained by the Kizil river which makes its way through the block-faulted northern ranges to the Black Sea. It has a deeply incised valley in the plateau surface. The western part of the plateau is drained by the Sakarya river but large areas of the plateau are endoreic, the rivers ending in salt lakes.

### *The Armenian mountains*

This region of eastern Turkey is a confused structure of fold mountains and narrow valleys. The main trend of the Taurus folding passes south of the town of Erzurum and north of Lake Van (1720 m) before turning southwards to join the Zagros mountains of Iran. The Kurdish Taurus follow a parallel trend south of Lake Van and Lake Urmia (1330 m). Vulcanicity has occurred in this area and lava flows have blocked valleys and built a barren upland plateau at 2000 m into which rivers are deeply incised. Several large rivers arise in this area, including the Tigris and Euphrates, fed by the melting winter snows and the glaciers during the summer. The Parrot glacier on the north flank of the volcanic cone of Mount Ararat (5165 m) is 2 km long, descending to 3600 m from the ice-filled crater. Lake Van has a fossil shoreline 55 m above its present water level and terraces can be traced from this level downstream from the lake.

*The Kurdistan uplands*

Emerging from the Armenian mountains the Euphrates and Tigris rivers cross an upland plain around the towns of Urfa and Diyharbakir. This area is 500 m above sea level and is semi-arid, and becomes drier further south as it merges into the Al Jezirah, the upper part of the Mesopotamian valley.

## The Indian peninsula

The extent of the Indian lithospheric plate is marked by the constructive margin along the mid-Indian oceanic ridge and the Carlsberg ridge on the south-western side. The west and eastern edges are formed by the conservative Owen fracture zone and the Ninety East oceanic ridge respectively. The northern boundary of the plate lies embedded in the continent to continent collision zone between the Indian and Asian lithospheric plates. The lands which lie south of the Himalayan mountain ranges have an area of almost 5 million km$^2$ and include the states of India, Pakistan, Burma, Bangladesh and Sri Lanka. In terms of a broad physiographic analysis the major geomorphological divisions are:

The Indian shield
The Indo-Gangetic plain
The Himalayan mountain ranges
Oceanic provinces of the Indian plate

Each of these major divisions may be further subdivided into a number of morphostructural provinces.

## The Indian shield

Triangular-shaped peninsular India has remained a stable area throughout the greater part of geological time and is composed of Archean or Proterozoic rocks. Folding has not influenced the shield since the end of the Pre-Cambrian, but faulting has occurred. The peninsula is largely formed from highly metamorphosed Pre-Cambrian gneisses with a foliated or banded structure interspersed with granitic intrusions; rocks that can be seen at the surface all the way from Cape Cormorin in the south to the Aravalli hills of Rajastan in the north.

In part contemporaneous with the Archean rocks and partly overlying them, a group of highly metamorphosed sedimentary and igneous rocks, the Dhawar system, were laid down. These rocks occur in the Aravalli hills, the Chota Nagpur plateau and in the western Ghats in Mysore. The goldfield of Kolar is in the Dharwar rocks and manganese deposits have been mined from these rocks in Madhya Pradesh, Orissa and Madras states. Following the episode of folding which affected the rocks of the Dhawar system, conditions of stability allowed the accumulation of the sedimentary Cuddapah and Vindhyan rocks, thought to be Protozoic and comparable in age to the Torridonian sandstones of Scotland. The Cuddapah and Vindhyan strata consist of quartzites, shales and limestones of marine origin which crop out in several localities in the northern part of the Indian shield, overlying the Archean and Dhawar rocks in the Aravalli hills.

Younger sedimentary rocks occupy only small areas of the Indian peninsula. As occurred with the Karroo rocks of South Africa, deep faulting has allowed the preservation of rocks of the Godwana system (Palaeozoic to Jurassic). These rocks are important to the economy of India as the lower sandstones and shales are followed by further massive sandstones and limestones which contain coal seams. These were preserved in deep troughs in the Archaen rocks of the shield; the trend of these fault troughs is followed by the Godovari and Madanadi rivers which cross the plateau to enter the sea in the Bay of Bengal. Other small troughs occur on the flanks of the Chota Nagpur plateau in Bihar and West Bengal where bituminous coals are mined.

Speculation has occurred about the relationship of the Gondwana rocks of India and rocks of similar (Permian) age in Australia. The distribution pattern of the rocks suggests that the two areas may have been in close proximity during that period of geological history. However, the deposition of the Indian Gondwana sedimentary rocks in deep troughs or basins, and the different pattern of the tillites (interbedded glacial deposits) of the two areas suggest the match may be coincidental. Erosion has removed traces of this Carboniferous episode of glaciation in southern India, but striated boulder beds may be seen in central and northern parts of India.

Outcrops of Mesozoic rocks on the Indian shield are limited to the edges with incursions only into ancient downfaulted areas. Jurassic rocks crop out in the Kathiawar and Gujerat areas and Cretaceous strata occur in Madras and extend up the Narbada valley north-east of Bombay.

An area of 320 000 km² in the north-west part of peninsula India is covered with outpourings of basalt which have subsequently solidified into the Deccan Traps. These volcanic eruptions began towards the end of the Cretaceous and continued during the Eocene. Where valleys have been cut in the lavas, they have a distinctive stepped appearance which gave them the name of trap. Although the greater part of the basalt plateau is composed of a uniform augite basalt, there are ashes and acid rhyolitic lavas as well.

The greater part of the Indian shield is an undulating plateau which has an elevation of between 200 and 1000 m above sea level. The plateau surface has been peneplained but inselbergs remain and the streams have incised their valleys below the general plateau level. On the highest parts of the plateau, King has traced fragments of the pronounced Gondwana surface, formed before the rifting of Gondwanaland took place. These remnants indicate a summit level which is inclined from south to north upon the Cardamon hills (2930 m) and the Nilgri hills (2920 m). Near Ootacamond some levelling suggests a post-Gondwana surface at 2100 m which may be related to summit levels in the eastern Ghats.

An early Tertiary surface is claimed to exist in many parts of peninsular India between 300 and 700 m above sea level. It post-dates the Deccan lavas as these are trimmed by it, so it must be middle or late Tertiary in age. This surface is capped by a massive laterite (duricrust), indicating its considerable age as a soil surface and it is cut into by a younger surface which has a less prominent lateritic development.

The eastern-flowing streams of the Indian plateau, with broad valleys and slight gradient, are claimed by Wadia to be mature landforms in the Davisian tradition. Many of these rivers rise in the western Ghats and flow eastwards, eventually to discharge into the Bay of Bengal. The nature and form of these rivers emphasises the long-continued stability of the Indian plateau. In the case of the Cauvery river, this smooth gradient is interrupted by waterfalls resulting from faulting.

The provinces into which the shield may be divided are as follows and are shown in Figure 6.17.

The Chota Nagpur plateau
Central plateau
Bundelkund
The Aravalli hills
The Chattisgarh plain
The Godavari valley
Hyderabad and Mysore plateau
Kistnar and Penner valley plains
Carnatic coastal plain
Sri Lanka
The Eastern Ghats
The Western Ghats
The Deccan Traps
The Thar desert
Kathiawar and Gujerat lowlands

*The Chota Nagpur plateau*

The north-eastern portion of the exposed Indian shield lies in the southern part of the state of Bihar and the northern part of Orissa. An extensive erosion surface is developed over Archean rocks in the north and the Cuddapah and Vindhyan rocks in the south and down-faulted graben have preserved the coals of the Gondwana system. This surface lies between 500 and 1000 m and an older summit level exceeds 1000 m above sea level.

Drainage of the plateau in the north is to the Ganges and in the east rivers such as the Damodar flow into the western part of the Ganges delta. In the south the rivers are tributary to the Mahanadi which flows directly into the Bay of Bengal.

*Central plateau*

This section lies to the west of the Chota Nagpur plateau and south of the upper course of the river Son. The landforms of the central plateau are developed over rocks of the Gondwana series but the general appearance is similar to the Chota Nagpur plateau, rising southwards from the Son river to heights of just over 1000 m. Drainage is all northwards to the Son and eventually the Ganges.

*Bundelkund*

Between the Son and the Ganges this plateau slopes gently towards the Ganges and has an elevation of between 300 and 700 m. It is developed upon rocks of the Cuddapah and Vinghyan systems. It is alternatively known as the central Indian foreland and it forms the western part of the State of Vinhya Pradesh, south-west of Allahabad.

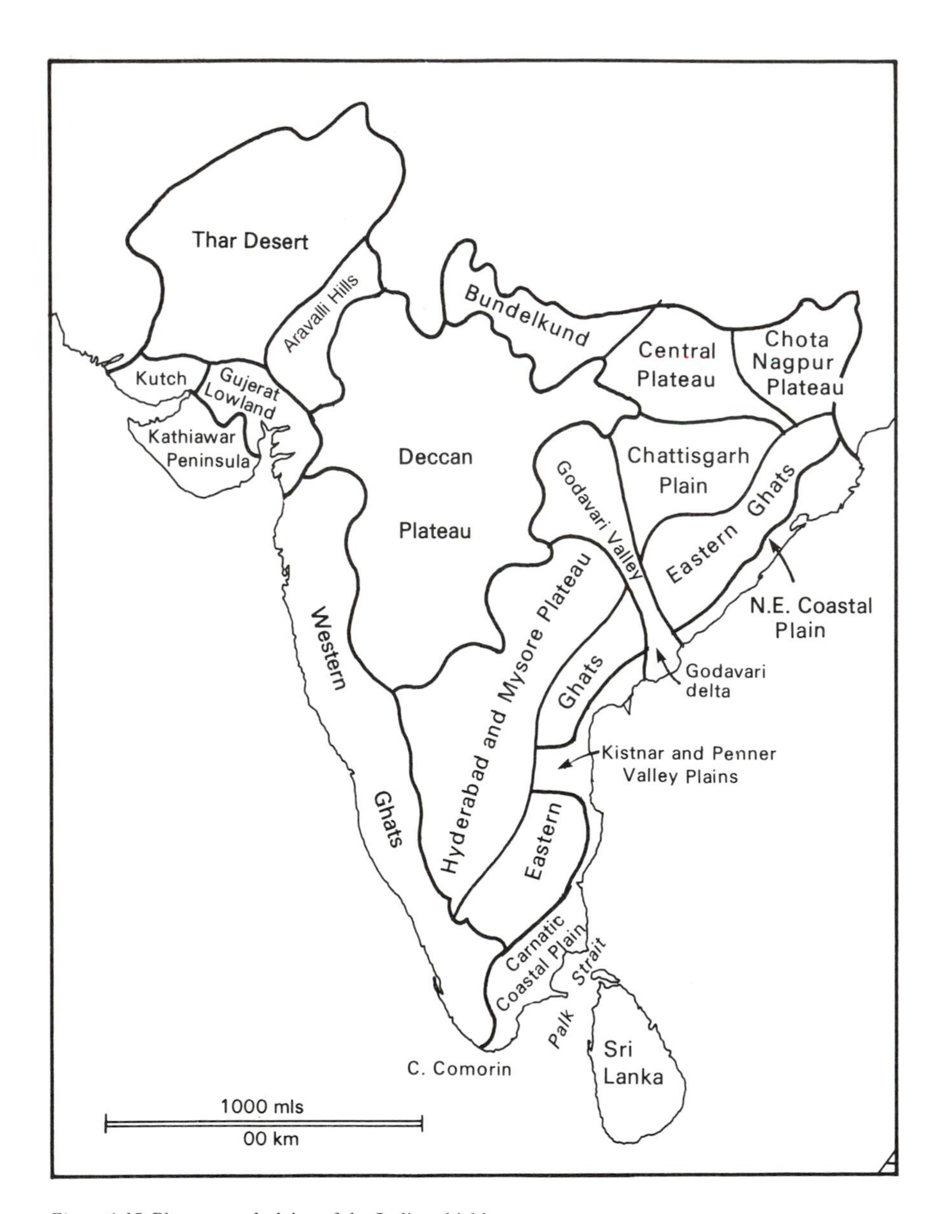


*Figure 6.17.* Plateaux and plains of the Indian shield.

*The Aravalli hills*

This line of hills has a north-east–south-west alignment in the north-western part of the shield from Jaipur to Udaipur in the State of Rajasthan. The Aravalli hills are formed from the Dharwar rocks, phyllites, quartzites and dolomitic limestones with manganiferous beds, which were folded and metamorphosed before the end of the Pre-Cambrian. These hills rise to a height of 1150 m but are mostly about 700 m above sea level where there is evidence of a peneplained surface situated above the dissection of the valleys. Its age is uncertain.

*The Chattisgarh plain*

South of the Chota Nagpur and central Indian plateaux the upper reaches of the Mahanadi river form a plain of lower elevation in the State of Madhya Pradesh around the town of Raipur. The plain is a lowland of 130–300 m elevation.

### *The Godavari valley*

A similar plain has been opened out by the headwaters of the Godavari river around the town of Nagpur in central Madhya Pradesh. The floor of the Godavari plain is between 200 and 500 m above sea level and the river passes over a series of rapids as it crosses the line of the Eastern Ghats. At the coast, the river has formed a delta which has built out seawards and laterally until it joins the Kestra delta in the south.

### *Hyderabad and Mysore plateau*

These areas of plateau country lie east and south-east of the Deccan lava plateau. The underlying rocks are Archean gneisses or granitic intrusions and the landscape is one of rolling plains slightly above 700 m, with scattered inselbergs rising above the general level. The surface, which slopes very gently eastwards, is attributed to peneplanation during the early Tertiary and there is a partial covering of laterite on the interfluves where current erosion has not removed it. The goldmines of Kolar exploit auriferous quartz veins in the Dharwar rocks to depths of 3000 m below the surface.

### *Kistnar and Penner valley plains*

The rivers Kistnar and Penner in northern Madras have eroded extensive valley plains west of the Eastern Ghats. These plains lie at an elevation of 200 to 700 m above sea level. Around the town of Cuddapah the rocks of that system have not been folded and give rise to flat-topped, mesa-like hills. Along the coast itself, lowlands occur where the major rivers reach the sea. In the north the delta of the Mahandadi has brought sufficient sand down to create a sand bar which encloses Chilka lake.

### *Carnatic coastal plain*

South of the city of Madras, the coastal plain becomes wider but isolated masses of the crystalline Pre-Cambrian rocks protrude through the alluvial sediments as steep-sided hills. In the Tiruchirapalli district upper Cretaceous and Tertiary limestone crop out. On the coast, just north of Madras, Pulicat lake is a lagoon enclosed by a sandbar.

### *Sri Lanka*

The large island of Sri Lanka (65 610 km$^2$) is separated from southern India by the shallow Palk Strait, 80–100 km wide. About half of Sri Lanka is plateau-like with an elevation of 400–500 m above sea level. It has been suggested that this plateau is of similar late Tertiary origin to the Indian plateau. Some higher, flat-topped hills and ridges at 800 m may be relicts of an early Tertiary surface. The strike of the Pre-Cambrian rocks in Sir Lanka is different from that in southern India which suggests the island may have moved independently of the main Indian lithospheric plate. The beaches of Sir Lanka show evidence of longshore drift, particularly on the western side of the country.

### *The Eastern Ghats*

The eastern edge of the Indian plateau is slightly more elevated than the central plateaux and many physical geographers have recognised a separate physiographic province in the Eastern Ghats. Unlike the Western Ghats, the eastern margin of India has several major rivers crossing to reach the sea. The increased elevation of the shield along this faulted coast may be caused by the upward flexing of the crust following rifting when Gondwanaland broke up. Maximum elevations of 1500 m are reached and an extensive planation surface is evident at 600 m above sea level. Drainage of the Indian plateau is from the Western Ghats eastwards, so it follows that the uplift of the Eastern Ghats is younger than the drainage pattern.

### *The Western Ghats*

The southern part of the western Ghats is developed upon the Archean rocks which form a major fault scarp extending without a break for 1500 km from Cape Cormorin in the south to the Tapti river 200 km north of Bombay. This major scarp is similar to the Eastern Highlands scarp in Australia and the Great Escarpment in South Africa. The Western Ghats are highest in the south, in the Nilgri and Cardomon hills, both of which rise to over 2900 m to form a summit plain which is possibly the Gondwana surface found on other parts of the former Gondwana continent. The north-western Ghats are formed by the edge of basalt flows of the Deccan Traps which overlie the Pre-Cambrian rocks.

The scarp edge rises sharply from the narrow coastal lowland to the edge of the Indian plateau which stands at between 900 and 1500 m above sea level. It has many fast-flowing streams which dissect the scarp face but two major streams, the Tapti and the Narbada which drain to the Arabian Sea, both occupy down-faulted troughs in the plateau surface. These valleys pre-date the Cretaceous as rocks of this formation are found within the troughs and they are covered by alluvial sands and gravels, some of which lie in rock basins 150 m above the present valley floor. This suggests that relatively recent faulting has occurred and is confirmed by the presence of the Jabalpur falls on the Narbada river.

At the foot of the Western Ghats on the Malabar coast there is only a narrow coastal plain but the streams have difficulty in flowing out to sea because of the beach ridge behind which a series of lagoons has formed. Offshore, the belt of abrasion is much wider than on the east coast, indicating the greater efficacy of the onshore winds in producing a wave-cut shore platform.

### *The Deccan Traps*

The lavas which erupted from late Cretaceous to Eocene times flowed out on to an irregular surface which has since been further disturbed by epeirogenic movements. Originally covering between 500 000 km$^2$ and 800 000 km$^2$, erosion has reduced the area of these basalt flows to 320 000 km$^2$, and from a total thickness of 3000 m to the present maximum of 2000 m. All previous landforms were covered and new plateau landscapes created by these lava flows which on average were about 5 m thick, although individual ones may be up to 33 m thickness. Some lava flows are separated by thin, discontinuous sedimentary layers, possibly soils, which break the succession into identifiable flows. In contrast to the infertile soils of the rest of peninsular India, weathering of the lavas has produced black fertile soils which can be used for a variety of crops, but particularly cotton. Locally known as regur, these black soils are widely known as vertisols.

### *The Thar desert*

From the 1000 m elevation of the Aravalli hills, the land gently slopes northwards to the valley of the Indus. The climate progressively becomes more arid and along the international border between India and Pakistan lies the Thar desert (Figure 6.18). In this region the Pre-Cambrian rocks of the basement complex crop out in only a few places and the area has a covering of Pleistocene sands. It has been reasonably stable tectonically and, with little rainfall to effect erosion, it is not strongly dissected by wadis. The sands of the Thar desert are mainly in the form of sief dunes, 15–30 m in height and partly stabilised by vegetation. The dunes were emplaced during the Pleistocene and their extent in the desert was formerly much greater as fossil dunes may be found as far east as Delhi and as far south as Ahmadabad. Within the dunes fossil soil formation and the presence of stone age implements indicates the sand mass has been stable for the last 3000 years. The fossil soils also show calcification and leaching at different periods suggesting that there has been climatic change. Where rainfall is less than 200 mm the crests of the sief dunes are of loose sand and between them small hummocky dunes 1–2 m high occur.

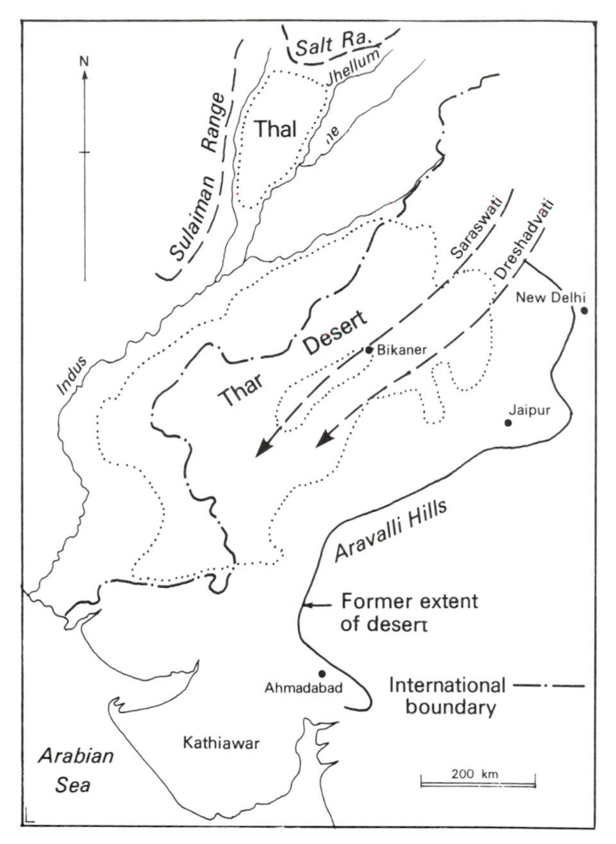


*Figure 6.18.* The Thar desert.

Were it not for the influence of the waters of the Indus river, the Thar desert would be much larger as many of the irrigated areas of its valley have insufficient rainfall. Desert conditions also occur on the interfluve between the Indus and Jhelum rivers in the Thal desert. A Pleistocene river terrace with sandy deposits has been worked upon by the wind in arid conditions to produce dunes 5–15 m in height. These occur in a complex pattern of longitudinal and transverse dunes.

### *Kathiawar and Gujerat lowlands*

Separated from the rest of peninsular India by the Gulf of Cambay (Khambat), the Kathiawar peninsula is generally of lower elevation. The Girner hills which rise above basaltic lava plains are of gabbro and in the area of Kutch there are many dikes and sills intruded into the country rock. Mesozoic and Tertiary rocks crop out around the periphery of Kathiawar and in the Gujerat lowlands. Jurassic sandstones occur in the north and a limestone, the Pobander stone, has been used extensively as building material in Bombay.

## The Indo-Gangetic plain

The area north of the advancing Indian lithospheric plate was occupied by the sea known as Tethys. Lying between the two land masses of Eurasia and Gondwana, the Tethys was gradually infilled with sediment eroded from the continents. Extensive deposits of carbonate rock were precipitated on the shelf seas during the Cretaceous. With the onset of the Alpine orogeny, the sediments of Tethys were uplifted and folded to form the Himalayas but the area lying to the south of the mountains was sharply depressed where the Indian shield was pushed below the Eurasian landmass. As the uplift proceeded this depression was infilled with debris eroded from the rising mountains. Isostatic readjustment continues, so erosion is still providing large quantities of sediment which is being deposited on the flood plains and deltas of the Indus and the Ganges rivers.

The Indo-Ganges lowlands cover over 1 380 000 km$^2$, varying in width from 140 km to 480 km. The watershed between the Indus and the Ganges is less than 300 m above sea level, 1600 km from the Ganges delta at Calcutta or 1000 km from the Indus delta at Karachi. These extensive lowland plains have been slightly incised leaving river terraces bordering the present-day streams with their contemporary alluvial deposits.

It is estimated that 5000 m of Tertiary, Pleistocene and Holocene sedimentary deposits have been laid down in the Indo-Gangetic depression. Authorities are divided in their opinions about whether this geological feature is a geosyncline or simply a foredeep of compensatory depth compared with the uplift of the Himalayas immediately to the north.

The Indo-Gangetic lowlands (Figure 6.19) may be considered in six provinces:

The Punjab plains
The middle Indus valley
The Indus delta
The Ganges valley
The Ganges–Brahmaputra delta
The Lower Brahmaputra valley

### *The Punjab plains*

The 'land of the five rivers' lies mainly in Pakistan and is sharply defined in the west by the Sulaiman ranges and the Salt range to the north. The eastern watershed between the Indus and the Ganges is less obvious, there being no relief obstacle which can be identified as the hydrological divide. The rivers Indus, Jhelum, Chenab, Ravi and Sutlej all emerge from the mountain front and proceed to cross the plain, eventually becoming confluent to form the Indus. The surface topography is made up of Pleistocene river terraces which can be traced as far downstream as Multan on the Jhelum–Chenab rivers. The easternmost of these interfluves has been brought under irrigation by the construction of canals which distribute water by gravity. The Pleistocene alluvia tend to be sandier than the present alluvium and in the case of the interfluve between the Jhelum and Indus, have been blown into dunes 5–15 m in height in the Thal desert.

Throughout geologically recent and historical times, the rivers of the Punjab have changed their courses on many occasions. The enormous load of sediment which the Himalayan rivers bring down from their upper reaches is deposited on the plains, raising the river beds above the general level of the plain so when floods occur the changes can be dramatic. The Chenab and Jhelum formerly joined the Indus 80 km further upstream than at present and the Beas river also abandoned its channel for a new one about 200 years ago. The headwaters of the Jumna formerly flowed south-west to join the Sutlej and Indus system and may have flowed across the Thar desert landscape.

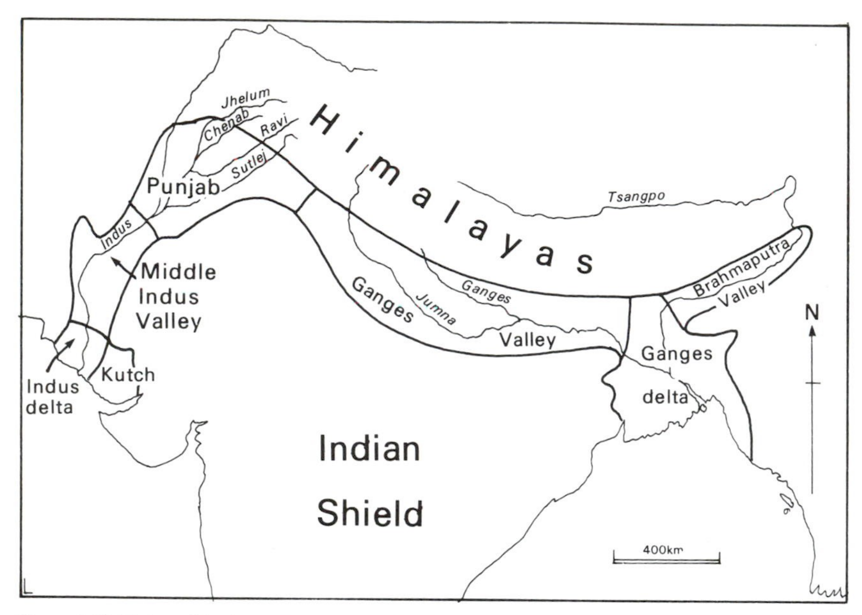


*Figure 6.19.* Plains of the Indo-Gangetic lowland.

### *The middle Indus valley*

The middle Indus valley is bounded on the west by the Kirthir and Sulaiman ranges which form the edge of the Baluchistan plateau. To the east lies the Thar desert. The alluvial plain of the Indus extends north-west towards Quetta in a pronounced embayment of the mountains which has been the site of strong earth tremors.

After the Indus leaves the Punjab plains, it flows through a rocky gorge where the Sukkur barrage was built in 1932 to provide irrigation water for 3 million ha downstream. Gravity canals convey water as far south as Hyderabad, 320 km from the barrage. Outside the limits of the irrigation area, many parts of the Indus lowland are as dry as the Thar desert immediately to the east.

### *The Indus delta*

Eighty kilometres from the sea the Indus begins to spread its volume of water into the distributaries of its delta. The delta has not been used extensively for agriculture in the past but local and world pressures for food production may require its increased utilisation in the future. The alluvium brought by the Indus becomes finer with distance from the headstreams, so by the time it reaches the delta it is clayey. In addition to this fine texture, salt is present in the soils of a considerable area near the coast.

Former channels of the Indus, now dry, once took water into the Rann of Kutch. This flat, low-lying area has gradually been silted up by the Indus floods to become what is now an arid plain, effectively an extension of the Thar desert. Formerly an arm of the sea, the Rann of Kutch still becomes flooded during the monsoon season. Salts which are left behind when the waters evaporate are subsequently blown inland as dust where they cause salinisation of soils. Although sedimentation is the main cause of the Rann of Kutch becoming infilled, some tectonic activity has affected this coast and uplift appears to have contributed to the conversion of this area from an arm of the sea into (almost) dry land.

### *The Ganges valley*

The River Ganges is 2510 km in length, with a catchment of 1.05 million km and an average rate of flow of 12 500 $m^3$ per sec. The watershed between the Indus and Ganges rivers lies immediately west of Delhi, separating the Sutlej river flowing south-west and the Jumna flowing eastwards. As in the Punjab, Pleistocene terraces accompany the river from the mountain front downstream until they are submerged in alluvia of more recent date. Locally these older alluvia are known as Bhangar and the newer alluvial plains as Khardar. The watershed between the Ganges and Indus is poorly defined and the headwaters of the Jumna previously flowed south-westwards. It is recorded in Hindu literature that a river, the Saraswati, flowed towards the town of Bikaner, but subsequent diversion caused the water to flow eastwards. The Jumna eventually becomes confluent with the Ganges at Allahabad where the gradient of the Ganges across the plains is less than 10 cm/km. Along the length of the Himalayas, rivers emerge from the mountain front, make a sharp turn to the east and, after flowing parallel to the Ganges for many kilometers, eventually join it. These left bank tributaries include the Ghaghara, Gandek, Kosi and many others. Fewer rivers join the Ganges from the plateau country of the shield but include the Son, Betwa and Chambal.

Terraces, formed by the Pleistocene alluvia are described as being eroded gradually by the meandering river channels. The older alluvia contain concretions of calcium carbonate called kankar, which in some districts amount to as much as 30% of the alluvial material. The younger Holocene alluvium contains less concretionary calcium carbonate.

### *The Ganges–Brahmaputra delta*

The delta of these two rivers is one of the largest in the world, the first distributaries beginning 320 km from the sea, shortly after the Ganges makes its turn southwards. The older part of the delta lies on the western side, and has changed little since it was first surveyed in the 1770s. The landscape is uniformly low and large areas have been drained for rice production. They are still subject to flooding but the main effect of the current sedimentation is to extend the eastern part of the delta seawards. Deltaic deposits cover 130 000 km$^2$ and comprise a sequence of clays, sands, marls, peats, lignites and forest beds. The swamps along the coast, forested with mangroves, are called the Sundarbans.

The major interfluve separating the Ganges and Brahmaputra is known as the Barind, but further migration of the rivers eastwards is prevented by the slightly higher land upon which Dacca is situated. Unlike the calcareous alluvia of the Ganges, that of the Brahmaputra is non-calcareous. Many changes have been recorded in the distributaries of the rivers in the delta region since 1750 and, as a rapidly developing geomorphological feature, it will change in the future when large floods occur.

### *The lower Brahmaputra valley*

Alternatively known as the Assam valley, the lower Brahmaputra valley is a tract of country 800 km long and about 80 km wide. The river tends to be braided and occupies a broad flood plain, the more elevated parts of which are used for rice cultivation. Oil occurs in four small fields where reserves in the mid-Miocene Tipam sandstone and upper Barail beds of the Upper Eocene are tapped. Lower Tertiary strata also contain coals which have been exploited from the Eocene Jainita beds.

## The Himalayan mountain ranges

The most impressive mountains of southern Asia, and indeed of the whole world, are the Himalayas. These mountain ranges, lying to the north of the Indo-Gangetic lowlands, form a major division of the Asian landmass with many peaks over 8000 m. They form a natural physiographic boundary to northern India, passing through the states of Kashmir, Nepal, Bhutan and Sikkim. After making an abrupt turn southwards the mountains eventually reach the sea on the southern coast of Burma (Figure 6.20).

The impact of the Indian lithospheric plate into the Asian plate has resulted in the strong folding and uplift of Mesozoic sediments from the Tethys into the ranges of the Himalayan mountains. It is argued that the depth of continental material beneath Tibet is much greater than normal, so there may be a double thickness of lithospheric crust where India has been thrust beneath Tibet. It is not thought that subduction is occurring in a deep Benioff zone as there is an absence of volcanic activity in the Himalayas.

The Himalaya mountains are drained by tributaries of the Indus in the west and the Ganges in the centre. The eastern part is drained by the Brahmaputra. It has long been assumed that these rivers were superimposed on to the geological structures, but recently it has been proposed that they began by utilising transverse synclinal structures across the ranges. The thick flysch deposits of Upper Cretaceous age, some of which are still preserved in Ladakh, are thought to have been the means of allowing streams to make the crossing of the ranges.

Although the greater part of the Himalayas is composed of Mesozoic sediments, older rocks have been caught up in the folding and metamorphism. Deep-

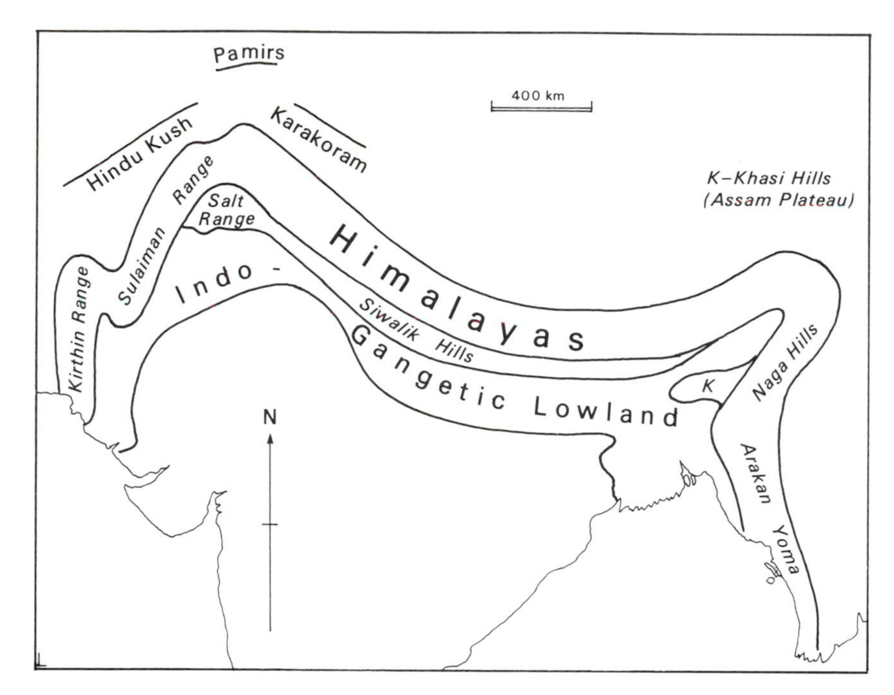


*Figure 6.20.* The mountain ranges of Southern Asia.

seated intrusive activity characterised the central parts of the ranges. Uplift began in the Jurassic and continued at intervals until the beginning of the Pleistocene.

The name Himalaya (abode of the snows) is used to describe all the mountain ranges which lie on the northern borders of Pakistan and India. However, the complexity of this major landform division of the Earth's surface is such that sub-division is necessary. In purely geographical terms, people speak of the Punjab Himalayas, the Kumaon Himalayas (between the Sutlej and Kali rivers), the Nepal Himalayas (between the Kali and Tista rivers) and the Assam Himalayas. Even these regional terms include a wide range of relief, structure and drainage. In terms of their geomorphology the Himalayas may be thought of in the following sub-divisions:

The Siwaliks
The middle Himalayas
The great Himalayas
Karakoram range
The Assam plateau

### *The Siwaliks*

Lying at the foot of the Himalayas from Pakistan to Assam is a belt of comparatively low hill country between 100 and 1500 m above sea level. The sediments which compromise the Siwalik hills are derived from alluvial materials eroded from the Himalayan ranges from the Middle Miocene to Lower Pliocene. They resemble the present alluvial deposits of the Ganges, except where they have been folded and uplifted in the most recent of the Himalayan orogenic episodes. Although relatively recent in geological terms, these strata have yielded an interesting mammalian fossil fauna, two-thirds of which are extinct. The Siwalik hills are made up of a series of anticlines and synclines, dissected into a sequence of steep escarpments by the subsequent streams of the strike valleys (duns). The northern boundary of the Siwaliks is an over-thrust reversed fault (the main boundary thrust) which marks the beginning of the main Himalayan ranges. Some of the larger river valleys have late Pleistocene terraces which may be traced throught the river gaps in the Siwalik escarpments and out on to the Indo-Gangetic plain.

### *The middle Himalayas*

The second group of mountain ranges in the Himalayas is the middle or lesser Himalayas. These have an average elevation of 4000 to 5000 m and consist of an intricate system of ranges. They are mostly composed of crystalline and metamorphic rocks of Pre-Cambrian

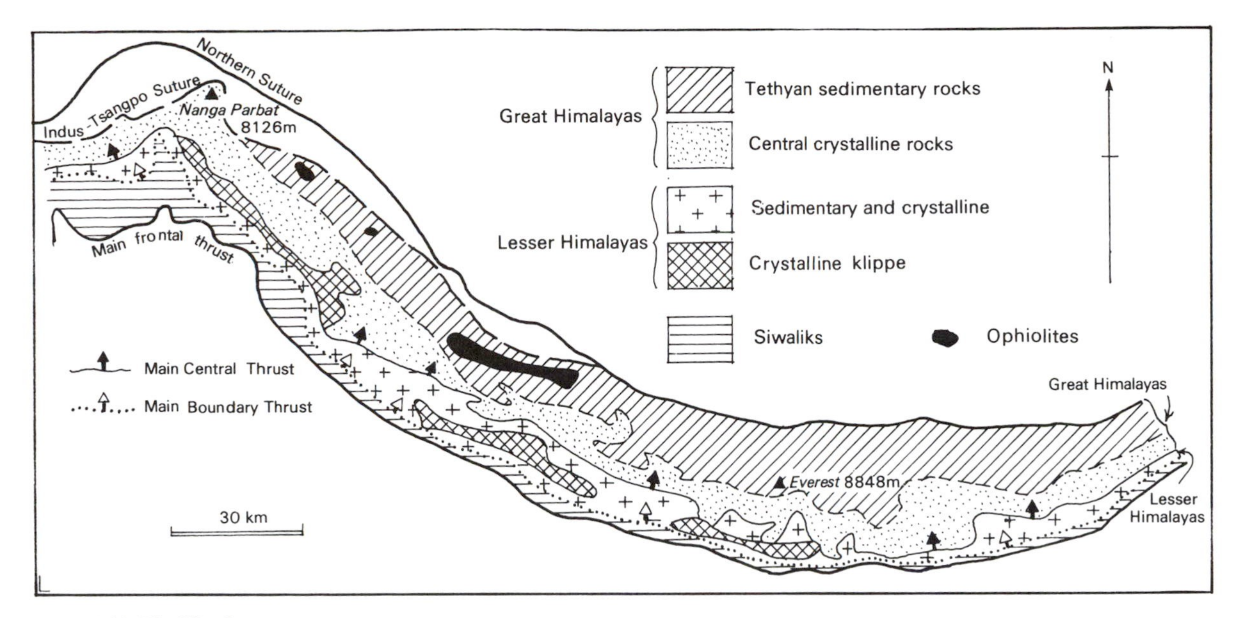


*Figure 6.21.* The Himalayas.

age, thick deposits of Palaeozoic rocks but only thin representatives of the Mesozoic. The major feature of these mountains is the thrust-faulting present. Crystalline gneissic rocks have been pushed or have slid southwards so that patchy remains of these crystalline rocks (klippen) remain to show that erosion has removed the greater part of the thrust material (Figure 6.21).

The relief of the middle Himalayas is such that the majority of the landscape is under steep slopes, often unstable with many landslides. The stability of the slopes is not improved by the extensive deforestation which is occurring.

### *The great Himalayas*

The highest Himalayan mountains lie on the northernmost ranges where the average height is in the region of 7000 m and many peaks exceed 8000 m. Many of the world's highest mountains are included in these ranges including Everest, known in China as Mount Qomolongma (8848 m) and Kangchenjunga (8600 m). In the western province of Gilgit in Kashmir and the southern parts of Tajikistan several of these large mountain chains converge into a knot of mountains, the Pamir knot. These higher peaks are well above the level of perpetual snows and permanent glaciers occur.

Uplift of the Himalayas began towards the end of the Eocene, following the deposition of the Nummulitic limestones, and continued until the Oligocene, elevating the central axis of ancient sedimentary and crystalline rocks. A further orogenesis of greater intensity took place in the Miocene when the multiple over-thrusting of sedimentary rocks took place and the intrusion of granites, now visible in the Brahmaputra valley, occurred. Finally, earth movements took place in Pliocene–Pleistocene times which raised the Himalayas to their present elevation. Erosion then cut the deep valleys and glaciers sculpted the impressive peaks. Pleistocene folding also affected the Siwalik beds which were themselves formed from the sedimentary debris eroded from the mountains.

Folding and thrusting in the Himalayas is directed towards the south and several major thrusts have been identified. The earliest thrusts recognised were known as the Krol and Garwhal thrusts, but at the present time geologists refer to the main frontal thrust, the main central thrust and the main boundary thrust. They are piled one on another so that a nappe containing Permo-Carboniferous rocks lies upon Lower Tertiary rocks, in places almost completely covering them (Figure 6.22). Subsequent erosion has revealed windows of younger rocks lying beneath the older rocks of the nappes. The rocks from which the great Himalayas have been formed were formerly Tethyan sediments. The crest of Mount Everest is formed from a massive sandy limestone, believed to be of Carboniferous age, but the roots of these mountains are highly metamorphosed and

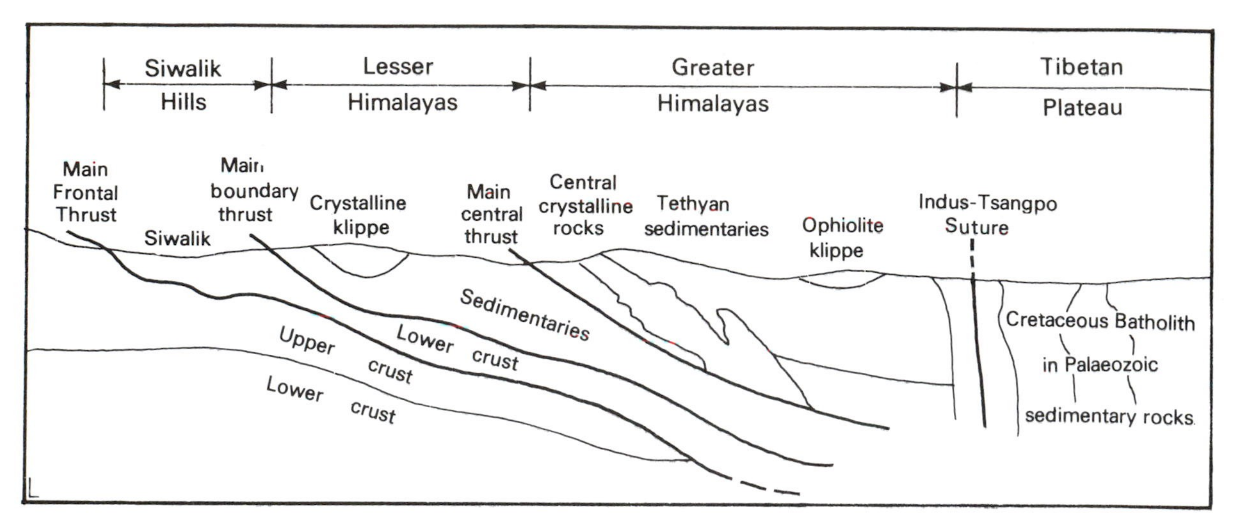


*Figure 6.22.* Section through the Himalayas.

intruded with plutonic rocks. Some authorities have suggested that the crystalline core of the Himalayas represents part of the frontal edge of the Indian shield which has been caught up in the mountain building activity. Evidence for this is the presence of typical Gondwana beds, including the tillites which have been recognised in the Darjeeling district. Other evidence includes the presence of slices of ocean-floor material (ophiolites) which were incorporated into the folding. Some of these rocks have been thrust up to 80 km southwards from the Indus–Tsangpo suture, where the Himalayan foldings have been welded onto the trans-Himalayan crustal material of the Tibetan plateau.

In such a newly-elevated landscape where erosion predominates, it is unlikely that many features remain of previous erosion cycles. However, broad ancient flat-floored valleys at 1000 m above sea level occur in the north-west Himalayas, and terrace gravels occur on the shoulders of some of the gorges but with epeirogenic movements of the magnitude which have created the Himalayas it is often difficult to assess the relationships of these remnants.

### *The Karakoram range*

The Karakoram range contains some of the largest glaciers outside Greenland or Antarctica. The Siachen glacier is 72 km long and 120–300 m thick at the present time, and was much larger during the Pleistocene. Many Himalayan mountain valleys that do not contain glaciers today have lateral and terminal moraines, ice-scoured basins and outwash terraces.

### *The Assam plateau*

The Assam plateau has an elevation of about 1000 m and reaches 2000 m in its highest parts in the Khasi hills. It is an upland area, sloping northwards, which lies between the valley of the lower Brahmaputra and the delta of the Ganges. Its rocks include ancient gneisses intruded by granites and thinly covered by Cretaceous sandstone. Accordingly, this area is thought to be part of the Indian shield. The older rocks are thought to underlie the Brahmaputra valley to the north, but on the southern side the Cretaceous sandstones thicken rapidly and pass beneath the Tertiary sediments of the upper delta region. The Assam block is a fractured horst which has become caught between the Naga hills and the main Himalayan ranges.

## Oceanic provinces of the Indian plate

As the terrestrial mass of the Indian plate migrated northwards, new sea floor was created between it and Antarctica. India probably moved northwards between the 90°E fracture zone and the fractures which are evident along the eastern side of Madagascar and across the Arabian basin. At the present time, the Carlsberg ridge and the mid-Indian ocean ridge are taken as the southern boundaries of the Indian plate; its northern margin is buried beneath the Himalayan mountains. After considering the broad features of the ocean floor, it is possible to sub-divide it into two: the Arabian basin, and the more extensive mid-Indian Ocean basin and the Bay of Bengal.

*The Arabian basin*

Although there is little coastal plain on the western coast of India, the continental shelf gradually widens northwards from Cape Comorin until at Bombay it is up to 250 km wide. Along the Baluchistan and Oman coasts the continental shelf is very narrow. The northern part of the Arabian basin is occupied by a large sedimentary fan of material brought in by the Indus. This fan, which is trenched by submarine distributaries, decreases in thickness southwards to the abyssal plain at about 15°N. The eastern margin of this basin is marked by the line of volcanic peaks from the Chagos to the Laccadive islands.

The western part of the Arabian basin is crossed by the Owen fracture which separates the smaller Oman basin from the larger Arabian basin. The southern margin of both basins is the Carlsberg ridge.

*The mid-Indian basin and Bay of Bengal*

This basin is much larger, extending as far south as the Tropic of Capricorn, with its deepest part lying east of the Chagos Islands. Its floor is at about 6000 m below the sea surface and several lines of volcanic seamounts occur, but only the Nikitin group of seamounts, at − 154 m, approach the sea surface.

Virtually the whole of the Bay of Bengal is occupied by the large submarine fan of sediment brought down by the Ganges. As in the case of the Indus, a number of undersea distributaries have been found crossing the fan. Except for the head of the Bay of Bengal, there is only a narrow continental shelf along the east coast of peninsular India, although Sri Lanka is only separated from India by the shallow Palk Strait. On the eastern shore of the Bay of Bengal, the Andaman and Nicobar islands lie on the edge of the south-east Asian plate.

# 7

# Australia

The continent of Australia has a land surface area of 7 674 518 km² lying between latitudes 11°S to 44°S and extending from 113°E to 154°E longitude. It has a generally low to moderate relief, average elevation 305 m, in the form of extensive plateau and broad sedimentary plains. The highest elevations occur in the Snowy mountains of the south-east, near the New South Wales–Victoria border and culminate in Mount Kosciusko (2442 m). The sedimentary plains of the interior are the result of downwarping of the basement complex and in some instances this has resulted in the land surface being below sea level; the surface of Lake Eyre is −12 m to −13 m below sea level.

The boundary of the Australasian lithospheric plate may be traced along the south-east Indian oceanic ridge from Amsterdam and St Paul Islands to a point south of New Zealand (Figure 7.1). It then passes east of New Zealand, the Kermadec Islands and Tonga. The northern boundary then extends from the vicinity of Fiji to the Bismarck archipelago and New Guinea. It continues north-westward along a line, clearly marked by the Java trench, as far as the Nicobar Islands. The western edge of the Australian plate is the Ninety East oceanic ridge, which separates it from ocean floor associated with the northward movement of the Indian plate.

Figures given by Butzer indicate that Australia has the greatest area of depositional plains of all the continents, 28%; and erosional plains cover 24% of the area of landmass. The ancient crystalline shield covers 31%, old mountain belts 10% and young mountain belts 7%. Despite their common occurrence throughout Austra-

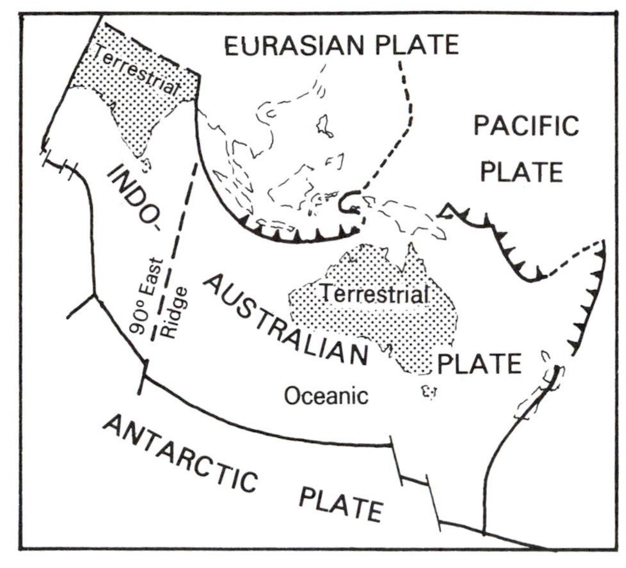


*Figure 7.1.* The Australian plate.

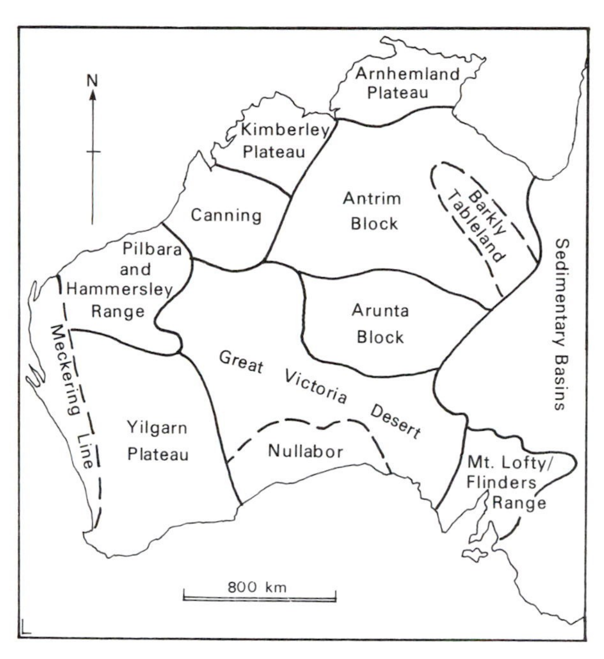


*Figure 7.2.* The Australian shield.

lia, volcanic plains and plateaux do not amount to more than 1% according to the available figures.

## Geological history

The oldest rocks in Australia occur in the continental crust of Western Australia. The Pilbara craton of Western Australia is more than 3000 million years old and possibly 3800 million years old. The Yilgarn craton of south-west Australia is almost as old at 2500 million years. North Australia became part of the continental block by 1850–1600 million years ago and the Gawler craton of South Australia 1450 million years ago. Some controversy surrounds the incorporation of the Arunta–Musgrave block of central Australia as it has been subjected to several periods of metamorphism. However, it appears to have been linked to the adjacent older cratons by 900 million years ago and carries the remnants of peneplain surfaces of considerable antiquity.

Australia was the last fragment of Gondwana to move away from Antartica. The final split came towards the end of the Cretaceous as the circumpolar spreading centre extended eastwards to make the separation. During the Tertiary, the Australian plate moved steadily northwards, although not so fast as the Indian plate from which it is separated by the Ninety East ridge on the floor of the Indian Ocean. This northward movement eventually brought the Australian plate into collision with the Banda island arc and New Guinea where orogenic activity has occurred from Miocene times. On the eastern side of the Australian plate, a fragment was rifted away and now forms the undersea plateau which surrounds New Zealand.

## The Australian shield

It is possible to sub-divide the major physiographic division of the Australian shield into ten provinces, based on structure and morphological criteria. These are shown in Figure 7.2 as:

The Yilgarn plateau
The Pilbara and Hammersley ranges
The Kimberley plateau
The Arnhemland plateau
The Canning basin
The Great Victoria desert
The Arunta block
The Antrim block
The Mount Lofty–Flinders ranges
The Nullabor plain

### *The Yilgarn plateau*

Archean rocks are exposed at the surface over an area of 1 032 000 km$^2$ in Western Australia where they form an extensive plateau between 400 and 1000 m above sea level. The surface of this block had been brought to low relief in the early Tertiary and drainage was integrated westwards. Deep weathering took place resulting in duricrust formation which has helped to preserve evidence for this 'Australian' planation surface. In the late Tertiary, differential uplift took place along a line parallel to the west coast and approximately 320 km inland, called the Meckering line. This disrupted the drainage pattern and the increasing aridity resulted in many playa lakes. In the coastal area erosion continued so that valleys were cut below the level of the Australian surface which was left isolated on plateaux and ridge crests. Break-up of this surface has provided detrital sands which become increasingly thick in the Gibson and Great Victoria desert where the Archean rocks are covered by a sand plain. East of the Meckering line deep weathering profiles can be seen but, nearer to the coast, erosion has removed the evidence.

The structural trend of the Archean rocks near Kalgoorlie is in a north–south direction and differential erosion has produced a ridge and valley topography. Mineralisation of the Pre-Cambrian rocks has led to the development of mining, especially for gold.

West of the Yilgarn block, and separated from it by the Darling fault are coastal lowlands which developed following the break-up of Gondwanaland. Consequently, only sedimentary rocks which are post-Jurassic age occur. The arching of the coastal area in the late Tertiary increased the erosive power of the coastal streams from the Darling scarp, resulting in deeply incised valleys developing during the Quaternary and destruction of large areas of the Australian surface.

### *The Pilbara and Hammersley range*

The Archean rocks of the Yilgarn block pass beneath the (Pre-Cambrian) sedimentary rocks of the Hammersley ranges, which lie on the surface of the Pilbara block. Except for uplift they have been very little disturbed. Between the two stable cratons some folding took place but was complete by 1800 million years BP.

### *The Kimberley plateau*

This is an area of dome-shaped relief, the highest parts of which rise to 530 m. Faulting has raised a few areas such as Mount Hann (850 m). Drainage is radially disposed with the northern rivers forming the Drysdale; the eastern rivers include tributaries of the Ord and Durack which drain into Joseph Bonaparte Gulf, and the southern drainage is collected into the Fitzroy and Isdell rivers. Nearly all the rainfall occurs between November and March.

Almost all the rocks of the Kimberley plateau are Upper Pre-Cambrian (Nullagine series) consisting of a thick series of sandstones and volcanics but Lower Cambrian volcanics crop out in the southern part of the province. The King Leopold sandstones form a maturely dissected plateau ranging from sea level to 600 m with the valleys incised 200 m below the surface. The volcanic rocks carry the remnants of a Tertiary surface, 150 to 700 m above sea level, capped with a laterite. In the east of the province and along the coast, scarp features are developed on the sandstones of the Warton beds. The coast is abruptly cliffed with many flat-topped offshore islands and peninsulas.

### *The Arnhemland plateau*

In structural terms, the Arnhemland plateau has much in common with the Kimberley plateau. The older Archean rocks has been brought to a plain of low relief before younger Pre-Cambrian rocks were laid down. Uplift then occurred to give these interbedded sedimentary and volcanic rocks a plateau form of relief, sloping down from 300 m in the south to pass below sea level in the north. The main component of the younger Pre-Cambrian rocks is the Kombolgie formation, dominated by a quartz sandstone which forms a pronounced scarp along the western edge of the plateau north of Katherine. Erosion has dissected the quartzite and cut through into the older rocks beneath, leaving little of the early Tertiary surface remaining but extensive areas of stripped quartzite rock crop out. The rivers of this province have a strongly seasonal regime resulting from the monsoon climate, with 60% of the precipitation falling between January and March.

### *The Canning basin*

The Canning basin lies between the Kimberley plateau and the Hammersley ranges. It is an area of the Western Australian shield which has continuted to subside after Pre-Cambrian times, so it contains Devonian and Permian rocks. The present surface is a plain, mostly below 100 m and covered with sand, which has been moulded into long sief dunes with an east–west alignment.

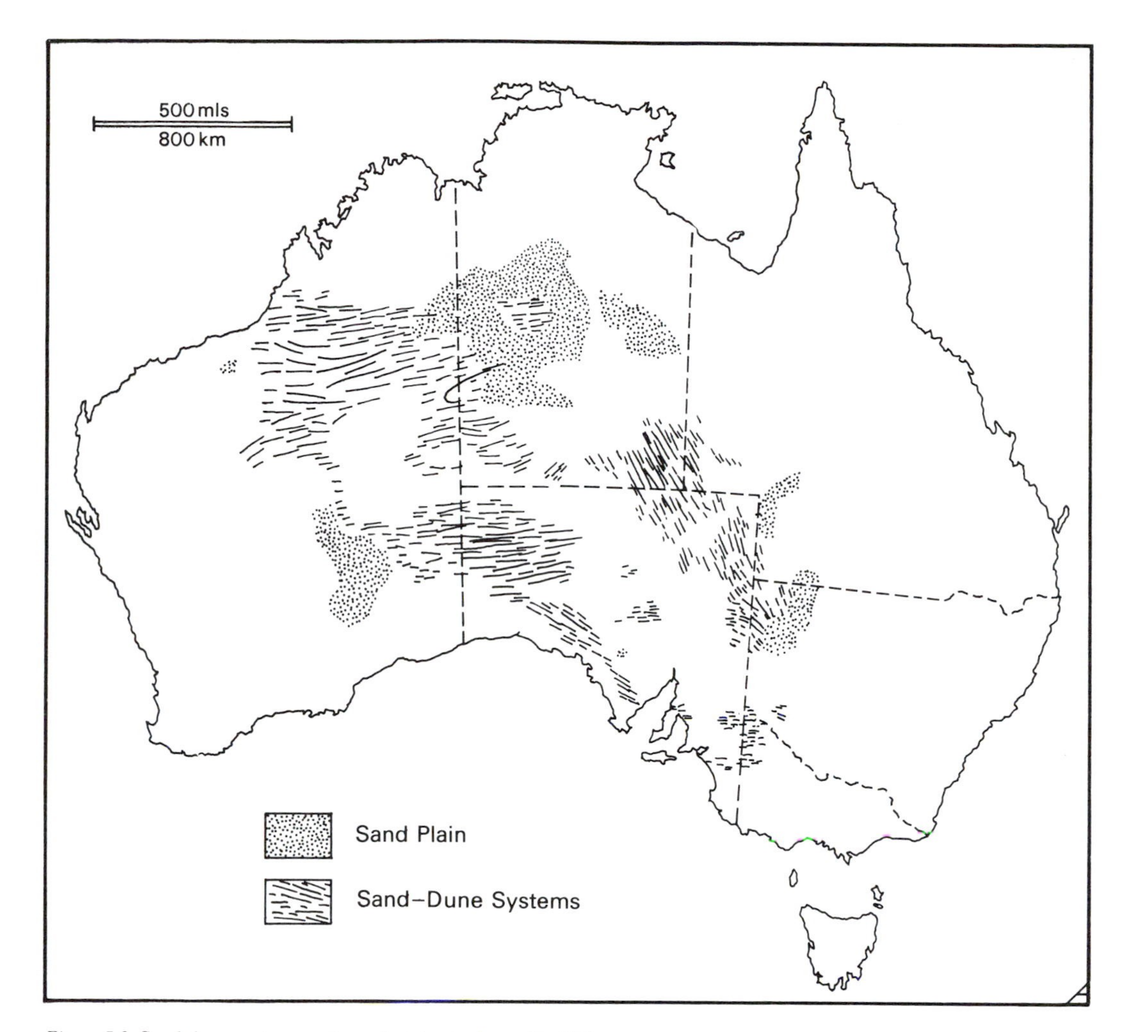


*Figure 7.3.* Sand dune systems ands sand plains in Australia. (After Leeper, 1960.)

### *The Great Victoria desert*

East of the Yilgarn and the Pilbara cratons, the Australian shield is extensively covered with sand dunes. 1 806 000 km² of Australia are estimated to have either sand ridges or sand plains as their characteristic landform. The sand dunes are sief ridges, extending for 80 to 320 km and are about 10 to 30 m in height. These linear dunes lie about 200–300 m apart, so there are about four or five ridges per km. Although predominantly on the old rocks of the shield, these sand ridges and sand plains extend onto the sedimentary basins of east central Australia, but they are most conveniently described in this section. In Western Australia, these sands form the Great Sandy desert, Gibson desert and the Great Victorian desert. It is thought that the alignment of these dunes resulted from the movement of wind around an anticyclonic sub-tropical high pressure cell, because in Western Australia their direction is east–west, whereas in South Australia and in Northern Territory they assume a north-west–south-east alignment (Figure 7.3). It is in these last two areas that the sand ridge country overlaps into the sedimentary basins of the Simpson desert and the Mallee country to the north-east of Adelaide.

The sand forming these linear dunes has a characteristic bright red coloration caused by a haematite coating on the sand grains. Individual grains are normally less than 1.0 mm in diameter and are formed of quartz. Over most of their area of occurrence the crests of the dunes

*Figure 7.4.* Ayers Rock and the Olga mountains in the distance.

are still active but in South Australia and the Mallee country the dunes are fixed by vegetation and no longer active. Presumably during a more arid episode in their past, their area of occurrence was extended and in the present, more humid, conditions a vegetative cover was established.

The drainage system over virtually all this sand ridge country is uncoordinated with ephemeral channels ending in salt lakes. Some of the major channels leading into the area from outside are well coordinated, as can be seen from the air.

### *The Arunta block*

The sand ridge country of the Great Victoria desert covers the northern margin of the Eucla basin, but on the northern side the Pre-Cambrian basement crops out again in the central Australian Musgrave ranges, Macdonnell ranges and Mount Doreen range. Between these outcrops, the Archean rocks are inter-cratonic troughs, such as the Amadeus trough in which 7000–10 000 m of sediments accumulated or the Ngalia trough with 3000–7000 m. The older Archean rocks are highly metamorphosed, tightly folded and intruded with granite batholiths and dolerite dykes. Sedimentation took place during the Proterozoic and continued until the Devonian; the rock succession includes sandstones, shales and limestones. The quartzite which forms the escarpment in which Heavytree Gap is cut, north of Alice Springs, is assigned to the Upper Proterozoic, together with conglomerates of a similar age derived from glacial debris of a Pre-Cambrian glacial period. Ayers Rock and Mount Olga are composed of coarse sandstones of Upper Devonian age (Figure 7.4).

A marine transgression affected large areas of the sedimentary basins of eastern Australia in the Cretaceous, but by the Eocene, uplift had caused the sea to withdraw and terrestrial weathering occurred to considerable depths producing duricrust. The presence of these highly weathered deposits is recognised in the 'grey billy' of siliceous deposits and laterite of iron-rich materials. Late Tertiary doming of these deposits led to the initiation of erosion and development of residual hills capped with duricrust. On the older rocks of the Macdonnell ranges remnants of the Older Australian surface are preserved.

Other Quaternary features of the landscape include the clays and evaporites in the playa basins and calcium carbonate concretions and crusts in alluvial materials. Active and ancient sief dunes, the latter fixed by vegetation, also occur in this province. Although rainfall is insufficient to maintain a permanent flow, the Finck river system is complex with anastomosing channels on an alluvial plain 2 km wide. Only from an aeroplane or a satellite photograph can the whole alluvial complex be appreciated.

### *The Antrim block*

South of the Roper river, and forming the central part of the Northern Territory is an extensive plateau called the Antrim block. This is an area with an elevation of between 150 and 300 m above sea level. The underlying geology is of Lower Palaeozoic sediments resting on the Pre-Cambrian shield. The sequence included flood basalts, probably of Cambrian age, which today cover 35 000 km$^2$ and originally may have covered over 300 000 km$^2$. The Antrim plateau was land throughout most of the Mesozoic but in the Cretaceous sandstones, shales and mudstones were laid down. Gentle uplift at the end of the Mesozoic permitted deep weathering to occur and lateritic soils were developed during the Tertiary. A further uplift in the late Tertiary has left mesas capped by laterite interspersed with alluvial tracts and partly covered by low dunes, especially in the southern part of the area.

The drainage of the western part of the Antrim plateau is integrated into the Ord and Victoria river systems. Parts of these valleys are deeply incised and the waters of the Ord have been exploited for irrigation at Kunnurra.

The eastern margin of the Antrim plateau is poorly marked physically as it gently inclines towards the Gulf of Carpentaria where the older strata are covered with Cretaceous and Tertiary strata. However, in the Mount Isa–Cloncurry district of the Barkly tableland strong folding occurs in the basement rocks which gives the relief a north–south trend. Cambrian rocks overlie the basement here and, although Cretaceous sediments were deposited, they have been extensively stripped in the planations of the Tertiary and the present land surface was subsequently duricrusted. The surface of the Barkly tableland is between 400 and 500 m and drainage is either to the Carpentaria basin or to the inland Lake Eyre basin.

### *The Mount Lofty–Flinders ranges*

This geomorphological province is structurally related to the Adelaide geosynclinal deposits and orogeny which occurred during the late Pre-Cambrian period of geological history. Extensive block-faulting in the Tertiary has thrown the area around Adelaide into a series of horsts and graben. The Mount Lofty ranges reach 883 m, the Flinders range 1300 m, and the less elevated Barrier ranges of western New South Wales are the upthrust features. The sea-filled graben of Spencer Gulf and Gulf St Vincent, and the ephemeral Lake Torrens, are examples of the downfaulted areas.

Eyre peninsula and the Gawler range, lying west of Spencer Gulf, are part of the Archean shield and include important iron ore deposits. This area acted as the western foreland during the Adelaide orogeny when the original folding of the rocks took place. It is claimed that early and late Tertiary erosion surfaces are preserved on the Mount Lofty ranges. The Australian erosion surface cuts across Cretaceous, Eocene and basement complex in the vicinity of Woomera. Coober Pedy on the Stuart ranges is famous for its opal mines.

### *The Nullabor plain*

Inland from the Great Australian Bight on the south coast of Australia is the 193 000 km$^2$ Eucla basin. Although sedimentation occurred in the period from Carboniferous to Tertiary, this is an area which is essentially still part of the Australian shield. It is better known as the Nullabor plain, an area of very low relief, ranging from 30 m elevation at the coast to 200 m inland. The surface is formed by a calcreted Miocene limestone which has been subjected to karstification in a pluvial phase of the Pleistocene. Conventional doline features are uncommon but circular shafts, 11 m deep and 2 m in diameter occur together with large subterra-

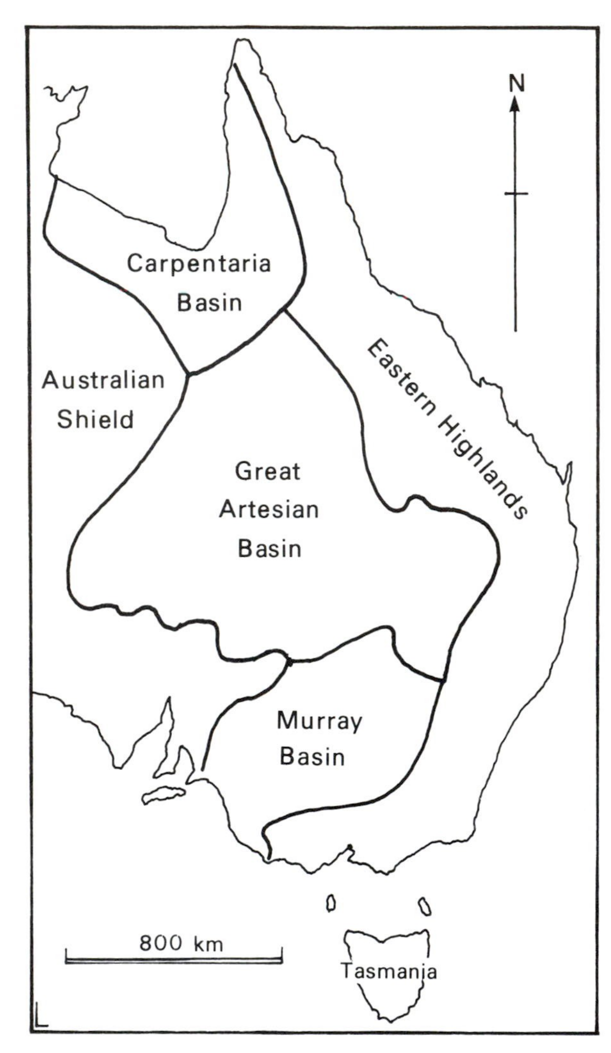


*Figure 7.5.* Sedimentary basins of Australia.

nean systems formed when the water-table was lower and the water more aggressive. The Mullamullang cave has 5 km of passages. The northern limit of the plain occurs where the dune sands of the Great Victoria desert begin to cover the land surface.

## The sedimentary basins

Three provinces comprise this major geomorphological division, all of which occur in eastern Australia (Figure 7.5). In this zone the emphasis in geological history has been one of gentle subsidence, accompanied by sedimentary infilling. The subsidence has not been uniform throughout the whole of the major relief division, and this has led to the three provinces being separated by sills which are more stable tectonically. The three provinces are:

The Carpentaria basin
The Great Artesian basin
The Murray basin

The landforms of these sedimentary basins are monotonous plains over which rivers flow, often with meandering courses in poorly defined valleys. Drainage of these rivers is to the Gulf of Carpentaria in the north and the Southern Ocean in the south. In the interior, though, the drainage is to Lake Eyre (−16 m), the lowest part of Australia.

### *The Carpentaria basin*

The least well-known of the major sedimentary basins of Australia, the Carpentaria basin lies between the Barkly tableland and the eastern highlands. The basement complex is near or at the surface in the Barkly tableland and Selwyn range but it dips gently northwards so that at the coast there is 1000 m thickness of sediment overlying the Archean rocks. Throughout the Mesozoic, intermittent deposition and planation occurred and an early Tertiary uplift resulted in the removal of Cretaceous sandstones and shales which were probably responsible for the superimposition of the rivers onto older structures beneath. Further uplift took place in the late Tertiary and some warping continued into the Quaternary.

The Carpentaria basin is drained by the Gregory, Leichart and Flinders rivers which rise on the Barkly tableland and flow to the Gulf. The divide between the northward-flowing rivers and those which turn south to the Lake Eyre basin is very indistinct. The area is subject to a monsoonal climate, so the regime of the rivers is strongly seasonal and channels are extensively braided.

### *The Great Artesian basin*

Artesian waters are present beneath the inland plains of Australia from the Barkly tableland in the north to the Cobar peneplain in central New South Wales and westwards as far as the Simpson desert and the Lake Eyre catchment. Crustal downwarping commenced in the Jurassic when lacustrine deposits were laid down. These sediments were sands and now form the aquifers

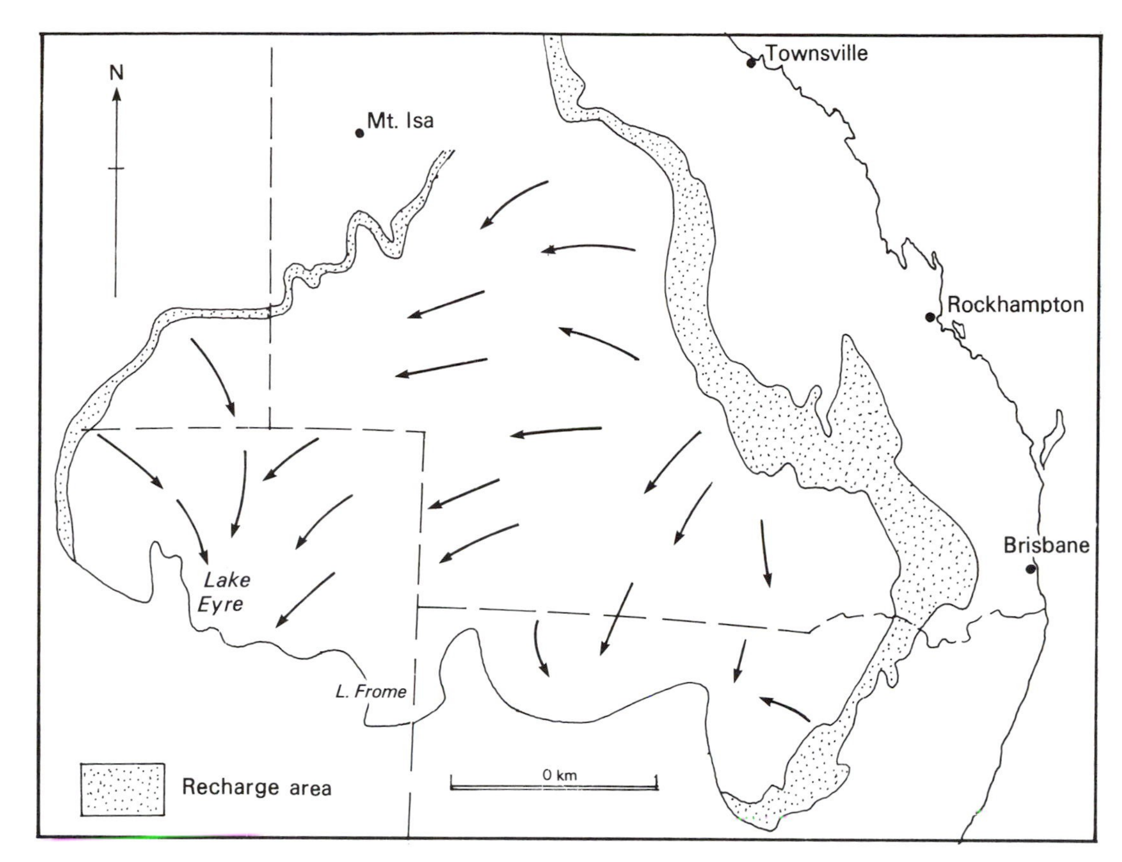


*Figure 7.6.* Underground water movement in the great artesian basin.

from which the artesian water is obtained (Figure 7.6). Subsequent sedimentation in Cretaceous times provided other aquifers and also impermeable shales which seal the water below and enable the artesian pressure to be built up. Lake sediments and alluvial materials accumulated until the Quaternary, but at the present a phase of erosion is in progress. The underlying basement complex is encountered at depths of over 3000 m below sea level.

The characteristic landform of the Great Artesian basin is a plain, usually at an elevation of less than 1000 m above sea level. The surface frequently has a duricrust upon it and where erosion has slightly incised the surface a gently undulating plain with low plateaux is developed. The duricrusting took place mid- to late Tertiary and was accompanied by deep weathering. Warmer temperatures are involved in these processes and, as a result of the migration of the weathering products within the weathering profile, the ferruginous or siliceous crusts were developed. Attempts to map the distribution of these features have indicated the boundary between ferrugineous and siliceous crusts roughly coincides with the catchment of the Lake Eyre drainage system. The siliceous crusts occur inland and the ferrugineous crusts near the coast, but their development is complicated by changing climatic conditions during Pleistocene and Holocene times. Breakdown of these siliceous crusts has resulted in 260 000 to 390 000 km$^2$ of plain covered with extremely stony soils. Removal of the finer fractions of the soil by wind leaves the stones at the surface as a desert pavement. These stony surfaces (hamadas) are known in Australia as gibber plains.

The drainage of the northern part of the Great Artesian basin is by the Georgina, Diamantina and Coopers Creek rivers towards Lake Eyre. These ephemeral rivers have braided or anastomising courses which have given the name 'channel country' to this area. The river courses have downstream gradients of only 0.02 to 0.05%. The catchment of Lake Eyre is

*Figure 7.7.* The clay plains west of Narrabri, New South Wales.

estimated at 1 164 000 km² with a rainfall of less than 45 mm and a potential evaporation measured at 225 mm per annum. In a similar manner to lakes elsewhere, Lake Eyre was more extensive in pluvial periods of the Pleistocene and a surface level of 60 m above sea level has been reported. This enlarged lake, referred to as Lake Dieri, was twice the size of Lake Bonneville, the predecessor of the Great Salt Lake of the USA. The drainage of the southern and eastern parts of the Great Artesian basin is taken via the Darling river and its tributaries into the adjacent Murray basin.

### *The Murray basin*

The Murray basin as a separate geological entity appears in the Eocene when an epicontinental sea spread onto the Australian shield. Currently, the Murray basin is choked with alluvial debris and there is evidence to suggest that the basement complex is slowly subsiding beneath the accumulated sediments. The basin receives drainage waters from the Darling river, the headwaters of which are outside the structural basin. These rivers such as the Warrego and the Paroo bring large quantities of sediment derived from the Cretaceous materials in their headwaters. The streams draining from the eastern highlands: the Namoi and the Macquarie to the Darling, and the Lachlan and Murrumbidgee to the Murray, also bring quantities of sediment.

Studies in the Riverina district have shown that several stream patterns have been overlaid on each other. These former stream courses have been given the name 'prior' streams and can be identified by the deposits they left behind. An early pattern of prior streams was disrupted by tectonic uplift near Echuca. This was responsible for deposition of sediment in a now dry lake east of the town, Lake Kanyapella, with the Barmal sandhills on its north-east side as its lunette. Many other lakes and former lakes have these new moon-shaped dunes on their leeward side, derived from sediment blown from the dried floor of the lake. Clays deposited by floods, dried and then broken into silt-sized particles by salt crystals, form a readily blown material called 'parna'. This former alluvial clay covers thousands of square kilometres in south-eastern Australia (Figure 7.7). The Murray river has a tortuous, meandering course and unlike many other rivers its rate of flow decreases downstream as water is lost by evaporation, irrigation and seepage. Its salinity also increases downstream.

## The eastern highlands

The elevated margin of the Australian continent, known as the Great Dividing range, appeared to the early settlers as a mountain range extending in elevation from 1500 to 2000 m, rising abruptly from the narrow coastal lowlands. Approached from the opposite direction, these 'mountains' are seen to be the edge of an uplifted plateau. From Cairns in northern Queensland to south of Sydney the eastern margin of Australia has a great scarp lying between 10 and 100 km inland. It has receded under the attack of numerous streams which flow directly to the Pacific Ocean, so it has a very irregular appearance. The physical geography of eastern Australia is dominated by the presence of this great scarp; east of it lie the coastal lowlands and west of it the elevated plateaux of the northern and southern tablelands of New South Wales and the Darling Downs and Atherton tableland of Queensland (Figure 7.8).

The geomorphical provinces which may be recognised from north to south along the eastern highlands of Australia are:

Normanby lowlands
Atherton tableland
Central Queensland uplands
Queensland coast lowlands
Darling Downs
Northern tablelands of NSW
Coastal lowlands of NSW
Southern tablelands
Western slopes
The Snowy mountains
Central Victorian uplands
Gippsland and West Victoria coastal lowland
Tasmania

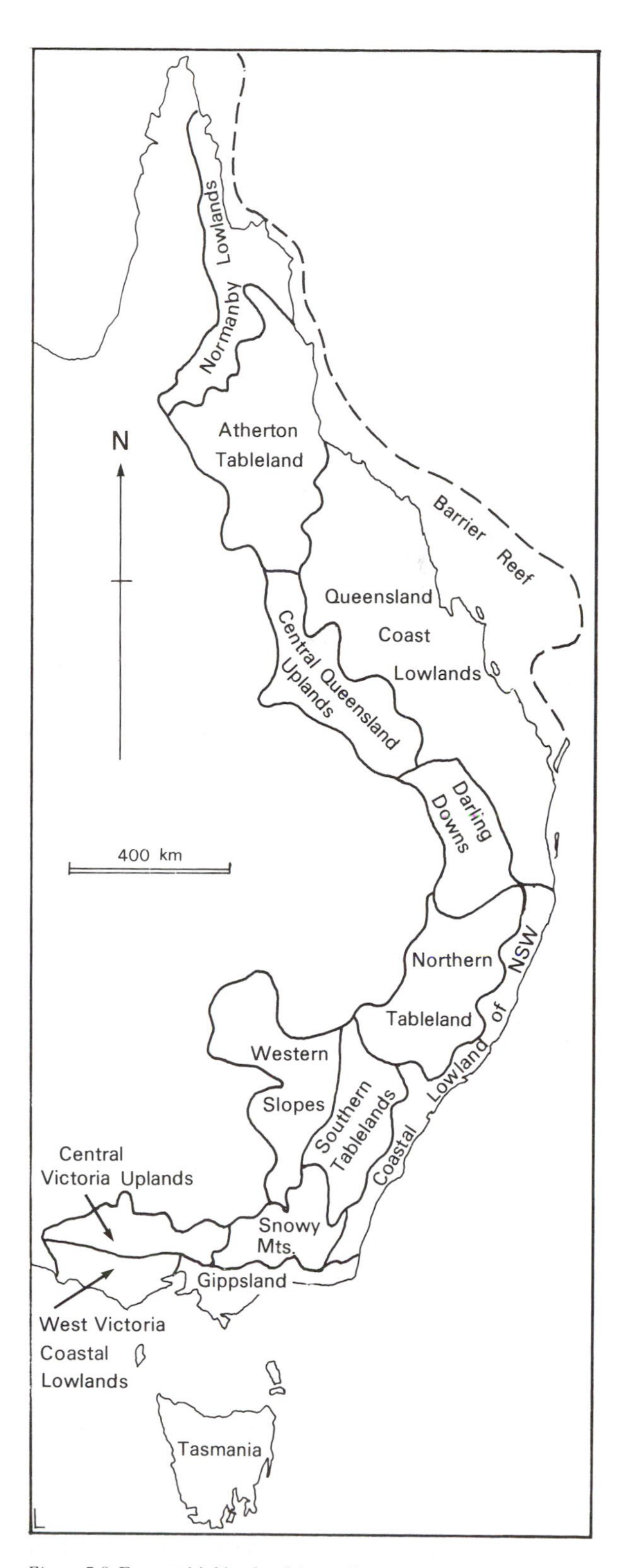


*Figure 7.8.* Eastern highlands of Australia.

### *Normanby lowlands*

The northernmost highlands of the Great Dividing range sub-divide the Cape York peninsula into three morphological sections. The mountains, mainly formed of granite-intruded Palaeozoic sedimentary rocks, include the McIlwraith range (900 m) in which gold was found, and the Iron range with Palaeozoic iron ores. East of the ranges the Normanby river drains a lowland developed over sandstones and mangrove swamps characterise its estuary. West of the mountains, in the Carpentaria basin, extensive bauxite deposits are mined at Weipa.

*Atherton tableland*

The eastern highlands come close to the coast north of Cairns, so the coastal plain is restricted. South of Cairns basaltic flows make up the Atherton tableland, a plateau consisting of three levels, the highest in the south around Millaa Millaa being just over 1000 m above sea level. Lake Eacham, south-east of Atherton is a volcanic crater lake. Near the coast, Mount Bartle Frere reaches 1600 m. Rivers have cut deep gorges and cascade over waterfalls on leaving the basalt plateau country.

*Central Queensland uplands*

The least spectacular part of the eastern highlands is no more than an elevated plateau of less than 900 m which acts as the continental divide for streams draining to the Coral Sea and those which flow to the interior. The importance of this area is that the Jurassic sandstones which crop out on the western slopes are the intake rocks for the artesian basin. Flow basalts occur north-east of Charters Towers.

*Queensland coast lowlands*

This diverse area includes the catchments of the Burdekin, Fitzroy and Burnett-Mary rivers. Geomorphological features are influenced by the underlying Permianrocks which have been folded and metamorphosed to give the country a broad north-west–south-east trend. Tertiary basaltic flows form plateaux at different elevations and provide an indication of the amount of erosion that has taken place. As the basalt flows originally must have been in the valleys, the present landscape indicates the considerable amount of erosion which has been accomplished (160–250 m has been estimated by this means). The mineral wealth of the area is in the coal mined at Moura and Callide; copper was mined at Mount Morgan and gold mined at Gympie.

*Figure 7.9.* Wollomombi Falls, Armidale, New South Wales.

Upland surfaces have been identified over the crests of the older, harder rocks, and river terraces of different ages occur along the major valleys. The coastline is a mixture of muddy estuaries with mangroves and beautiful sandy beaches. Some of the sands have been blown inland as in the Cooloola National Park to form dunes ranging from Pleistocene to Holocene in age. They have their axes at right angles to the coast indicating how they have been blown inland. The different ages of the dunes are reflected by the amount of soil formation and vegetation present.

*Darling Downs*

Extensive basaltic flows have contributed to the formation of the Darling Downs west of Toowoomba in southern Queensland. East of the town the edge of the plateau is marked by a 300–500 m erosion scarp. The basaltic flows have infilled former valleys and their thickness can be as much as 800 m. The present

*Figure 7.10.* The Blue mountains, New South Wales.

landsurface is a uniform plateau with a slight slope westwards to the upper valley of the Condamine river. Below the basalts Mesozoic sedimentary strata crop out in the steep east-facing escarpment, and granites occur in the Bunya mountains north of Toowoomba.

### *Northern tablelands of New South Wales*

An explanation of the landforms of the district known as New England has been given by Ollier. It is thought that the area was reduced to a peneplain by the end of the Cretaceous and that an uplift of about 500 m occurred to give an Eocene peneplain surface with a few monadnocks rising above it, remnants of the 'Australian' peneplain of King. Flow basalts were then erupted during the Palaeocene to early Oligocene, probably accompanied by some doming of the earth's crust. A Miocene surface was then cut and subsequently lateritised, now at 1030 m elevation. Further uplift gave the impetus for cutting a lower surface represented by the Dorrigo plateau at 760 m and the late tectonic disturbance known as the Kosciosco uplift occurred in Plio-Pleistocene times, the effects of which are seen in the lower river courses.

The slightly more elevated areas of the Liverpool and Nandewar ranges rise to 1300 m. Erosion has worked back from the adjacent coastal plain into the highlands and many streams descend from the plateau by spectacular waterfalls as at Wollomombi (Figure 7.9). In the south-west of the New England plateau, the trachyte plugs of the Warrumbungle mountains are partly exposed by removal of the country rock leading to speculation about the total amount of rock removed.

### *Coastal lowlands of New South Wales*

In the vicinity of Sydney, east of the Great Dividing range, the older sedimentary rocks have been folded into a basin structure which extends 800 m below sea level, the Cumberland basin. Coals of Permian age are included in the geological sequence and are mined in the Hunter valley. In the surroundings of Sydney, deltaic sandstones, the Hawkesbury sandstone, form a low plateau dissected by streams coming from the highlands. The well-known Blue mountains are the west-facing scarp of these sandstones (Figure 7.10). There are also basalt residuals on the sandstone plateau which provide areas of more fertile soils. The coastline is an attractive one of rocky headlands and sandy bays with estuaries which extend well into the Hawkesbury sandstone country. The accumulation of successive sand ridges at Woy Woy provides an interesting illustration of the influence of time on soil formation.

### *Southern tablelands*

Surrounding the Australian federal capital of Canberra is an undulating upland plateau at 600 to 800 m elevation. Its geological foundation is of Ordovician, Silurian and Devonian strata which have been thrown into a number of horsts and graben by block-faulting with a north–south alignment. These sedimentary rocks are mainly sandstones and shales with quartzites forming scarp features. They have been metamorphosed and intruded with granite. On this landscape, investigators have identified four erosion surfaces: the 1000 m Monaro surface, the 800–900 m Molonglo surface, the 700 m Yass–Canberra surface and the Yarralumla surface at 600 m. At elevations greater than 1000 m evidence of former periglacial activity becomes apparent.

The southern tablelands are drained by the Shoalhaven and Murrumbidgee rivers to east and west respectively. The Shoalhaven river has terraces which relate to sea levels at 7–8 m, 2.5–3.0 m and 2 m above present sea levels. Block-faulting has caused an internal drainage basin to form centred on Lake George. This 150 km$^2$ lake is no more than 5 m deep and occasionally dries out completely, but it is not saline. Former shorelines indicate the lake to have been deeper and more extensive in past wetter periods which may have corresponded to glacial episodes during the Pleistocene.

### *Western slopes*

The inland boundary of the Eastern highlands is less abrupt than that on the coastal side and the Palaeozoic rocks in this province have been extensively pedimented where they extend out into the plains on the Cobar block. Remnants of a duricrusted surface are preserved on the summits of residual hills at elevations of between 250 and 500 m.

### *The Snowy mountains*

The highest parts of Australia mainland lie astride the New South Wales border south-east of Canberra. Described as a boomerang-shaped area of mountains the Snowy mountains and the Australian alps have many crests over 2000 m, reaching the maximum height of 2440 m on Mount Kosciusko. Except for the crest of Kosciusko, true glacial features are absent, but the area has been strongly influenced by solifluction.

The underlying rocks of the Snowy mountains are Lower Palaeozoic sediments into which granitic intrusions have occurred. In Victoria these elevated areas are referred to as high plains and their crests are smooth even though the valleys are deeply incised. This upland surface is a remnant of an early erosion surface which has been uplifted to form the mountains. Mount Kosciusko sits upon this upland surface as a result of late Pliocene–early Pleistocene tectonic movement. Before descending into their gorges, the rivers of these highlands occupy broad, shallow valleys. At their headwaters small moraine-dammed lakes occur but the larger bodies of water, Lakes Eucumbene and Jindabyne, are man-made, being associated with the Snowy mountains hydroelectric scheme. Some level areas in the highlands are the result of Tertiary basalt flows as at Kiandra in the Snowy mountains or the Nunyang plateau in Victoria.

### *Central Victoria uplands*

West of the Melbourne–Benalla highway the elevation of the land decreases until it is a hilly upland of about 300 m with residuals such as Mount Macedon reaching just over 1000 m. The basis of the area is Lower Palaeozoic rocks intruded by granite, but late Tertiary basaltic lavas have covered parts of the upland around Daylesford, leaving basalt and scoria cones and explosion craters.

### *Gippsland and the west Victoria coastal lowland*

The landforms south of the Australian alps may be described as rolling country which reaches 300–400 m in the Strzelecki hills and slightly lower in the attractive Dandenong hills east of Melbourne. The area is founded upon Mesozoic and Tertiary sedimentary rocks which form the coast from Cape Howe westwards as far as Lake Entrance. West of that point is the sand dune and lagoon coast of Ninety Mile Beach.

Port Phillip Bay, at the head of which Melbourne is sited, is a shallow, flooded lowland of Tertiary sediments enclosed in a rim of Silurian rocks. Western Port, the inlet east of Port Phillip Bay is a block-faulted inlet and the south coast of Phillip Island has 'organ pipe' basaltic cliffs. West of Geelong, the coastal plain is dominated by the low lava plateau south of Ballarat, 23 400 km$^2$ in extent. Ninety-four vents have been recorded and there is evidence of deranged drainage as a result of the porosity of the lava, which often provides only arid pasture interspersed with stony lava masses where the lower layers of fluid lava flowed out from beneath an already solidified crust.

### *Tasmania*

The island of Tasmania is separated from the Australian continent by the shallow Bass Strait where there have been finds of oil and natural gas in Tertiary sedimentary strata. The island may be divided into four sections, the north-east and the west coastal regions of folded Archean rocks intruded by granites, a central block-faulted Triassic and Tertiary plateau, mainly above 300 m and a lower, block-faulted mixture of horsts and graben in the south-east around Hobart.

A major influence on the relief of Tasmania is the presence of dolerite igneous intrusions into the horizontal Triassic rocks of the central plateau. Faulting of these dolerites, which are resistant to erosion, has given the 1300 m cliffs of the Great Western Tiers, described as one of the most outstanding landforms in Australia.

The eastern part of the western highlands of Tasmania has the greatest amount of evidence for glaciation in Australia. Lake St Clair is a glacially gouged basin dammed by moraines and most of the highland zone was actively influenced by periglacial processes.

The west coast ranges of Tasmania and the central plateau were centres of glacial ice accumulation in the Pleistocene. The west coast ranges were also mineralised with ores of copper, zinc and tin. A wide range of coastal features is found around the periphery of Tasmania with deep, flooded valleys, sandy beaches and rocky cliffs.

## Oceanic provinces of the Australian plate

The Australian plate is considered to extend from the Ninety East oceanic ridge in the west to the Tonga–Kermadec trench in the east. In a north to south direction it extends from the Indian–Antarctic ridge to the Sunda trench, New Guinea and the complex of trenches and islands north of the Coral Sea. The major provinces are:

The west Australian (Wharton) basin
The south Australian basin
The Tasman basin
The Coral Sea
The Barrier Reef

### *The West Australian basin*

This basin lies between the west Australian coast and the Ninety East sea-floor ridge. The floor of the basin is almost entirely at the level of the abyssal plain, 6000 to 7000 m below the sea surface. Its northern boundary is the Sunda trench lying parallel to the coast of Sumatra and Java and its southern margin lies along the south-east Indian ridge (Broken ridge). There are numerous seamounts present but only two are of significance, forming the Cocos and Christmas islands.

The west coast of Australia has a narrow coastal plain, but on the north-west coast it is wider, and the continental shelf increases in width until it extends almost all the way across to Timor. Between Australia and New Guinea the Arafura Sea is also underlain by continental shelf. A deeper terrace of the coast of north-western Australia is called the Exmouth plateau. The southern margin of this province is problematic. It would seem to be along the Broken ridge as south of this ridge some authorities have identified the Ob and the Dimantina trenches which would form an obvious natural boundary.

### *The south Australian basin*

Terrestrial Australia has been forced northwards by the creation of new sea floor material along the Indian–Antarctic ridge. From this ridge, which is only 1100 m below the surface south of Australia, and breaks the surface in the St Paul (279 m) and Amsterdam (911 m) Islands in the Indian Ocean, the sea floor descends to the abyssal plain over 6000 m deep. This extends northwards to the continental rise and a continental shelf of

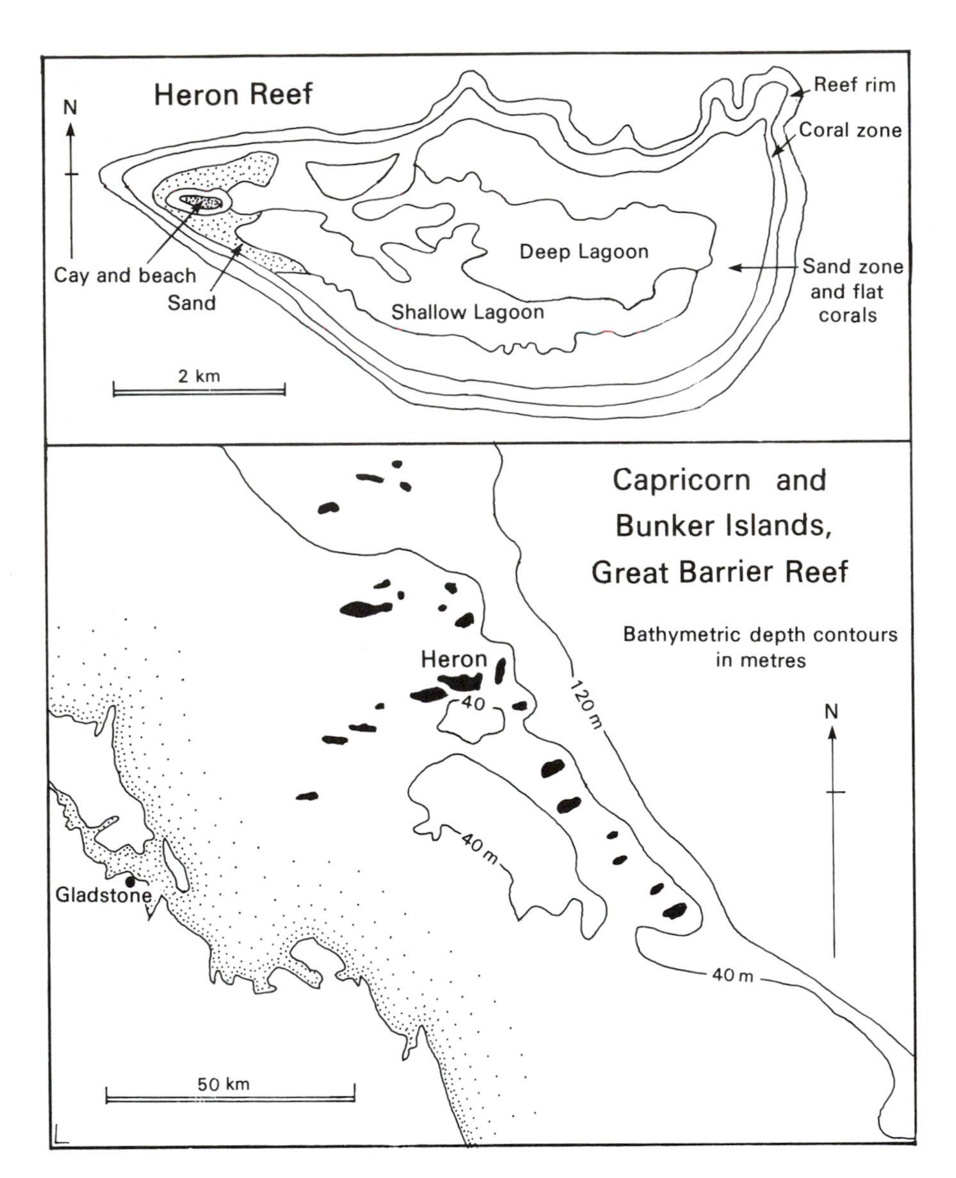


*Figure 7.11.* Heron Island.

up to 200 km occurs along the southern Australian coast. The submarine Ceduna plateau in the Great Australian Bight and the Naturaliste plateau off south-west Western Australia are thicker portions of the oceanic crust. South of Kangaroo Island, the Sprigg and Couedic submarine canyons cut deep gorges in the edge of the continental shelf.

*The Tasman basin*

The island of Tasmania lies upon the Australian continental shelf which forms the natural western boundary of the Tasman basin. The eastern margin is the Lord Howe Rise–Campbell plateau, 350 to 450 m below the sea surface upon which Norfolk Island and the islands of New Zealand rest. Sea-floor spreading has affected the Tasman basin because Lord Howe Rise was adjacent to the east coast of Australia until about 80 million years ago. The continental shelf of New South Wales is narrow, and north of the latitude of Sydney the Tasman basin gives way to the shallower Coral Sea.

*The Coral Sea*

The Coral Sea is a triangular area of relatively shallow seas, numerous islands and several deep sea trenches.

*Figure 7.12.* Heron Island on the Great Barrier Reef.

The eastern margin is the Tonga–Kermadec trench and the northern boundary the Solomons–New Hebrides trench. Associated with the trenches are arcs of volcanic islands beneath which subduction is taking place. However, the extensive underwater plateaux and numerous seamounts provide sites for coral growth north of the Tropic of Capricorn. The continental shelf along the Queensland coast north of Gladstone is much wider, supporting the Great Barrier Reef.

### *The Barrier Reef*

The corals of the Barrier Reef of Australia extend 1900 km from approximately 10°S to 25°S and at their greatest extent occur up to 200 km from the mainland. An earlier interpretation of the structure of the Barrier Reef suggested that it was a faulted, sunken block, the surface of which lay just below the sea, but recent studies have shown that many of the structures seen on land are continued onto the sea floor below the Barrier Reef. Geomorphological evidence points to the uplift of the Great Dividing range as warping the pre-existing erosion surfaces so that they declined in elevation on both east and west flanks of the mountains, so a simple downfaulted segment of the Earth's crust is not a complete answer to the formation of the Barrier Reef.

The coral polyp does not like living below 60 m depth, it does not like sediment-laden waters and it must have a water temperature which does not fall below 19°C. Most of the gaps in the reef may be explained by one or other of these conditions not being satisfied.

Heron Island is an example of a true reef island cay formed from calcareous coral debris (Figure 7.11). Cays are normally perched on the rim of the coral reef and are quite small compared with the submerged area of the reef. Heron Island is 830 m by 300 m and only rises out of the sea by 2 m. The island is underlain by bedded beach rock which crops out on the eastern side of the island at the foot of the beach. Many reefs do not have an island, only a raised rim with a central lagoon, a feature known as an atoll (Figures 7.12 and 7.13).

*Figure 7.13.* Corals and a clam on the Heron Island reef.

Offshore islands further north along the Queensland coast are not coral islands, but fragments of the mainland, although surrounded by coral reefs. The Hayman, Whitsunday, Magnetic and Palm islands come into this category. At the southern end of the reef Fraser Island and Moreton Island are immense sand dune systems, pushed shorewards during the Pleistocene and Holocene.

## New Guinea

The northern limit of the Australian lithospheric plate occurs along a line through Timor and the larger island of New Guinea. This is the only 'active' edge of the Australian plate and the only place where vulcanicity and mountain building is occurring at the present time. The northern edge of the plate was the site of a geosynclinal couplet throughout the Mesozoic and early Tertiary and the deeper eugeosynclinal part was affected by the eruption of basic marine volcanic rocks. Landform development began in the Miocene with the emergence of the present mobile belt and the general outline of the island of New Guinea became apparent by the late Pliocene. Uplift is continuing at a rate of about 3 mm per year.

From south to north the main geomorphological features of New Guinea (Figure 7.14) are:

The Digoel–Fly lowlands
The southern plains
The southern fold mountains
The highlands
The metamorphic ranges
Cape Vogel basin
The intermontane trough
The coastal ranges
The Timor Sea

### *The Digoel–Fly lowlands*

The lowlands of the southern part of New Guinea reach a maximum width of 400 km and are a northwards extension of the stable Australian shield. The basement complex of granites is seen in the Torres Islands and

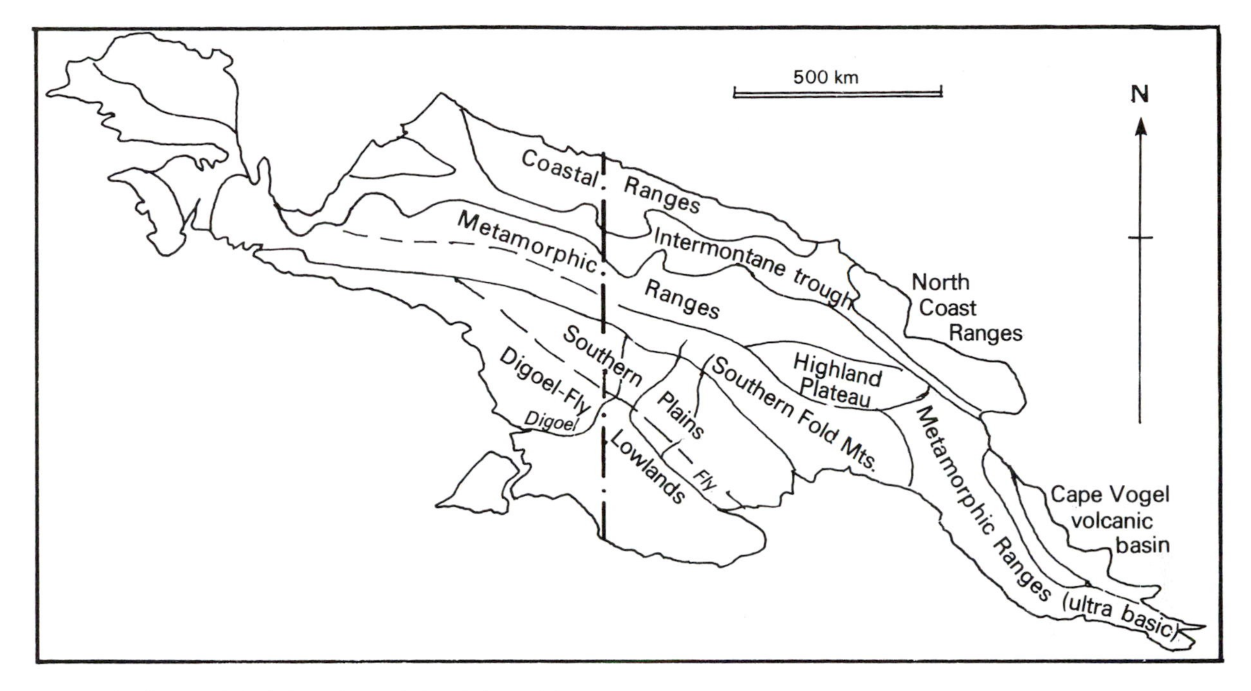


*Figure 7.14.* Geomorphological provinces of New Guinea. (After Loffler, 1977.)

near Mabaduan in New Guinea, but for the most part the shelf is covered by Mesozoic and Tertiary sedimentary strata and the surface of this area is covered by a Miocene limestone. South of the Fly river this limestone is covered by up to 300 m of Quaternary coral debris and alluvium. The Fly estuary is characterised by mudbanks and mangroves.

### *The southern plains*

North of the Fly river the limestone reaches the surface and karstic features occur north-west of Kikori where small roughly hemispherical or cone-shaped hills and stellate depressions occur, collectively named polygonal karst. The crests of all the interfluves in this area are all at the same level and it is thought this is a former piedmont alluvial plain, now being dissected following uplift. Fans of volcanic debris also occur in this north-east part of the lowlands.

### *The southern fold mountains*

There is strong lithological control of landforms in the southern fold mountains of New Guinea. The strata are folded and eroded so that steep escarpments face south to south-west, but elevations do not exceed 2000 m above sea level. The rocks include limestones, greywacke, siltstone and mudstones of Miocene–Pliocene age. Intense weathering conditions have resulted in the development of tower and cone karst where limestone crops out in the Kikori–Lake Kituba and Darai areas.

### *The highlands*

The central region of the New Guinea highlands is characterised by a series of upland valleys and montane plateaux upon which some large strato-volcanoes are built. These include Mount Giluwe (4368 m), Mount Lalibu and Mount Hagen, all of which rise 2000 m above the montane plateau of the highlands which ranges from 1300–1800 m.

A topic of major interest in the New Guinea mountains has been the effects of glaciation upon these tropical mountain peaks. Mount Giluwe had the most extensive ice cap, 188 km$^2$, which extended down to 3200–3500 m above sea level. On south and south-west slopes a broad band of moraine, 20–30 m thick was deposited but on north and north-east slopes only small valley glaciers pushed out from the ice cap.

*Figure 7.15.* Volcanic island, Auckland Harbour, New Zealand.

### *The metamorphic ranges*

To east and west of the highland province, mountains formed from metamorphic rocks occur. These have narrow-crested ridges with straight slopes to the valley floors where streams with steep gradients flow. Mount Wilhelm (4510 m) is the highest peak in Papua New Guinea. This area is dominated by erosion and no evidence of former surfaces survives. The eastern section of the metamorphic ranges is formed by the Owen Stanley range which forms the backbone of the 'tail' of New Guinea where Mount Victoria reaches 4073 m above sea level. Accordant summits are widespread in the Owen Stanley ranges especially on the main divides. They range from 200 m at the eastern end of New Guinea to a maximum of between 2800 and 2300 m. These are described as not being the remnants of former peneplain surfaces but the result of erosion into uniform ridge and ravine landforms.

The glaciation of Mount Wilhelm was accomplished by corrie glaciers which cut deep glacial troughs to give it a very Alpine appearance. Most corries are situated on the west and south-west of the summit. One argument for their location is that the west and south-west faces receive less insolation as the build-up of clouds during the afternoon limits this. North of the Owen Stanley ranges, and separated from them by a pronounced fault trough are the ultra basic ranges. These mountains are situated where the Australian plate is passing beneath the Pacific plate. Uplift occurred in the Eocene and now these ultra basic rocks have been revealed by erosion.

### *Cape Vogel basin*

In complete contrast to the landforms of the metamorphic rocks, the Cape Vogel basin, which lies along the northern shore of the tail of New Guinea, is volcanic in origin. Individual features include Hydrographer's Volcano, a deeply dissected extinct feature and Mount Lamington which erupted in 1951 with nuées ardentes

*Figure 7.16.* Thermal energy, Wairakei, New Zealand.

and ash showers. The line of small islands off the northern coast of New Guinea, Manam, Kar Kar and Bagbag, are all volcanoes emerging from the sea. The Entrecasteaux Islands are also volcanic and part of this geomorphological province.

*The intermontane trough*

The low ground occupied by the Sepic river of Papua New Guinea and the Mamberamo river of West Irian is a graben feature which continues eastwards to join the New Britain trench. The fault margins of this province are marked by coalescing alluvial fans and the rivers crossing them are braided and have constantly shifting channels. South of the Sepic river, the lowland is very swampy and the valley has the appearance of a drowned valley which is rapidly being infilled with alluvial sediment. The Sepic river meanders greatly and air photographs indicate many oxbow lakes.

*The coastal ranges*

There is no coastal plain on the northern side of New Guinea, the ranges rise from the sea to a maximum height of 4121 m in Mount Bangeta. This series of mountain ranges is in progress of being uplifted at a rate of 3 mm per 1000 years. Raised coral terraces on the coast at the Huon peninsula show this uplift in a striking fashion. The largest doline mapped in Papua New Guinea occurs north of Lae, also on the Huon peninsula.

*The Timor Sea*

Below the Timor Sea the Australian plate extends north-westwards until it meets the Timor trough. This submerged shelf is now regarded as a passive margin with Mesozoic and Tertiary sediments. However, the Timor trough is thought to be the site of a former subduction zone and the island of Timor an emergent mass of much faulted material in the 'melange' of rocks scraped up as the Australian plate was being subducted. Another theory suggests that Timor has always been part of the Australian plate which is being subducted beneath the Banda arc, hence the line of volcanic islands north-west of Timor.

## New Zealand

The islands of New Zealand are the emergent parts of a fragment of Gondwana, most of which is a submerged plateau. This submerged fragment has moved away from the Australian mainland since late Cretaceous times. The most extensive area, known as the Campbell plateau, lies south of South Island, but submerged continental areas continue northwards as Lord Howe Rise and Norfolk Island ridge.

Tectonic activity, known as the Kaikoura movements, gave New Zealand its basic structure in the Tertiary. This resulted in two ridges with a trough between. This trough became filled with volcanic lavas, ashes and material eroded from the ridges. Uplift continued until the Quaternary with local displacement amounting to 12 000 m. Off the east coast of New Zealand the Kermadec–Hikurangi trench indicates a subduction zone which some authors indicate passing onto the South Island in the vicinity of Christchurch. Other authors suggest that there is a foredeep with a recurved shape passing through North Island and following the line of the major slip-fault of South Island eventually swinging eastwards across central Otago. This slip-fault is interpreted as a plate boundary along which it is claimed up to 480 km displacement northwards has occurred since the Cretaceous.

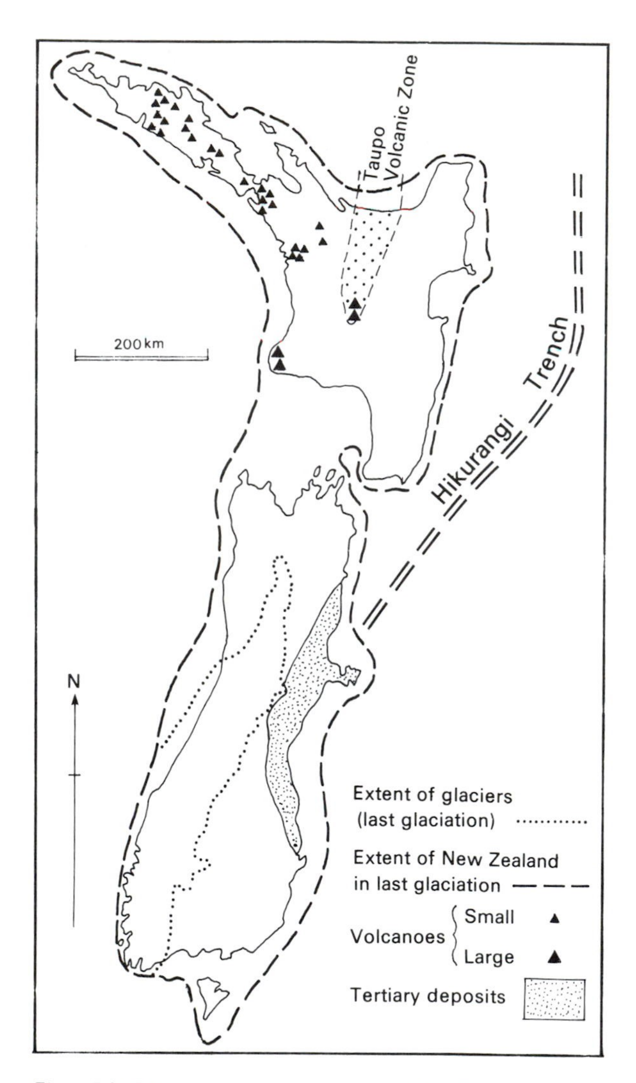


*Figure 7.17.* Major morphological features of New Zealand.

In North Island the emphasis has been on volcanic activity, especially during the Quaternary, when andesite lavas erupted to form Mount Ruapehu (2796 m) and Mount Egmont (2518 m). Basaltic cones occur in Auckland and in the northern peninsula. The plateau around Rotorua is formed of ignimbrite which was erupted violently in the early Quaternary. The central parts of North Island are covered by many layers of tephra. Active hot springs, mud volcanoes and fumeroles occur at several sites between Lake Taupo and the Bay of Plenty. A strong earthquake was experienced in 1987 in the Bay of Plenty area (Figures 7.15 and 7.16).

The South Island is dominated by the Southern alps, rising to 3764 m in Mount Cook with many other mountain peaks reaching over 3000 m. The massive slip-fault which passes through the southern alps has graywacke and schists lying to the east with granite and Tertiary sediments to the west. The greywacke and schists are being thrust upwards against the granite; this upward movement is estimated at 10–20 mm per year and is balanced by an equivalent amount of erosion to give a 'steady state' landform. East of the alpine range in Otago, the greywacke mountains tend to be flat-topped. These are remnants of a Tertiary peneplain which rises inland from the east coast. As it rises this old surface becomes increasingly dissected until the boundary with the Alpine peaks is reached.

During the last glaciation, sea level dropped by about 100 m and New Zealand became one island but with only a slightly enlarged outline (Figure 7.17). The cover of ice during this glaciation, 70 000–80 000 years ago, extended throughout most of the high ground of the South Island and small ice gaps also appeared on the volcanic mountains of the North Island, such as Mount Ruapehu. In the South Island ice from the highlands spread down existing valleys and created the fjordland of the south-west coast. Glaciers spreading eastwards brought masses of debris which has accumulated to form the Canterbury plains and other lowlands, choking the valley floors and resulting in anastomosing channels. As the forest cover returned after about 13 500 BP these alluvial fillings were incised to give the spectacular terrace sequences seen along the rivers of both North and South islands.

# 8

# Europe

Traditionally Europe is regarded as a separate continent although it is in fact only a peninsula of peninsulas extending west from Asia. The Ural mountains are the accepted eastern boundary of Europe which has an area of 10.4 million $km^2$. It extends 3500 km north to south from latitude 73°N on the shores of the Arctic Ocean to latitude 36°N on the Mediterranean Sea, and 4500 km east to west from 60°E to 10°W in longitude. The highest peak within Europe is Mont Blanc (4807 m) but on the boundary of Europe as considered here, Mount Elbrus in the Caucasus reaches 5642 m.

The boundaries of the lithospheric plate occupied by Europe are difficult to draw. The southern margin of the Alpine folding is a logical structural boundary, but this would include the Mahgreb lands of North Africa. The rest of the Alpine folding lies on the northern shores of the Mediterranean, but the boundary is complicated by the structure of the Mediterranean basin, the presence of a spreading centre south of Crete and the presence of the micro-plates of Italy–Yugoslavia and Turkey. The western boundary may sensibly be described as the mid-Atlantic ridge, a constructive margin which extends from the Azores northwards through Iceland and around Spitzbergen into the Arctic Ocean.

The mean elevation of Europe is 300 m above sea level and 31% of the continent is in the form of depositional plains, 30% erosional plains and plateaux in sedimentary rocks. Exposed crystalline shields account for 13%, old mountain belts 14% and young fold mountains 11%. Volcanic plateaux only account for 1% of the continental surface area (Butzer, 1976).

## Geological history

The geological history of Europe is complex but the same basic pattern of development may be traced as in North America or Asia. A stable Pre-Cambrian shield occurs in Scandinavia, forming the core of the continent and around which folding has taken place. The mountains resulting from this folding have been worn down by erosion, uplifted again and the sediment incorporated in the later mountain building. The Pre-Cambrian rocks of the shield were reduced to a peneplain before the Cambrian sedimentary rocks were laid down and throughout the Lower Palaeozoic extensive sedimentary accumulation took place on the western and southern margins of the shield. At the end of the Lower Palaeozoic, the Caledonian orogeny took place forming the mountains of north-western Europe.

Sedimentation continued through the Upper Palaeozoic during which period sediments ranging from desert sands to swamp forests accumulated, the latter forming important coal deposits. Again Europe was affected by mountain-building forces, this time of the Hercynian, affecting the central part of the continent. It appears this was not geosynclinal activity, rather faulting was accompanied by uplift of areas such as Brittany and the Rhine highlands whilst sedimentation continued in the depressed graben alongside.

Following this activity, a sea, the Tethys, had opened out along the southern side of the developing European continent. Sediments which were eroded and transported into this sea were subsequently involved in the Alpine orogeny, being folded and uplifted into the Pyrenees, Alps and Carpathian mountains as well as the other ranges in southern Spain, Italy, south-east Europe and north Africa.

Possibly the most significant event in the geological history of the continent for landform development took place in the Pleistocene when northern Europe, and many of the higher mountain ranges further south, were glaciated. Originally, on the evidence of deposits in the Alps, four major phases of glaciation were identified but in recent years it has been established that there were probably twenty or more glacial and interglacial stages during the two million years of Pleistocene. Those areas surrounding the ice sheet underwent a periglacial climate with solifluction occurring on slopes and much fine material was blown from outwash plains to form the accumulations of loess on the northern flank of the Alps and extending into the southern part of the Soviet Union. Post-glacial modification has occurred as the landmass rebounded isostatically after the ice melted, resulting in dissection of the unconsolidated Pleistocene deposits and aggradation in valleys.

A German geologist, Stille, proposed a four-way division of Europe in 1924 which is still valid today. The four regions are:

1. *Eo-Europe*: those areas which have not been disturbed by orogenesis since the end of Pre-Cambrian times.
2. *Palaeo-Europe*: those areas which have not been disturbed by orogenesis since the end of Lower Palaeozoic times.
3. *Meso-Europe*: those areas which have not been disturbed by orogenesis since the end of Upper Palaeozoic times.
4. *Neo-Europe*: areas which were affected by orogenesis in Tertiary and Quaternary times.

In broad terms these geological divisions relate to the major physiographic divisions of Europe which form the framework of the following discussion. These are the Fenno-Scandian shield, the Atlantic highlands, the European plain, the European scarplands, the central European uplands and basins and the Alpine highlands (Figure 8.1). Each of these divisions contains several geomorphological provinces, the landforms of which will be briefly outlined.

## The Fenno-Scandian shield

The Pre-Cambrian foundation of Europe crops out in parts of Norway, Sweden and Finland but it extends southwards beneath the north European plain where it is covered by younger rocks. Other exposures of Pre-Cambrian occur throughout Europe, notably in Wales and in the Ukraine. The exposures in northern Scotland belong to the North American province but have become attached to western Europe after the closure of the proto-Atlantic ocean (Iapetus). The oldest rocks of the shield, in excess of 2500 million years old, crop out in the Kola peninsula, Karelia and south-west Sweden. Proterozoic rocks, more than 1800 million years old, form the greater part of the lands surrounding the Gulf of Bothnia in Finland and central and north Sweden (Figure 8.2).

The rocks are metamorphic, gneiss and mica schist, with intrusions of granite which form a large saucer-

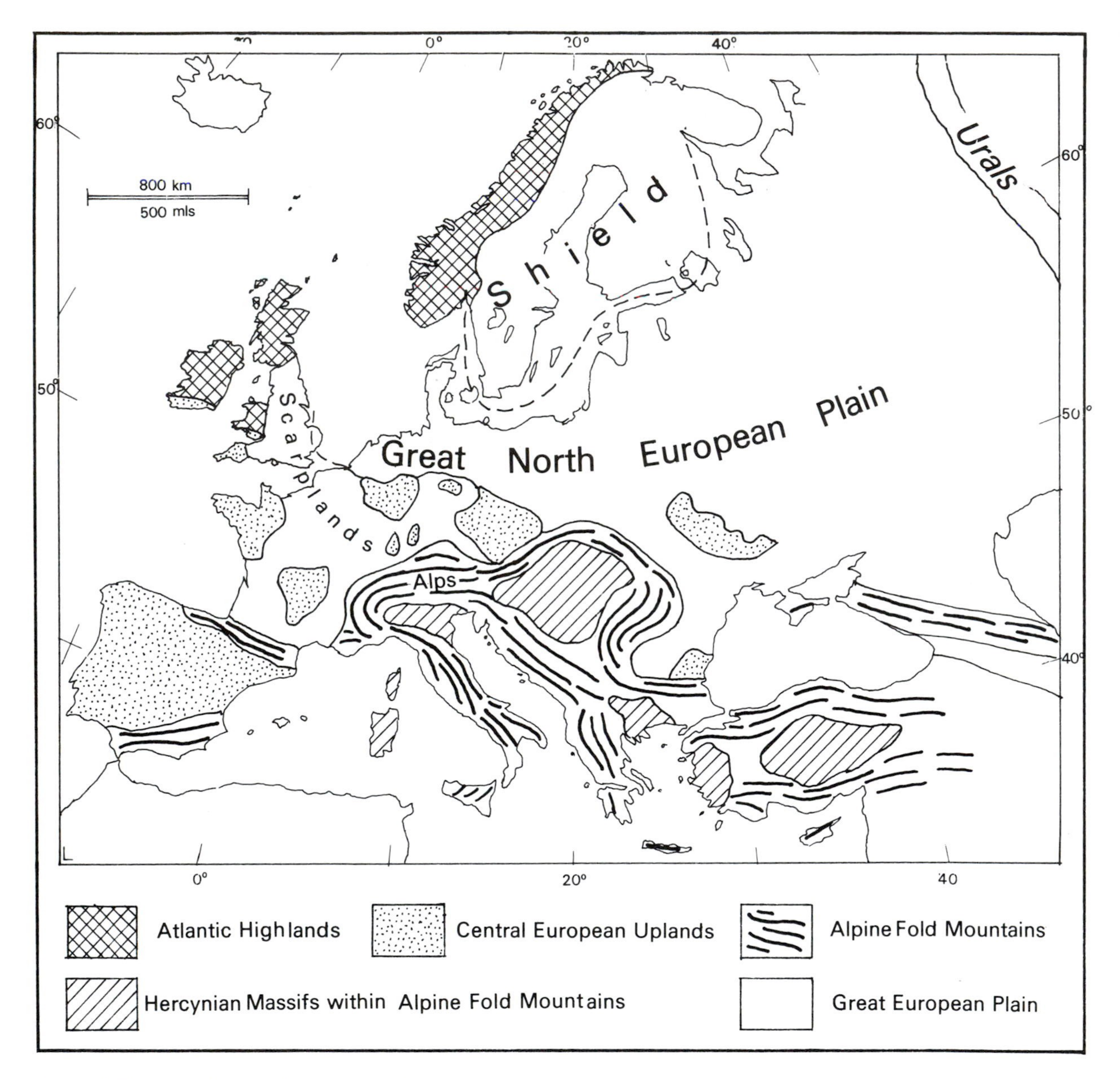


*Figure 8.1.* The major physiographic divisions of Europe.

shaped area, 900 km across, centred on the Gulf of Bothnia. Its western rim reaches 1500 m in the Norwegian mountains but the eastern rim only attains 350 m in Karelia, and much of southern Finland is 150 m or less above sea level. The Gulf of Bothnia is less than 200 m deep. The surface of these Pre-Cambrian rocks is relatively smooth with only occasional monadnock-type hills. It was eroded to a peneplain before the Cambrian rocks were laid on its surface and these still remain in southern Sweden, on the floor of the Gulf of Bothnia and across the Baltic Sea in Estonia.

Much of the present relief and landforms result from glaciation during the Pleistocene, when this area formed the centre of an ice sheet which covered Scandinavia and extended onto the European plain. When ice accumulated, the land surface depressed because of its weight. The Scandinavian ice sheet may have been 2000 m thick and the land surface was probably depressed by

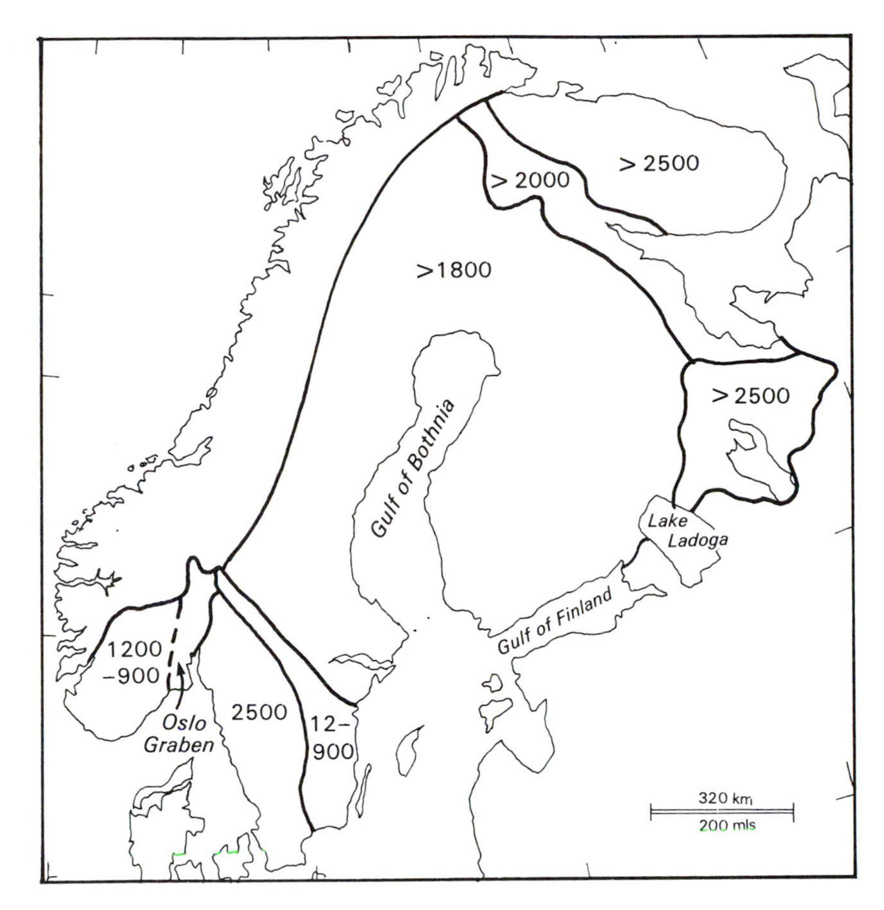


*Figure 8.2.* Age of cratons in the Fenno-Scandian shield.

650 m at the maximum stage of glaciation. When the ice melted, the weight was removed and the Earth's crust began to return to its former position (isostatic rebound). The situation was complicated as the water returned to the sea rapidly and resulted in a eustatic rise of sea level. At first the Baltic depression was a pro-glacial lake, but as sea level rose it caught up with the rising land and the sea flooded across southern Sweden to form the Yoldia Sea. Isostatic rebound then surpassed the sea level rise and the Ancylus lake was formed about 9500 BP.

Eustatic rise in sea level in the Flandrian transgression once more caught up and the sea again was able to link with the lake waters through the outwash channels between the Danish islands to form the Littorina Sea, the immediate precursor of the Baltic Sea as it is today. Uplift continues to affect the Gulf of Bothnia; at Oulu it is 1 m per 100 years, at Stockholm 50 cm per 100 years and equilibrium is reached in Denmark and northern Germany (Figure 8.3). The sections of the Fenno-Scandian shield, illustrated in Figure 8.4, are:

The premontane section
The Norrland section
The emergent section
The Baltic Sea section

### *The premontane section*

The western edge of the shield occurs in western Sweden at elevations of 500–1000 m where plateaux between rivers have been separated by deep incision and glacial scouring to form 'finger' lakes dammed by glacial debris.

### *The Norrland section*

This name has been given to the main extent of the shield surface in eastern Sweden and Finland. It is a

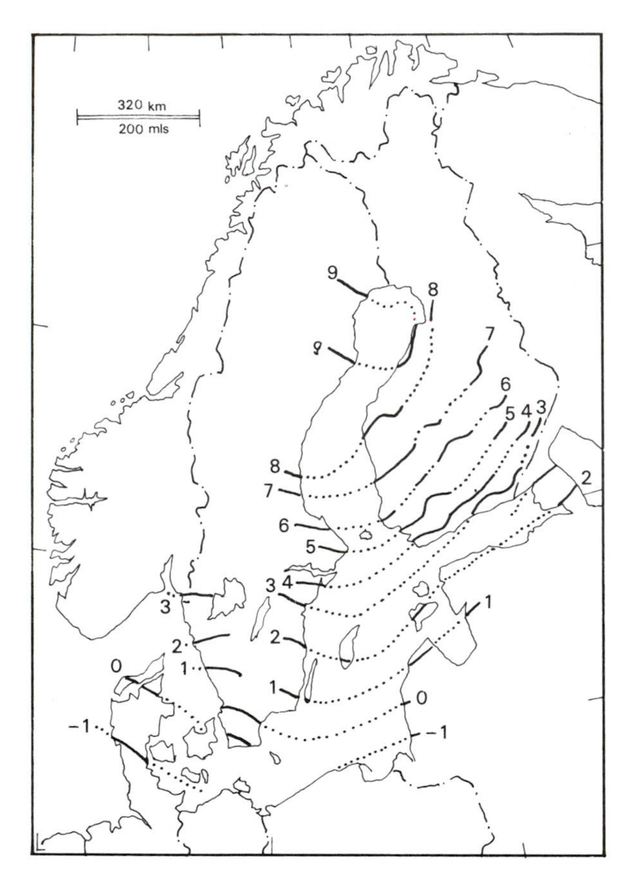


*Figure 8.3.* Contemporary rates of isostatic readjustment in millimetres per year. (After B. John, 1984.)

gently tilted, glacially scoured shield surface, 200–500 m above sea level, with occasional monadnocks, mainly covered with glacial and fluvioglacial deposits. Drainage tends to be complex with many lakes. Much of the glacial debris from the melting ice sheet was deposited in shallow, flowing water which left a deposit lacking in fine materials. The finer sediments accumulated on lake beds which when drained, provide level land.

Elsewhere the landscape is one of low relief resulting from the deposition of till and fluvioglacial deposits such as eskers and kames. Across southern Finland a large multiple recessional moraine, the Salpauselka, extends from the Russian border to the Baltic coast. Its northern flank is an ice-contact slope and outwash deltas extend south of the ridge. Chaotic dumping of glacial debris behind it has given rise to a very complicated drainage pattern with many lakes. Similar ridges of moraine can be traced around southern Norway as the Ra moraines (Figure 8.4). In the central lowlands of Sweden and in southern Finland, streams have cut through the glacial deposits and have been guided by faults in the granite beneath to make the fissure-valley landscape.

### *The emergent section*

As isostatic recovery has taken place the lower lands have emerged from below the level of the Baltic Sea as a result of uplift. Usually they are covered with glacial deposits into which streams have become incised. Raised beach deposits are a common occurrence, providing freely drained level sites. Protruding areas of bedrock are glacially scoured.

### *The Baltic Sea section*

The enclosed Baltic Sea is connected with the open ocean only through the narrow channels between the Danish islands and southern Sweden. It is extended north of the Åland islands by the Gulf of Bothnia and eastwards by the gulfs of Finland and Riga. Between Sweden and the soviet republics of Estonia and Latvia successive submarine escarpments occur on the floor of the sea, related to the south-eastwards dipping of Ordovician, Silurian and Devonian rocks. The major islands of Bornholm and Gotland are tectonic rises in the basement capped by Ordovician rocks. The northern coasts of the Baltic are all emergent but the Polish and German coasts lie south of the flexure in the shield and are constructed from Pleistocene sediments. Longshore drift has been active to produce the classic sand spit and lagoon (haff and nehrung) coast.

## The Atlantic highlands

The Atlantic highlands province extends from North Cape, Norway across Scotland, Wales and Ireland, where it abruptly ends at the Atlantic coast. It consists of a highland region reaching 2400 m in Norway and 1400 m in Scotland and is formed of Lower Palaeozoic strata. The line of the structures is continued in Newfoundland, New England and the Appalachian mountains. Sediments eroded from both the Canadian and Fenno–Scandian shields were deposited in a geosyncline lying between them from Cambrian to Silurian times. Lithological evidence points to previous rifting with the development of oceanic floor between the

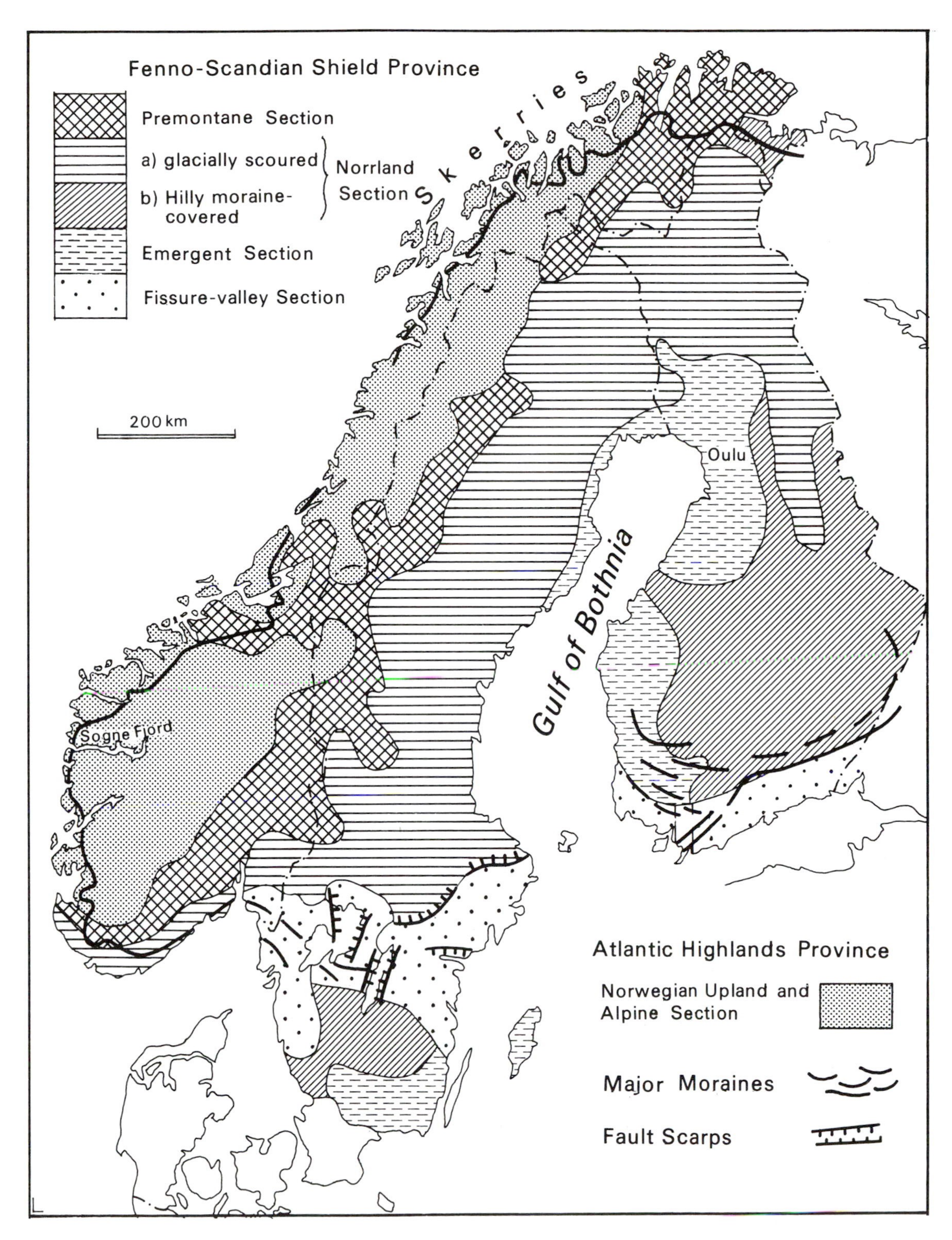


*Figure 8.4.* Geomorphological provinces of the Atlantic highlands and Fenno-Scandian shield.

European and American shields. Up to 10 km thickness of shales, sandstones, conglomerates and limestones with volcanic rocks accumulated on the trailing edges of the continental masses. When this ocean (called Iapetus) subsequently closed, the resulting movements produced a strongly folded mountain chain with overthrusts (nappes) which were pushed eastwards into the Swedish part of the mountain chain. Metamorphism of the sediments took place increasing in intensity from east to west. Erosion of these mountains took place throughout the Upper Palaeozoic and although affected by the Hercynian and Alpine orogenies, the results were confined to faulting and uplift which in Norway may have amounted to 1000 m.

The present landforms of Norway, west of the Scandinavian shield have largely resulted from erosion during the Tertiary when two upland surfaces were cut, a lower one lying between 700 and 1100 m above sea level and an older summit surface which has been attributed to the early Tertiary.

The elevated level surfaces produced during the Tertiary were ideally suitable for the accumulation of snow and its conversion into glacier ice. The adjacent North Atlantic Ocean provided plentiful moisture for precipitation. As the ice thickened, it first covered the highland surfaces and gradually extended to occupy the whole of the shield to the east.

Sub-division of the Atlantic highlands results in the following sections: Norway, Scotland, Wales, Ireland and the Irish Sea basin.

*Norway*

There seems to have been a strong structural control of drainage to the west coast of Norway, but the eastward flowing streams developed on the tilted surface of the shield with parallel courses towards the Gulf of Bothnia. However, the dominant influence on the present scenery was the Pleistocene glaciation.

The most dramatic landforms produced by glacial action in the Atlantic highlands are the fjords of Norway. The pre-existing valley pattern was used by the ice to flow from the plateau surfaces. As the volume of ice is roughly equivalent to its power to erode, the valley trough was deepened to below sea level (1308 m at the deepest point in Sogne Fjord), so that when deglaciation took place the sea flooded the deep valleys to produce the fjords. Near the coast, the ice became less constricted and could escape from the confines of the valley, and was floating on the sea water, so the power to erode was reduced. The valley floor was not eroded as deeply and so a threshold was left only 100–300 m deep at the mouth of the fjord (Figure 8.5).

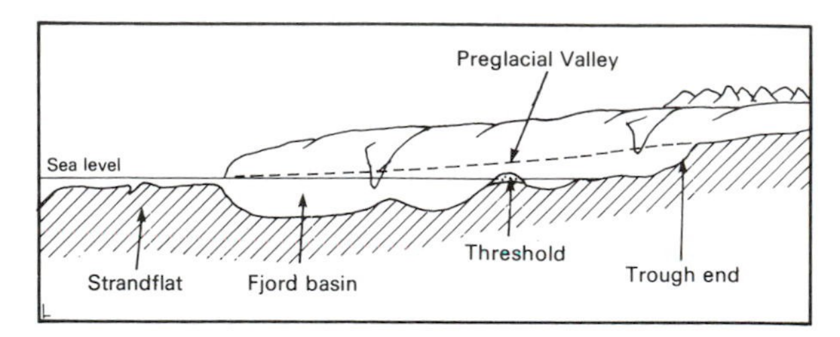


*Figure 8.5.* Section of a fjord valley.

The geomorphology of the Atlantic highlands in Norway can be summarised as alpine landforms, ice-scoured plateaux and the fjord coast. The alpine landforms include the higher peaks of the Norwegian mountains which retain some glacial activity. Steep peaks, arêtes and cirques occur, becoming progressively nearer to sea level further north. Many of the Norwegian mountains are really extensive, high, ice-scoured, rocky plateaux covered by thin drift and cut by glacial troughs and transfluent channels. The fjords, deeply cut glacial troughs, flooded by the sea, extend into the ice-scoured plateaux. The coast is characterised by the strandflat, a wide rock platform with islands (skerries).

*Scotland*

In the Hebridean district of Scotland, ancient shield rocks occur north of the Highland boundary fault. A consensus of opinion now accepts that these rocks are a fragment of the North American shield which became attached to Europe during the Caledonian orogeny. When America was rifted away from Europe in the Jurassic, this fragment of the northern shore of the geosynclinal trough was left behind. The rocks of the Outer Hebrides are gneisses of Archean age, representing a long period of repeated sedimentation, orogenesis, metamorphism and intrusion. Towards the end of the Proterozoic, the well-bedded, red Torridonian sandstones were laid upon the Archean rocks from a north-westerly direction, probably eroded from the mountains of the Grenville orogeny along the south-east margin of the Canadian shield.

Caledonian structures of the British Isles are of two contrasting rock types, those in northern Scotland have been strongly metamorphosed but in southern Scotland, Wales and Ireland metamorphic alteration has been slight. The metamorphic rocks between the Great Glen fault and the western edge of the Moine thrust were folded in the Ordovician and Silurian, metamorphosed and thrust westwards some 20 km over the Hebridean basement and Cambrian rocks. Between the Great Glen fault and the Highland boundary fault rocks of a similar nature continue in the Dalradian. These were originally sedimentary rocks, which are thought to include upper Pre-Cambrian and Lower Palaeozoic strata, but they have been metamorphosed and intruded by granites. Faulting along Caledonian trend lines downfaulted the central valley of Scotland (in which Upper Palaeozoic rocks accumulated) and raised the Southern Uplands. Sedimentation continued in basins such as the central valley of Scotland with Old Red Sandstones and Carboniferous sediments accompanied by the volcanic activity responsible for the Ochil and Sidlaw hills. A long history of erosional activity followed to produce the Tertiary landscape which was uplifted to form the basis of the present landforms. Substantial lava flows occurred on Skye, and Mull during the Tertiary.

The Highlands have an erosion surface between 760 and 880 m above which monadnocks rise to over 1200 m in Ben MacDui (1265 m), Cairn Toul (1247 m) and Cairn Gorm (1201 m). Partial planation features have been observed between 250 and 300 m. The upper surface has been referred to as the Grampian main surface. Glaciation has had a profound effect on the higher Scottish mountains, producing corries and glacially eroded U-shaped valleys cutting into the Tertiary erosion surfaces. At lower elevations the solid rocks are covered with glacial debris and certain parts of the west coast exhibit evidence of a post-glacial rise in sea level with raised beaches. The carselands of the Solway, Clyde, Forth and Tay estuaries include raised beaches and late glacial clays and silts laid down 6500 years BP.

### *Wales*

The north-east–south-west Caledonian structural trend is continued in North Wales but in South Wales it swings east–west to lie almost parallel to the later Hercynian folding. The sedimentary and volcanic rocks of the Cambrian, Ordovician and Silurian are weakly metamorphosed, sufficient to convert shales into slates in North Wales and parts of south-west Wales. The sandstones and shales of this 6-km thick succession have

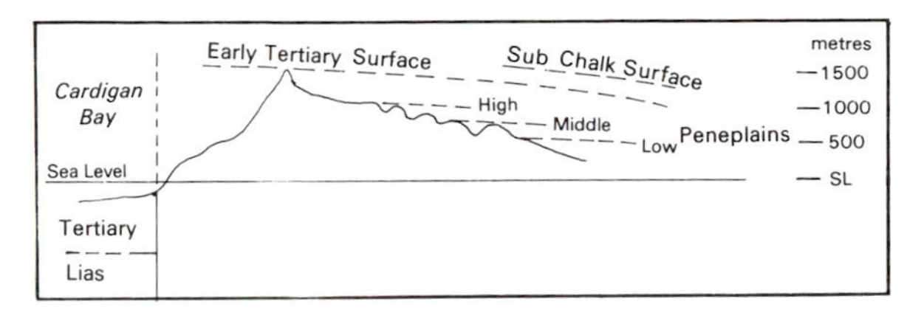


*Figure 8.6.* Erosion surfaces in Wales.

been uplifted to form plateaux into which the drainage is deeply incised. An upland surface lies between 580 and 600 m with monadnocks rising above it such as the Berwyn mountains (784 m), Plynlimon (752 m) and Cader Idris (892 m). Two partial planations occur at lower levels on the Welsh highlands apparently to a base level of about 370–490 m and 210–330 m above present sea level (Figure 8.6). Snowdonia, which is synclinal in structure, reaches 1085 m and has been strongly glaciated with excellent examples of glacial erosion including corries, arêtes, and U-shaped glacial troughs containing lakes.

Virtually the whole of Wales was glaciated, the mountains acting as a centre for ice dispersal during the Pleistocene glaciations. Not all of South Wales was glaciated during the Devensian glaciation, and those areas outside the glacial limits experienced a periglacial climate which is reflected in the landforms. Remnants of marginal lakes and former overflow channels occur in the area near Cardigan.

The rivers of Wales all appear to be superimposed upon the older, Palaeozoic rocks and geographers have been trying to account for their courses by invoking a cover of Mesozoic rocks upon which the drainage system could have been initiated. Linton advanced the hypothesis that the River Dee was formerly the headwaters of the proto-Trent and that the parallel valleys of South Wales were once left-bank tributaries of a proto-Thames. In both examples, these rivers have been dismembered by subsequent river capture and other geological events.

### *Ireland*

Except for the extreme south of Eire, rocks of the Lower Palaeozoic and structures of Caledonian origin also occur in Ireland. Their relief nowhere exceeds 852 m and Ireland has been described as a lowland of 60–90 m with isolated hills such as the Mourne mountains (852 m) and the Wicklow mountains (850 m). Although the

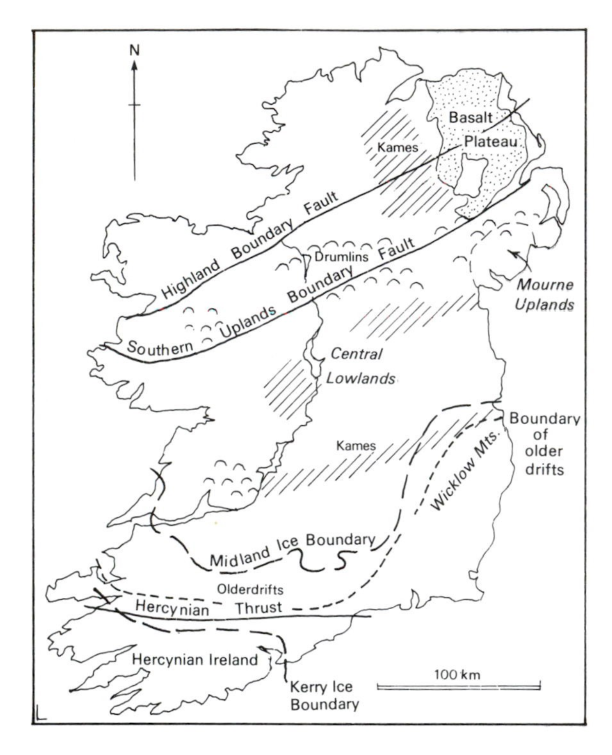


*Figure 8.7.* Geomorphological features of Ireland.

Caledonian structural trend may be seen in Northern Ireland and Eire, the Lower Palaeozoic rocks are neither metamorphosed nor strongly folded. A large part of central Ireland is formed by the Carboniferous limestone which is variably covered by glacial drift and, in places, peat bogs. The Burren on the west coast of Eire is one of the more extensive areas of limestone pavement and associated karstic features (Figure 8.7).

Early Tertiary vulcanism resulted in the outpouring of 800 m thickness of basalts in Northern Ireland forming a 400 km² plateau. Along the coast the famous Giant's Causeway provides sections in these volcanic flows. Lough Neagh is a downwarp in the volcanic plateau and is floored with Oligocene sediments. This volcanic activity and up-arching preceded rifting of the British Isles and the adjacent continental shelf areas below the sea.

### *The Irish Sea basin*

Between Ireland and Wales lies the downfaulted basin of the Irish Sea. Prospecting for oil has revealed that the submarine structures have the same Caledonian trend as seen in Wales, Scotland and Ireland. The basins in the Irish Sea contain Jurassic, Cretaceous and Tertiary sediments and the presence of Upper Cretaceous rocks in the St George's Channel basin suggests that a former cover of Mesozoic rocks over Wales is not as impossible as might be thought, although it may have been deposited around the Welsh highland massif without ever being deposited on the higher parts (see Figure 8.6).

## The north European plain

An extensive lowland extends from Britain, across the shallow North Sea, through the Netherlands, Denmark, Germany and Poland to the USSR where it expands greatly in width. This plain lies to the south of the Fenno–Scandian shield, and to the north of the central European uplands. From north to south it is between 200 and 400 km wide and the distance from East Anglia to the Urals is about 4000 km. Everywhere on this plain is below 300 m and parts of the Netherlands and northern Poland are below sea level.

The Pre-Cambrian basement of the Scandinavian shield continues beneath the north European plain but south of the Baltic it is covered by Mesozoic, Tertiary and Quaternary deposits. Overlying the basement rocks in southern Germany are Permian strata which include thick evaporites. Potash salts, 40 m thick, are mined at Stassfurt and greater thicknesses of anhydrite and rock salt are present. In western Germany and beneath the North Sea, these salt deposits and diapirs from them have trapped oil and gas. In the lowland embayments between the central European uplands the Tertiary rocks include lignite deposits. However, with all these sedimentary rocks, exposures are rare as the northern part of the plain has been covered by glacial deposits from the Pleistocene glaciations and the southern part by loess.

The Elster glaciation pushed furthest south and its deposits remain as scattered erratics on the northern slopes of the central European uplands. Saale drifts cover most of the plain west of the Elbe river; these have been decalcified and remain as sandy areas like the Luneburger Heide and the Geest further north. East of the Elbe, the Saale drifts are covered by the Vistula drifts (equivalent to the Wurm or Devensian). A terminal moraine (Brandenburg moraine) and two recessional moraines (Frankfurt and Pomeranian) have been recognised from this last glaciation. The ice sheet and the deposits it left behind strongly influenced the drainage of the plain. Rivers flowing north from the central European uplands and meltwaters were diverted westwards along broad spillways, now occupied by the smaller present-day rivers. These channels are called Urstromtaler. Landscapes outside the terminal moraines are characterised by outwash sands but inside their limits boulder clays were deposited and a landscape of lakes with hummocks formed by kames and eskers is typical. The Mecklenberg and Pomeranian lake districts of Germany and Poland are representative of this landscape. The cold climate of the glacial periods resulted in little vegetation and wind action upon the outwash deposits and led to accumulations of loess, especially in the embayments between the central European uplands.

The sections of the north European plain are: the North Sea basin, the central European plain, the Köln and Münster embayments and the Russian plain.

### *North Sea basin*

The western limits of the north European plain lie on the dipslope of the chalk escarpment of East Anglia. The chalk and Pliocene crag deposits are extensively covered by glacial drifts laid down in the 'Anglian' glaciation which reached almost to the Thames. In Norfolk subsequent ice advances produced the contorted drift, exposures of which may be seen in the cliffs of East Runton. In the Netherlands, the push moraines of the Arnhem–Apeldoorn district were contorted during the Saalian glacial stage. The Devensian (or Weichselian) glaciation just reached the north coast of Norfolk, depositing the Hunstanton boulder clay and eroding small marginal meltwater channels. Inland, glaciation reduced the height of the chalk escarpment and probably was responsible for erosion of the gap in the chalk scarp occupied by The Wash. In the Devensian glaciation cold conditions south of the ice front produced periglacial activity, now revealed in the Breckland and other areas as fossil features including patterned ground and ice wedge pseudomorphs.

On the western shore of the North Sea, the post-glacial rise in sea level, the Flandrian transgression, produced the conditions suitable for sand dune accumulation and behind the dunes, marsh development. Good examples of an offshore bar with sand dunes and coastal marshes may be seen at Scolt Head; Blakeney Point has extended westwards from the mainland at Weybourne and also has salt marshes on its sheltered, landward side.

Coastal erosion of glacial deposits occurs east of Weybourne on the Norfolk coast revealing sections in the Pleistocene Cromer forest bed, a deposit which preceded the first major glacial advance into East Anglia. East of the village of Runton excellent examples of contorted drift and large chalk erratics may be seen in the eroding cliffs. On the east-facing Norfolk and Suffolk coasts considerable erosion has occurred in historical time with the loss of the villages of Eccles and Dunwich.

The floor of the North Sea is a drowned part of the continental shelf. Following exploration for oil and natural gas it is now known that rifting and subsidence have occurred (associated with the line of the Rhine graben), which may account for the migration southwards of the lower Thames during the Pleistocene. Rifting has also been followed by an accumulation of over 3500 m of Tertiary sediment in the deepest part of the northern basin (Figure 8.8). Natural gas has been found in the southern North Sea basin in Permian sandstones capped by evaporites and to a lesser extent the Triassic Bunter Sandstone rocks. The Dogger Bank is associated with a buried salt diapir. In the northern North Sea basin, oil occurs in Jurassic and Lower Cretaceous rocks.

During the Pleistocene much of the North Sea basin was dry land and the continental ice sheet was able to spread from Scandinavia across a low-lying undulating plain to reach eastern Britain. As the ice melted this low-lying area was covered with boulder clay and outwash deposits. These have been washed by the sea as the post-glacial rise in sea level took place (Figure 8.9), and features such as the offshore bar of Scolt Head Island developed and possibly outwash gravels were modified and pushed onshore to form the core of Blakeney Point.

The eastern shore of the North Sea has developed under the influence of the Rhine delta. In the Pleistocene the river emerged from the faulted Rhine highlands and flowed across the low-lying lands of the North Sea basin. When the Saalian ice advance took place, the Rhine was diverted to a more westerly route. In Holocene time, the rising sea level inundated the lower part of the North Sea basin and produced 20–30 m dunes along the Dutch coast, overlying marine and fluviogla-

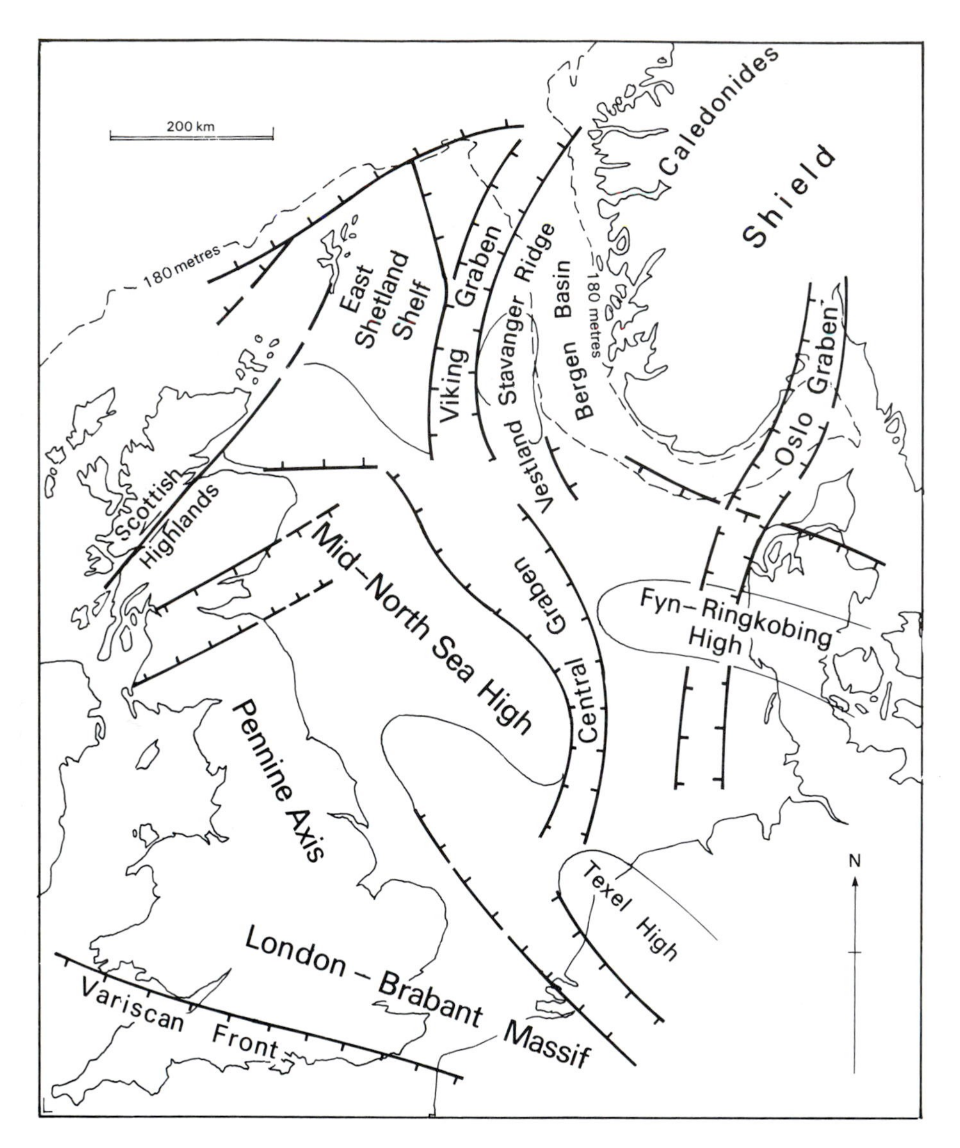


*Figure 8.8.* Tectonic features of the North Sea basin.

cial Pleistocene deposits. Behind the dune barrier, peat accumulation occurred. Following peat cutting for fuel, many of these former peat bogs have been converted by embankment and drainage into the polders, some of which are more than 6 m below sea level.

### *Central European plain*

The landforms of the north European plain in Denmark, East and West Germany and Poland may be subdivided conveniently into three parts: landscapes developed on the older glacial drift deposits, those on the younger drifts and those adjacent to the central European uplands which were not glaciated.

The most extensive stage of glaciation, the Elster or south Polish, covered the plain as far south as the slopes of the central European uplands. This early Pleistocene glaciation left drifts which have been weathered and eroded to form subdued landforms. Outwash sands and gravels often overlie the till deposits where they result in extensive heathlands as in western Denmark, and the Geest and Luneberger Heide districts of western Ger-

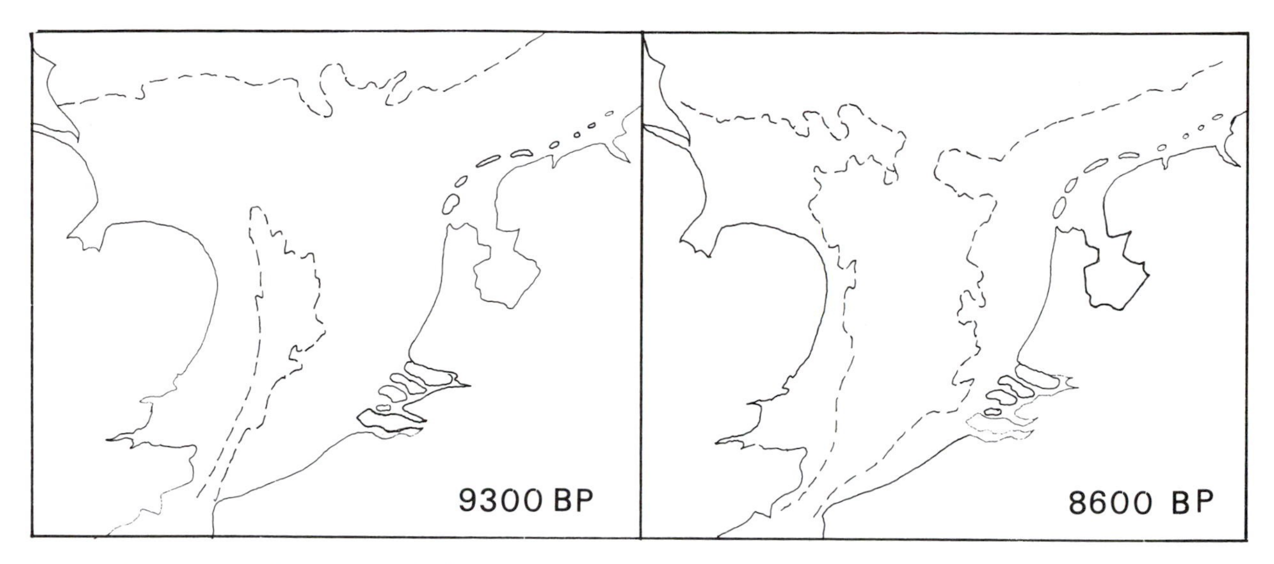


*Figure 8.9.* Expansion of the North Sea with the Flandrian transgression.

many. There are few lakes in the older drift areas. In Poland this stage of glaciation affected the country as far south as Cracow and Wroclaw (Figure 8.10).

The late Pleistocene Weichsel or north Polish glaciation did not extend so far south, only reaching the vicinity of Berlin and Warsaw. In the west it reached as far as the river Elbe, its margin curving northwards through Denmark. In Poland, the deposits from this glacial stage are known as the Vistula drifts. In Germany, the terminal moraine from this stage is known as the Brandenburg moraine. Inside the limits of the terminal moraines, boulder clays were deposited and a landscape of hummocks and lakes formed by kames, eskers and kettleholes is typical. The Mazurian and Pomeranian lake districts are representative of the young glacial landscapes.

The deposits left by these ice sheets strongly influence the drainage of the plain. Glacial meltwaters were diverted westwards along broad spillways between moraines, now occupied by smaller, present-day rivers. In Germany, these channels are known as urstromtaler and in Poland, pradoliny. The cold climate of the glacial stages did not encourage vegetation and wind acted upon the outwash deposits, blowing the silt-sized material further beyond the areas affected by glacial action to accumulate as loess, especially in the embayments between the central European uplands.

### *Köln and Münster embayments*

Two major lowland embayments occur in the central European uplands. The Münster embayment lies between the industrial Ruhr Valley and the ridge of the Teutoburger Wald and is drained by the Ems river. It was also an area of subsidence in the Tertiary and it acquired a mantle of loess overlying older drift deposits during the Pleistocene. The northern boundary of this section is ill-defined, merging with the glaciated lowlands of Lower Saxony and the Netherlands.

The Köln embayment is crossed by the Rhine after it emerges from the Rhenish highlands. The Rhine has a steeper gradient resulting from subsidence associated with deep faults. These lie deep below the surface alluvia and Tertiary deposits and are related to the line of downfaulting associated with the Rhine rift valley as it continues northwards across the floor of the North Sea. The Tertiary deposits include lignite which is mined by opencast methods. Restoration is made easier by the presence of a covering of loess which makes a deep, easily workable, fertile soil.

West of the Köln embayment the level of the land rises to the low (150–200 m) loess-covered plateau of Limburg. The Meuse, which in France occupies the Lias clay vale, has been superimposed on the western Ardennes which it crosses by two gorges. At Namur the

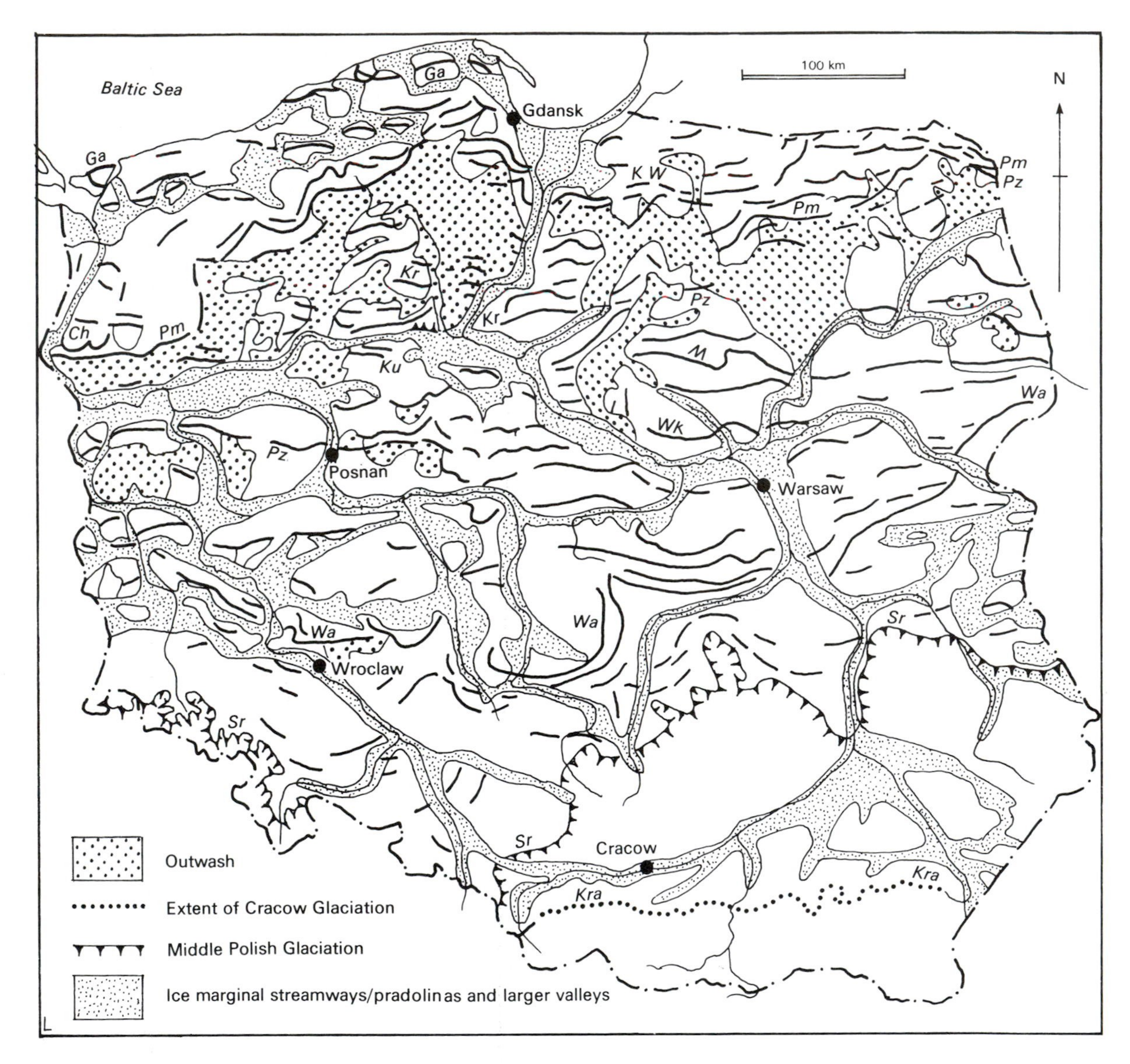


*Figure 8.10.* Glacial features of the north European plain in Poland. (After Embleton, 1984.)

river turns abruptly and follows the Hercynian structural trend to Liège where it begins to cross the Limburg plateau passing through Maastricht in an incised valley, providing the Netherlands with some attractive, hilly country. North-west of Maastricht, the Kempen plateau is a former alluvial fan of the Meuse and is composed of sands and gravels (Figure 8.11).

### *The Russian plain*

The basement rocks of the Fenno-Scandian shield continue at depth beneath the Russian plain but crop out again in the Ukraine massif. Between these two exposures it is known that there is a sequence of horsts and graben in the Pre-Cambrian basement over which has accumulated Palaeozoic, Mesozoic and Tertiary rocks of varying thicknesses. For example, between Kursk and Voronezh a horst brings the basement rocks close to the surface, but between this horst and the exposed shield in the Ukraine there is a deep graben in which the Donbas coal measures have been preserved.

As with the western and central parts of the European plain Pleistocene glaciation has occurred and had a profound effect on the landforms. There is evidence of

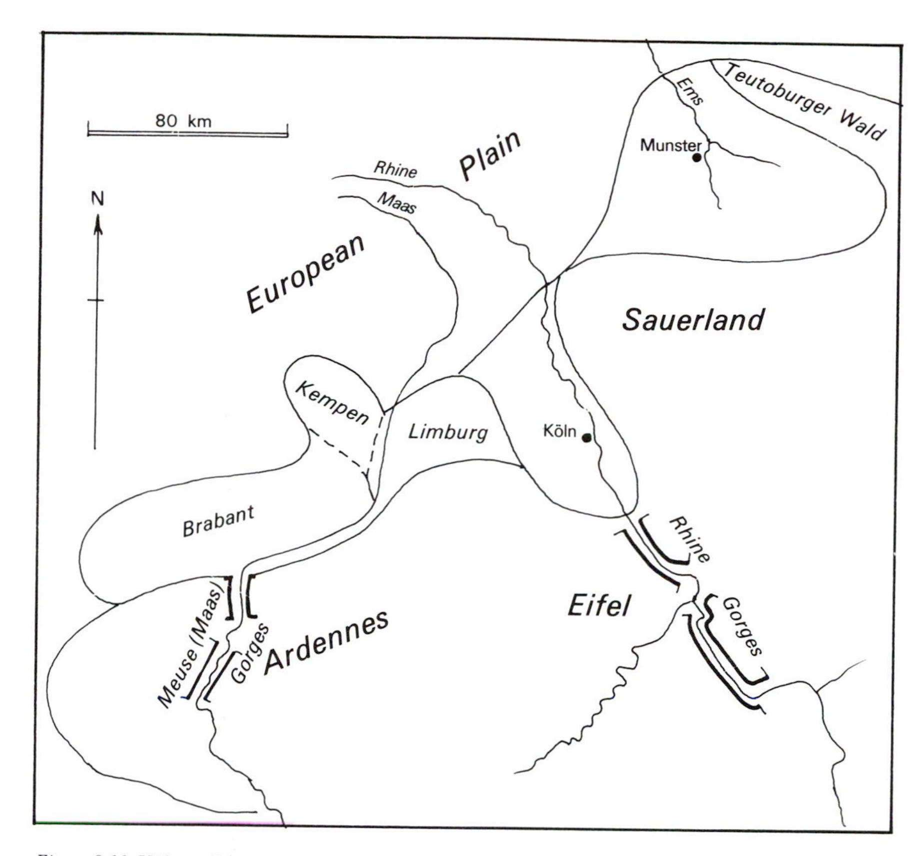


*Figure 8.11.* Köln embayment and the Kempen plateau.

at least two major ice advances across the plains of European Russia; these are referred to as the Dnepr–Don and Valday glaciations. The Dnepr–Don glaciation, probably contemporaneous with the Saale glaciation of northern Europe, pushed tongues of ice down the Dnepr and Don valleys to within 200 km of the Black Sea. The Valday, equivalent to the Vistula and Weichsel of the central European plain and Devensian of Britain, did not spread so far south, leaving terminal moraines along a line from Minsk to Smolensk and Moscow (Figure 8.12).

The geomorphology of the Russian province of the European plain may be discussed in eight sections: the Arctic lowlands, the Baltic lowlands, the Moscow–Oka plain, the Ukrainian plateaux and lowlands, the Volga heights, the trans-Volga plains, the Black Sea lowlands and the Caspian lowlands (Figure 8.13).

*The Arctic lowlands*

There are three sub-sections to the Arctic lowlands: the lands north of the latitude of Leningrad, east of Lake Onega, and in the west, the basin of the Dvina river which flows into the Barents Sea at Arkhangel. The landscape is one of a plain at 180–250 m above sea level which has been glaciated in its western parts and which has received fluvioglacial deposits elsewhere. As part of the Weichsel glaciation, the landforms it left are still fresh and there were proglacial lakes as the ice dammed the northward-flowing Dvina. The 300–400 m Timan ridge of uplifted Palaeozoic rocks was glaciated earlier, not at the Weichsel stage. It separates the Dvina from the Pechora plain. The Pechora plain, drained by the Pechora river, lies mainly below 100 m and is underlain by a considerable thickness of Tertiary and Quaternay sediments. It was glaciated earlier in the Pleistocene but not in the Weichsel glaciation, when it was exposed to periglacial conditions. Patches of permafrost still remain.

*The Baltic lowlands*

South of the Gulf of Finland, Lower Palaeozoic strata crop out and the Valdai hills, formed by an escarpment of Silurian limestone, are capped with recessional moraines of the Valdai glaciation. These low hills (300–325

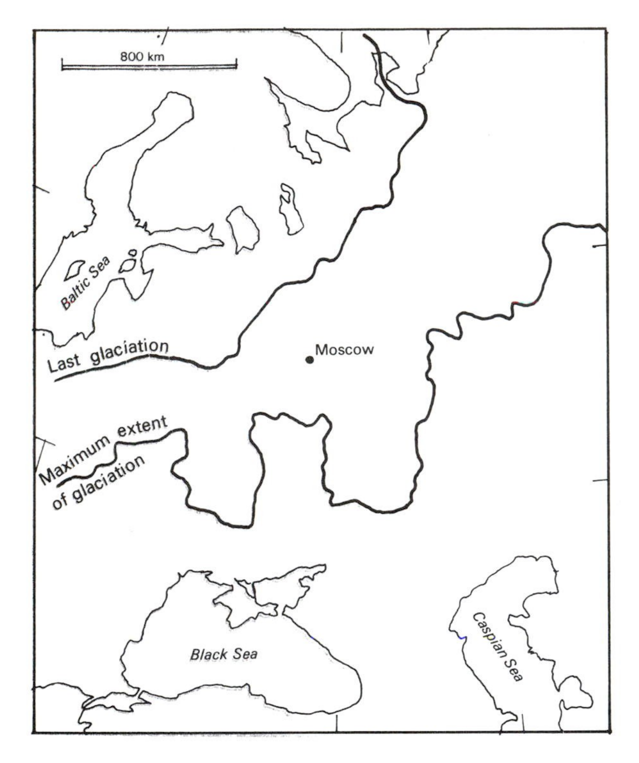


*Figure 8.12.* Glacial limits on the Russian plain.

m) form a major watershed on the plain between rivers flowing to the Baltic and those flowing via the Volga to the south. North of the Minsk–Smolensk–Moscow line of moraines, the Baltic lowlands landscape is one of fresh-featured landforms and disorganised drainage with many lakes, such as Lake Ilmen and Lake Beloye. The land subjected to the older Don–Dnepr glacial tongues has been exposed longer to erosion and so deposits are thinner and features more subdued. Additionally, outwash from the later Valdai glaciation spread sands along the slightly incised courses of the major rivers. Thus, where outwash streams scoured the landscape it is split into low plateau interfluves and broad valley plains. Where glacial outwash accumulated there may be extensive and poorly drained lowlands, as in the Pripyat Marshes.

*The Moscow–Oka plain*

This physical region extends from around Moscow and the Valdai hills in a north-easterly direction gradually increasing in elevation from just over 100 m to 150–180 m on the Klin–Dimitrov plateau. The landform is an early Tertiary peneplained surface which has been covered by glacial deposits in mid-Pleistocene times, especially fluvioglacial material, which is now being eroded to give complex detail to the landscape.

*The Ukrainian low plateaux and lowlands*

This sub-section extends from the Carpathians in the west to the Volga in the east and includes most of the Ukraine. The crystalline basement complex is broken into several extensive blocks, some of which have been uplifted and others depressed to give a sequence of low plateaux with lowlands between (Figure 8.14). In the west the Dnestr plateau at 450–500 m is the most elevated and rivers are deeply incised. Near to the Carpathians, the effects of the Alpine earth movements resulted in the strata being tilted and these have since been eroded to become west-facing escarpments. Planation has affected this area as these escarpments are reported to have their crests bevelled. The area has also been mantled by loess.

The Denpr–Donetz (130–140 m) and the Don–Oka (100 m) lowlands are both tectonic depressions, lying on either side of the central Ukrainian plateau. This low horst was peneplained in the Eocene, uplifted in the Miocene and subsequently dissected. It ranges in height from 250 to 280 m and is badly affected by gulleying of the loess mantle on its surface. The lowland depressions on either side were partly infilled with Tertiary sediments and their northern parts overrun by ice in the mid-Pleistocene glaciation. Fluvioglacial deposits up to 100 m thick characterise the southern parts of these lowlands. The disruption of the basement complex brings to the surface the crystalline rocks at several places between Kiev and the heights overlooking the Sea of Azov. However, the shield rocks are mostly covered with Mesozoic of Tertiary strata and the surface has been peneplained to give an elevation of 200–300 m above sea level. Glacial deposits only occur on the northern slopes, but loess is widespread on its surface. The shield plateau reaches it maximum height of 324 m in the Podolsk–Volyn uplands near the Sea of Azov where an old erosion surface with tors and monadnocks occurs.

*The Volga Heights*

The easternmost of these faulted blocks underlying the Russian plain lies between the Volga river and the Don–Oka lowland. A steep fault-scarp faces eastwards overlooking the river but the western side slopes gently from a maximum height of 370 m. A similar history of

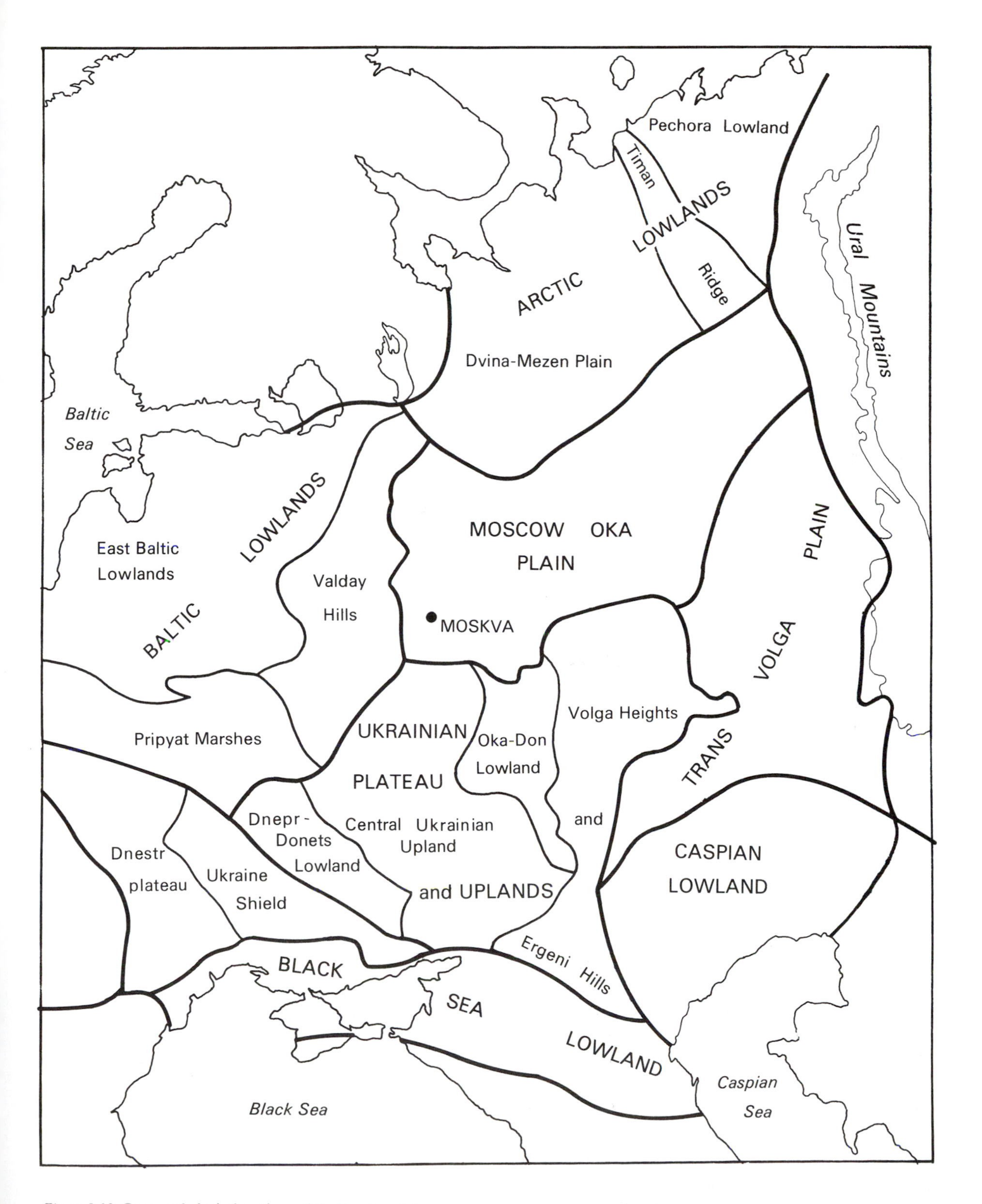


*Figure 8.13.* Geomorphological sections of the Russian plain.

*Figure 8.14.* The Russian European plain, Ukraine.

planation during the Eocene followed by disruption in the Miocene has occurred. A loessial cover occurs and there are problems with erosion. The southern part of this sub-section is the Ergeni hills, a 160–220 m planation surface.

*The trans-Volga plains*

East of the Volga plains continue at about 300 m above sea level but the underlying geology is different. The plains rest on a structural foredeep lying to the west of the Urals which has been infilled with Mesozoic and Tertiary sediments. The northern part of this section was glaciated in the mid-Pleistocene and the southern part received a loess cover. Oil and natural gas reserves occur in the Mesozoic structures beneath the slightly dissected planation surface which is preserved on the gently folded strata.

*The Black Sea lowlands*

Tertiary rocks deeply covered by loess form the surface of the Prichernomorsk plain adjacent to the shores of the Black Sea. It is extremely flat and the loess may have been reworked by the sea and subsequently uplifted.

*The Caspian lowlands*

The lowest part of the Russian plain occurs around the Caspian Sea. Its surface ranges from 28 m below sea level to 50 m above. It is a downfaulted area with the crystalline basement complex at great depth. Sediments have accumulated since the Permian and some folding is associated with marginal deep-seated faulting; minor relief features are caused by salt diapirs rising up through the extensive lacustrine deposits. The Volga has laid down a delta after crossing the undissected semi-arid lowland with areas of sand dunes and, in places, barchans.

## The European scarplands

Escarpments form distinctive landscapes in several parts of Europe where sedimentary strata are slightly folded and subsequent erosion preferentially lowers the clay vales, leaving the more competent sandstones, limestones or conglomerates as scarp-formers. The scarplands of Europe are formed from Upper Palaeozoic strata in northern England and Wales and Mesozoic strata in southern England, France and Germany. The scarplands do not often exceed 1000 m in elevation, and although the scarp faces may be steep, the dip slopes are gentle and even plateau-like. Provinces of the European Scarplands are:

British scarplands
The Paris basin
German scarplands

### *The British scarplands*

This province may be divided into a number of smaller but distinctive sections in which the cuesta landform is dominant. These include the Pennine scarps, the Welsh scarps and the English scarps. The position of these districts and the major scarps are shown in Figure 8.15.

The Pennine scarps are formed from slightly folded and faulted Upper Palaeozoic rocks of the Pennine anticline in northern England. Erosion has revealed the Carboniferous limestone Alston and Askrigg blocks in Yorkshire and the Derbyshire dome, an upland plateau encircled by inward-facing escarpments of the Millstone Grit and coal measures. On these rocks the highest surfaces have an elevation of 700 m on the peat-covered Kinder Scout. Most escarpments and the limestone plateau do not exceed 500 m, and there is a pronounced bevelling of the escarpments of the Low Peak in Derbyshire at 330 m and a minor incision of this Peak District upland surface at 300 m. The character of the river pattern suggests that superimposition has occurred and there are places where river capture has taken place. There are many karstic features in the Pennines, including limestone pavements, swallow holes, underground caverns and resurgences. Below the upland surfaces, a sequence of knick points and associated river terraces may be traced at several different levels along valleys. The history of events is complicated by glaciation during the Pleistocene and major valleys have fluvioglacial terraces as well as Holocene alluvium.

The Welsh scarps are also formed from Upper Palaeozoic rocks, folded in the Hercynian orogeny to form the syncline of the South Wales coalfield. The Brecon Beacons are a steep north-facing escarpment rising to 900 m overlooking the Towy and Usk valleys. Wales was twice subjected to glaciation during the Pleistocene and this high escarpment was also subject to late glacial ice action in small corries on its northern scarp face. Post-glacial solifluction has smoothed the dip slope of these uplands, redistributing their cover of boulder clay and forming solifluction terraces in the valleys. In South Wales the Carboniferous limestone does not make a pronounced escarpment, nor does the Millstone Grit but karstic features occur as rivers cross the limestone outcrop, especially in the Neath and Swansea valleys.

Another prominent escarpment is formed by the Pennant sandstones of the middle coal measures below which the productive coal seams lie. The Pennant escarpment can be traced all round the coalfield but it is highest in the north where it reaches a maximum height of 600 m. The northern-facing slopes were also subject to late-glacial ice action in corries. The escarpment forms a high plateau the dip slope of which has been disected by the parallel valleys of the Ebbw, Rhymney, Rhondda, Taf, and in the west, the structurally controlled Neath and Swansea valleys. The glacially oversteepened valley sides are the site of numerous landslips which are a hazard to the local communities. Recessional moraines and kame terraces lie on valley floors aggraded with fluvioglacial debris.

The English scarplands are developed on Mesozoic rocks, specifically the Jurassic limestones and the Cretaceous chalk. The Jurassic rocks form a pronounced north-west-facing escarpment called the Cotswold hills (346 m), which extends northwards to become Lincoln Edge and the Cleveland hills of the Yorkshire moors. Slightly to the east, the Cretaceous chalk forms the dissected plateau of Salisbury plain (100–150 m), and the Chiltern hills (240 m) before becoming very subdued in Suffolk and Norfolk. The Wash breaks the escarpment which reappears as the Wolds in Lincolnshire. South-east of London, the Cretaceous rocks form the anticline of the Weald where escarpments of chalk form an elliptical outcrop around the Gault and Greensand with a smaller inlier of Jurassic at the centre of the anticlinal structure near Battle.

The crests of the chalk escarpments have been bevelled by erosion but where they rise above the bevelling the escarpments carry deposits of clay with flints; a residual deposit of weathered chalk or remains of stripped Tertiary sediments. The general accordance of summit levels across southern England led Wool-

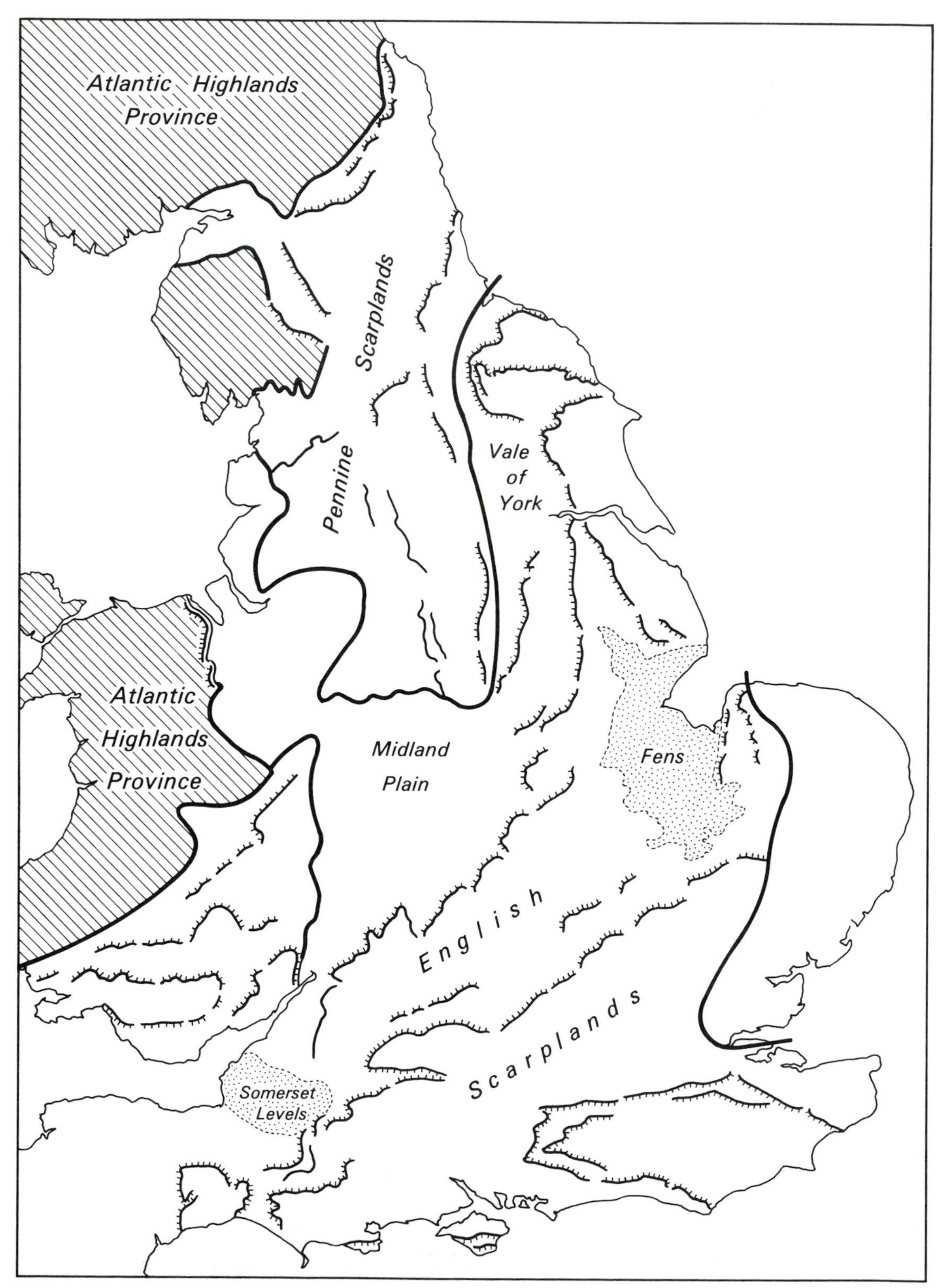


*Figure 8.15.* The British scarplands.

dridge and Linton to postulate a Mio-Pliocene peneplain surface from which much of the relief of the English scarplands has developed. On the North Downs a pronounced bench at 200 m carries sands which indicate a Pliocene incursion of the sea. Below this feature the most significant facet in the landscape is the stripped sub-Eocene surface. There is an extensive literature on the Weald and scarplands of southern England where greater detail may be obtained.

Eastward-draining consequent rivers were postulated by Linton as the predecessors of the present Thames and Trent. These were disrupted by river capture into the present river systems. Below the Pliocene bench a flight of up to nine terrace features have been recognised including the fluvioglacial gravel trains and river terraces alongside the Thames. Clapham Common and Wanstead are on the Boyn Hill terrace; Twickenham, Hyde Park and Holborn are on the Taplow terrace; Fulham and Chelsea are on the flood plain terrace; Rotherhithe and Wapping are on the floodplain alluvium.

As far south as the line of Severn–Thames the English scarplands have been glaciated. The glacial drift removed from the Jurassic and Cretaceous limestones is calcareous and the clay vales between have provided clay material for till deposits. Sand and gravel outwash deposits frequently occur along valleys which have partly removed the drift deposits to reveal the underlying solid geological deposits. Some river valleys were blocked by the ice and lakes were formed; a good example occurs in the English scarplands in the Vale of Pickering which was flooded to form Lake Pickering during the Devensian ice advance.

South of the extent of ice cover, a periglacial climate led to the development of many features of interest to geomorphologists. Examples of patterned ground, layered slope deposits, large-scale landslips and cambering of competent strata over clays has occurred. In the Chiltern hills asymmetric valley forms resulted from more rapid solifluction on west-facing slopes than on east-facing ones.

Ripples from the Alpine folding affected the British scarplands, producing the major synclines of the London and Hampshire basins as well as the anticline of the Weald, notably the vertical folding of the Isle of Wight.

### *The Midland plain*

The central part of England, developed mainly on level-bedded Triassic sandstones and marls, covered with glacial deposits, is a low-lying undulating plain. It extends on either side of the Pennines into the Vale of York and the lowlands of Cheshire and Lancashire. South-westwards the plain follows the River Severn and passes into the Vale of Glamorgan, the largest area of lowland in Wales. The northern part of this plain is drained mainly by the River Trent and the southern part by the Severn and its tributary, the Avon.

During the Wolstonian stage of the Pleistocene, much of the Midland plain of England was occupied by a proglacial lake, called glacial Lake Harrison, the waters of which were trapped between the ice to the north and west and the Jurassic escarpment to the south-east (Figure 8.16). Overflow from this body of water took place through gaps in the escarpment to the Nene and Thames drainage systems. Other glacial lakes formed at different times and overflowed, cutting gorges such as the Ironbridge gorge in Salop, subsequently used by the Severn in post-glacial time.

Glacial drift landscapes are particularly significant on the Cheshire plain and in the Vale of York. The City of York is built upon moraines left by the last major ice advance, the Devensian. Alongside the River Trent fluvioglacial terraces run parallel to the river at three different levels above the present floodplain. The path of the highest gravel train shows that the river once reached the sea through a gap in the Jurassic escarpment at Ancaster, near Sleaford, 70 km south from its present outlet in the Humber estuary. Fluvioglacial terraces similarly follow the Severn to the Bristol Channel.

Two areas of alluvial sedimentation occur on the British lowlands and are most conveniently described at this point, the Somerset Levels and the Fens. Both are low-lying areas which became marshy as the sea level rise of the Flandrian transgression took place and peat accumulated. In the Fens, almost all the peat has been oxidised through intensive agriculture following drainage in the early seventeenth century but peat is still extensive on the Somerset Levels which have been used more for grazing.

### *The Paris basin*

The Paris basin lies between the uplands of Brittany to the west, the Massif Central to the south, the Vosges to the east and the Ardennes to the north. It is approximately 400 km across from east to west and a similar distance from north to south. At its centre is a plateau of

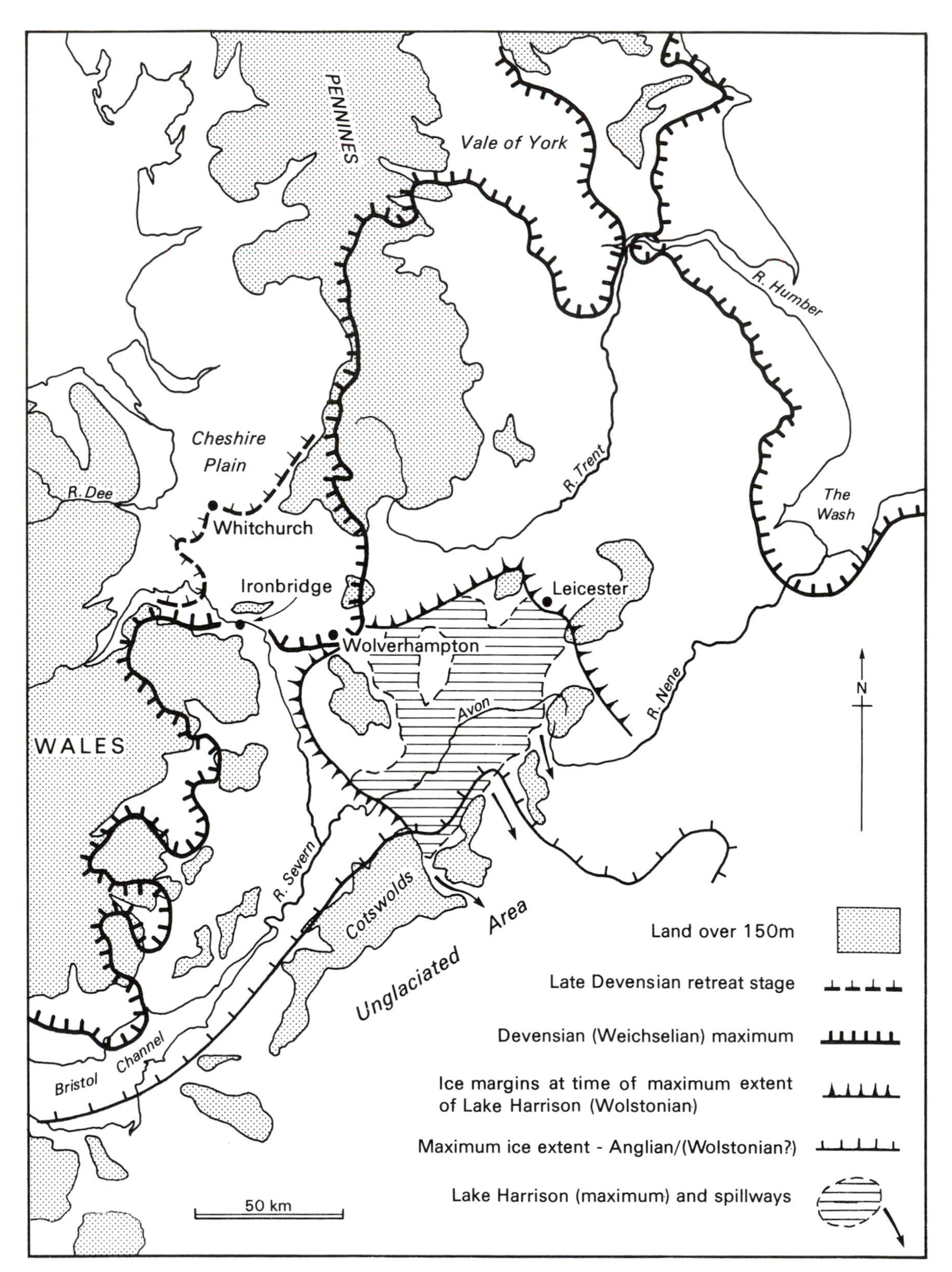


*Figure 8.16.* Glacial features of the Midland plain of England.

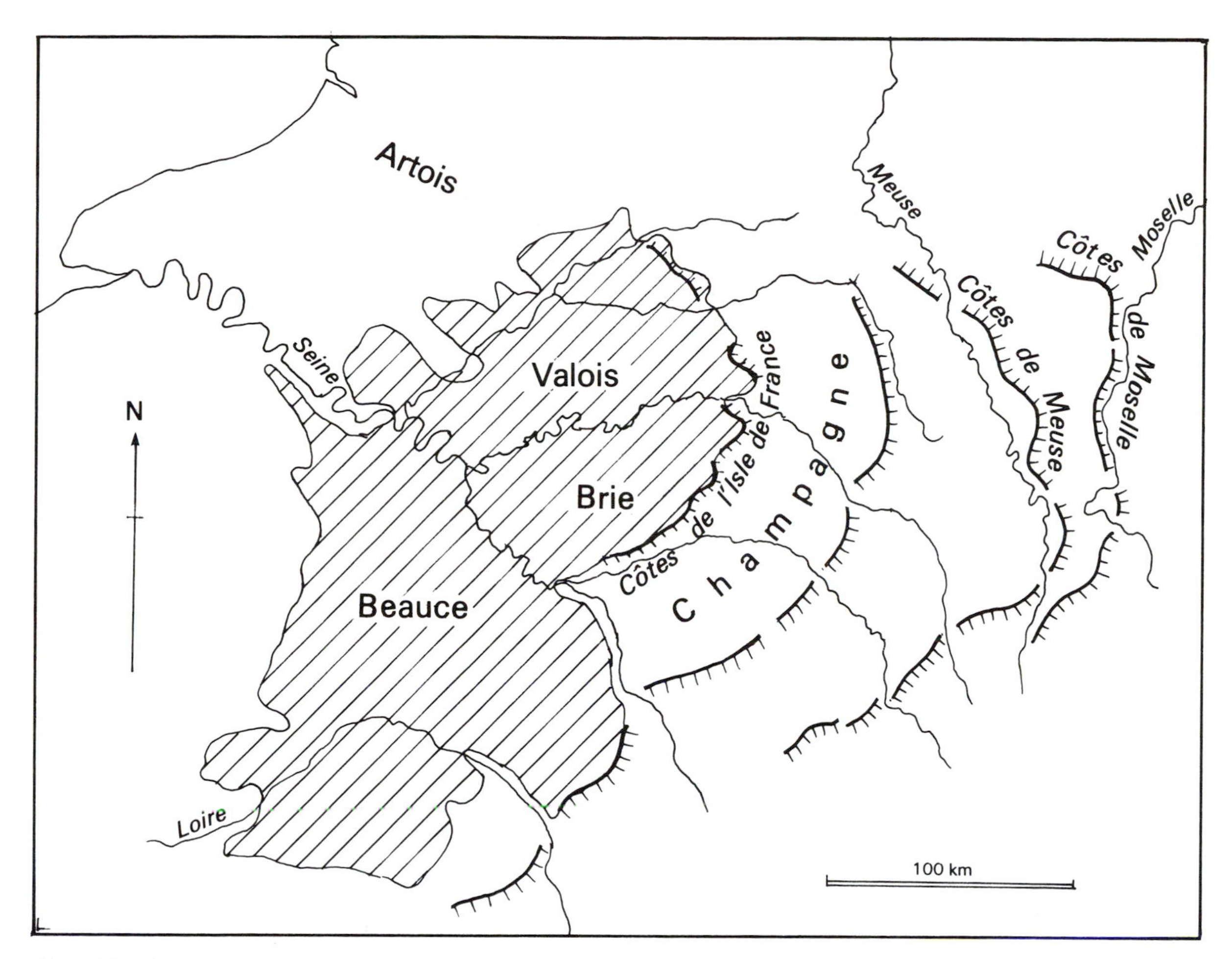


*Figure 8.17.* The French scarplands.

Tertiary rocks, elongated north-east to south-west. This central plateau which is 120 to 150 m above sea level is bounded on the east by the escarpment of the Côtes de l'Isle de France (Figure 8.17). The rocks which form this central plateau are varied, giving rise to the pays of Brie, Beauce and Valois. Silcretes were developed in sands and gravels since disrupted to form 'meuliniers' similar to the sarsen stones of southern England.

East of Paris the Cretaceous rocks emerge from beneath the Tertiary escarpment and form the chalk plateau of the Champagne Sèche. The eastern edge of this famous pays is formed by the chalk escarpment of the Côtes de Meuse along the western bank of the Meuse river. Emerging from beneath the chalk are clays which cover part of the Jurassic dip slope to give the pays known as the Champagne Humide. The crest of the Jurassic escarpment, known as the Côtes de Moselle rises to a maximum height of 545 m. Before the Tertiary sands and gravels were laid down, the chalk surface was peneplained to a low relief, and so the chalk escarpment is bevelled on its higher parts. North and west of Paris, the escarpments are not as pronounced as in Lorraine, but there are localised breached anticlines with inward-facing escarpments in Picardy and Boulonnais. Subsequent planation during the Tertiary has given plateau surfaces on the chalk of 200 m in Normandy, but somewhat less elevated in the Artois plateau, 180 m above sea level. Higher parts of the chalk carry a covering of clay with flints (argile avec silex). Periglacial conditions affected the Paris basin during the Pleistocene leaving solifluction deposits and screes of talus, known as 'grèzes'. There is also a loessial cover of variable thickness throughout the basin.

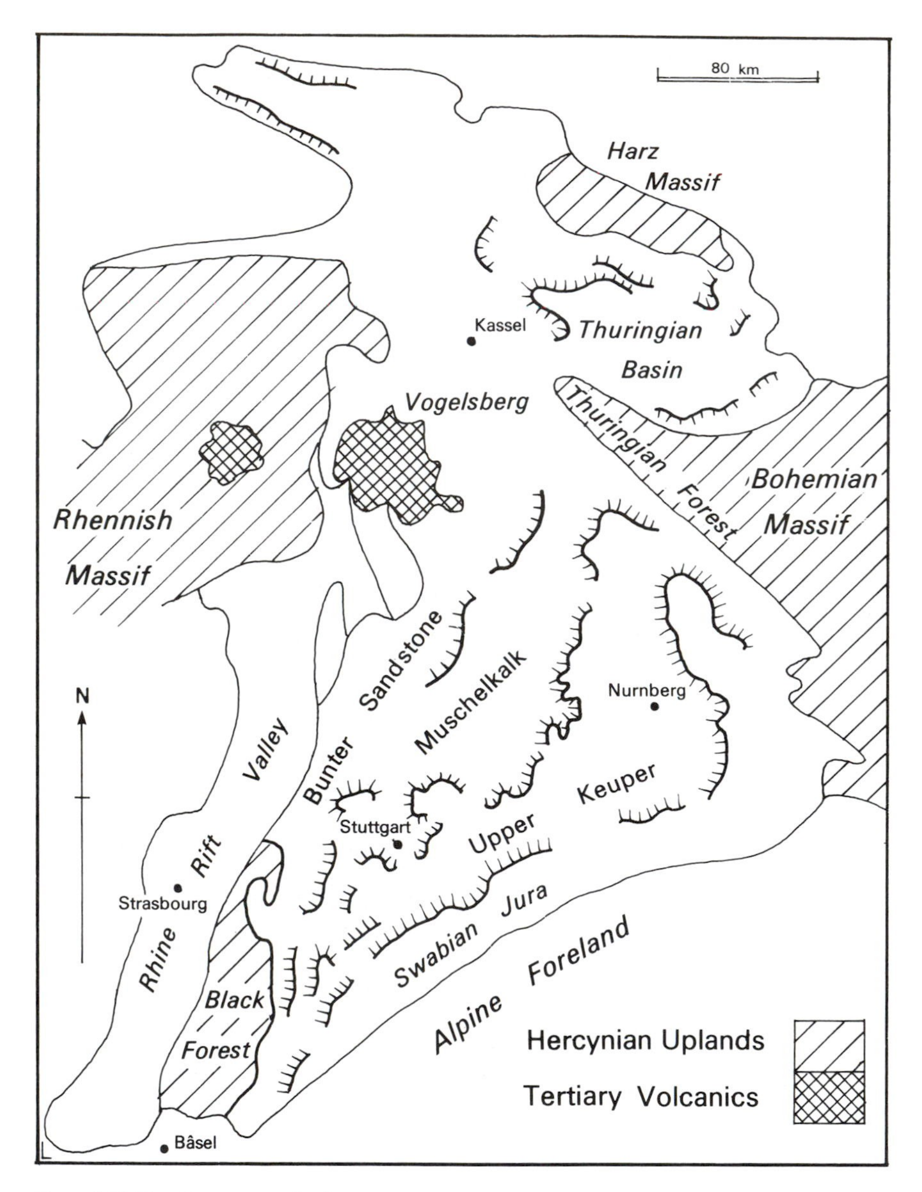


*Figure 8.18.* The German scarplands.

The River Seine, which rises on the plateau of Langres is joined by centripetal tributaries of the Aube, Loing and Yonne south-east of Paris, the Marne in Paris and the Oise downstream of the city. Presumably, these major streams originated on a warped constructional surface and proceeded to incise their valleys and in so doing cut across the different geological strata. River terraces have been laid down along the major river valleys. As a result, the landforms of the Paris basin are polycyclic with some surfaces inherited from the Tertiary or earlier and later facets cut by rivers in the Pleistocene and Recent times.

### *The German scarplands*

The Bavarian area of Germany, lying north of the Danube and between the Black Forest uplands to the west and the Bohemian highlands to the east is an Hercynian graben in which Triassic, Jurassic and Cretaceous rocks were deposited. Slightly tilted towards

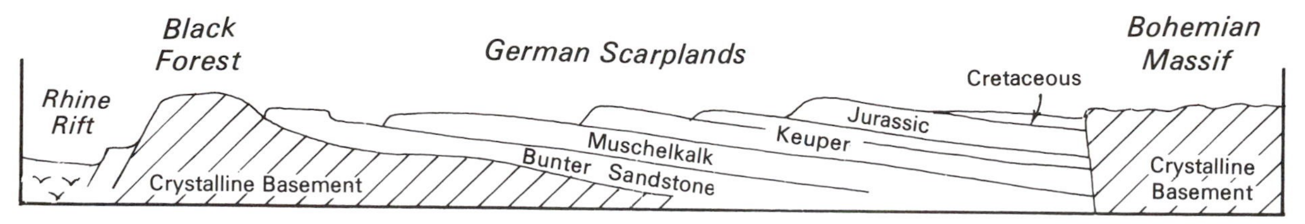


*Figure 8.19.* Section of the German scarplands.

*Figure 8.20.* Jurassic scarp, near Fridingen, West Germany.

the south-east, it is an area where cuestas have developed with a relief range of 200 to 1000 m above sea level. The southern part of this area is drained by the rivers Necker and Main and the northern area is drained by the headwaters of the Elbe (Figure 8.18).

The oldest rocks of this area, the Triassic Bunter Sandstones, crop out in the west to form plateaux on the flanks of the Black Forest and the Oldenwald uplands. Eastwards, these sandstones pass beneath the Muschelkalk and the Lower Keuper, both of which make north-west facing scarps (Figure 8.19). The succeeding Upper Keuper makes a major escarpment with an extensive plateau on the dip slope. This also dips eastwards and passes beneath the Jurassic scarp which extends from Schaffhausen on the Rhine to north of Nürnburg (Figure 8.20). In the south-east it reaches its maximum height, 1000 m above sea level, in the Swabian Jura. East and north of the Reis, where there is a 29 km diameter meteor crater, the escarpment continues as the Franconian Alb at the lower elevation of 500–600 m. Karstic features occur in the limestones and there are a number of small volcanic cones, basalt flows and crater lakes present.

## THE PLATEAUX AND LOWLANDS OF CENTRAL EUROPE

Extending across central Europe from the Atlantic to the Black Sea is a series of upland plateaux with intervening lowland basins most of which lie between the north European plain and the Alps, but some have become caught up in the Alpine folding. These areas have not undergone orogenic activity themselves since the end of the Palaeozoic era but they have experienced Mesozoic sedimentation and have been influenced by the neighbouring tectonic activity in the Alps. In general they appear to have acted as resistant massifs around which later sediments accumulated and have been crumpled. It is possible to identify northern and central groups of uplands, the rocks of which have been both strongly folded and metamorphosed. A southern group of massifs became involved in the Alpine orogeny and are surrounded by fold mountains. Unlike the Caledonian and Alpine orogenies, where extensive thrust-faulting and nappe structures developed, some authorities would interpret these Hercynian massifs in terms of vertical tectonics alone. The lack of former sea-floor rocks (ophiolites and greenstones) tends to support this interpretation and vulcanism is associated with the block margins rather than the margins of tectonic plates. Amongst these massifs are scattered lower-lying, downfaulted areas containing younger rocks but with a complementary history of development. The geomorphology of the floor of the western Mediterranean Sea also consists of sunken blocks including the Tyrrhenian Sea and the Malta–Gozo plateau.

There is no completely satisfactory method of arranging the sections of this province, so for convenience they will be described as a northern group including:

Southern Eire
South Wales
South-west England
The English Channel
The Ardennes
The Rhenish highlands
The Mittelgebirge
The Harz massif
The Thuringian basin
The Polish uplands

A central group includes:

The Armorican massif
The Aquitaine basin
The Massif Central
The Rhone–Saône valley
The Vosges, Black Forest and the Rhine valley
The Bohemian massif

A southern group includes:

The Meseta
The Guadalquivir plain
The basin of Old Castile
The central Cordillera
The plain of New Castile
The Iberian cordillera
The Catalonian cordillera
The Ebro basin
The Portuguese lowlands
The Maures and Esterel massifs
Corsica
Sardinia
The Po plain
The Hungarian plains
The Trans-Danubian uplands
The Apuseni mountains
The Rhodope massif

### The northern group

The southern part of Eire and the south-west peninsulas of England and Wales are structurally distinct from the land to the east and north. All three areas were strongly affected by the Hercynian earth movements and represent the westernmost of the faulted blocks which comprise the central uplands of Europe. In his essay on the delimitation of morphological regions, Linton referred to this area, together with Brittany, as part of the oceanic uplands province of Europe. The geological map shows the relationship of these areas to the Caledonian structures of the Atlantic highlands (p. 000).

#### *Southern Eire*

The folded and northward thrust Lower Carboniferous rocks of southern Eire are in marked contrast to the central lowland of Ireland where rocks of similar age lie relatively undisturbed on the Lower Palaeozoic basement. The greatest elevations are in Macgillycuddy's Reeks (1041 m) but the majority of this area lies between 100 and 400 m above sea level. An extensive erosion surface has been identified across the interfluves between the 250 m contour and the coast. The pattern of folding can be seen clearly in the morphology of the south-west peninsulas with the rias of Dingle Bay, Kenmare river, Bantry Bay and Dunmanus Bay between them. Attention has been drawn to the differences in dissection of the Old Red Sandstone ridges between their western and eastern ends. This may partly be a result of the late Pleistocene glaciation of this area by a separate ice cap on the Kerry mountains.

### *South Wales*

Strong folding and faulting of the Devonian and Carboniferous rocks have preserved the synclinorium of the South Wales coalfield, the northern edge of which is marked by the steep, north-facing escarpments of the Pennant Sandstones (Craig-y-Llyn 600 m) and the Old Red Sandstones (Pen-y-Fan 886 m) considered in the section on the Welsh scarplands (pp. 000–000).

Below 180–200 m around the coast of Wales coastal platforms have been identified by accordance of summit level and a clear break in slope profiles below the higher erosion surfaces. These marine abrasion surfaces are particularly well developed in South Wales and occur at 180–200 m, 120–130 m and 60–80 m above present sea level. The possibility exists that these may be pediment surfaces exhumed from beneath a Triassic cover rather than being of marine origin. No marine sediments have ever been found upon these platforms which have been attributed as Pliocene or early Pleistocene in age, but there is evidence of a former Triassic cover.

There are no doubts about raised beach deposits at 10 m above sea level in the Gower peninsula of South Wales. These contain rounded pebbles and fossils, notably *Patella vulgata* which gives the name Patella raised beach to this feature, reliably dated as Ipswichian, 125 000 BP.

### *South-west England*

The south-west peninsula of England, comprising the counties of Cornwall and Devon, constitutes another horst or massif of the central European uplands. The lower Palaeozoic rocks of the area have been strongly deformed by Hercynian earth movements and granite intrusion occurred. Erosion later revealed these granite bosses which form Dartmoor (600 m) and Bodmin Moor (450 m) as well as the less elevated St Austell and Land's End uplands. Exmoor (550 m) still has a cover of Devonian sedimentary strata. The granitisation was accompanied by mineralisation, especially tin, and subsequent kaolinisation of the granite provided the source material for the china clay industry of St Austell moor.

A former covering of Mesozoic sediments has been hypothesised to account for the complex drainage pattern which appears to have been superimposed upon the Palaeozoic rocks. Geological studies of this area have drawn attention to a Pliocene erosion surface at about 130 m which carries patches of gravel and a 'flight' of partial erosion surfaces has been identified by geomorphological investigations in both Devon and Cornwall at 330 m, 250 m, 180 m, 130 m, and 60 m having an accordance of coastal summit levels similar to those seen in Wales.

### *The English Channel*

The English Channel is a downwarped area, with Mesozoic and Tertiary sediments which infill the basin and overlap onto the buried Palaeozoic platform on either side. During the Alpine orogeny these sediments were thrown into a series of folds, one of the largest can be seen in the Isle of Wight, where the Cretaceous beds are thrown into a vertical position. After submergence again during the Pliocene, this low-lying area between England and France became dry land during the Pleistocene and Britain remained joined to the mainland of Europe until about 7500 years before present.

An extensive river system drained the low-lying land, its tributaries included the Solent and the Hampshire Avon on its northern bank and the Somme, Seine and Orne on the southern bank. The famous 'white cliffs of Dover' and the Falaise of France have been formed by cliff retreat as the sea attacked the foot of the rising land to form a cliff and a shore platform. Elsewhere the rising level of the sea in the Flandrian transgression resulted in such features as Chesil Beach in Dorset and Dungeness in Kent as gravel and sand banks were driven on shore and sorted by longshore drift. On the floor of the Channel, near the Straits of Dover sand ridges 7—10 m high and tens of kilometres in length occur, moulded by the force of tidal currents.

*The Ardennes*

The Ardennes of eastern Belgium is composed of folded Palaeozoic rocks which form an upland of less than 500 m above sea level. It is a deeply dissected sequence of synclinoria in which coal measures have been incorporated (the Belgian coalfield) along the line of the Meuse at Namur. This is a complex area in which it is claimed that evidence of several former landsurfaces is preserved at 550 m, 450 m and 400 m and on the northern side only at 300 m above sea level. The summit level represented by elevations of over 650 m is a smooth, peat-covered moorland, the soil of which contains flints implying it has been developed not far below the Cretaceous surface. Silcrete has been found on this surface near Jalhay, probably derived from Oligocene sands and clays now eroded from the plateau. Surfaces attributed to the Miocene occur at 340 to 400 m and two late Tertiary surfaces form the interfluves of the Ourthe, Amblève and Moselle rivers. Strong solifluction occurred during the Pleistocene and redistribution of the erosion products has given rise to river terraces.

*The Rhenish highlands*

The area of the Rhenish uplands includes the uplands of Eifel, Westerwald, Hunsrück and Taunus. Lying immediately to the north is the upland of Sauerland with the Ruhr coalfield on its northern margin. The rocks of this area are mainly slates of Devonian and Carboniferous age which were folded in the Hercynian orogeny, worn down to a peneplain and then uplifted to give the plateaux of 700–800 m above sea level. Some rocks of Triassic age are present on these uplands but Tertiary vulcanity produced peaks such as the Hohe Acht (2447 m) on the Eifel. Effusion of pumice, referred to as 'bims', especially in the Neiuweid basin, gives rise to rather more fertile soils than those on the slate. The Rhine crosses this upland area through a rift valley which is discussed elsewhere, but the presence of a low base level adjacent to these upland areas has resulted in deep dissection of the plateau by the rivers.

The Taunus and Hunsrück are elevated plateaux with quartzites forming ridges extending in a north-east–south-west direction. These plateaux lie on either side of the Rhine rift valley, the Taunus to the north-east and the Hunsrück to the south-west. The southern margin is an abrupt fault scarp which overlooks the Main valley near Wiesbaden (Figure 8.21).

The Westerwald lies between the Lahn and Sieg valleys, north-east of the Rhine gorge and is also an upfaulted horst with a plateau morphology. Tertiary volcanic activity has resulted in a basaltic plateau surface at about 650 m above sea level. South-west of the Rhine rift valley is the Eifel, which is the most varied part of the Rhine highlands. Bunter sandstone and Muschelkalk partly cover the metamorphic Lower Palaeozoic rocks which together form faulted plateaux surfaces between 500 and 700 m such as the Schnee-Eifel and the Hohes Venn. Eruptive materials from over 200 Tertiary volcanic cones and explosion craters has provided a varied and attractive landscape including the Laacher See and the Totenmaar.

The Siegerland and Sauerland plateaux continue the Rhine highlands north-eastwards as far as the Ruhr valley and the Münster embayment. As has been discussed in the case of the other Rhine uplands, the upland surface is considered to have been developed during the Tertiary with the quartzites and Tertiary volcanic rocks standing above the plateau as monadnocks.

*The Mittelgebirge*

North of the German Scarplands is an area drained by the headwaters of the Weser the structure of which is a series of horsts and graben formed from the Lower Palaeozoic rocks and faulted in the Hercynian orogeny. Tertiary deposits occur in the graben and vulcanicity occurred in the Vogelsberg (772 m), the basaltic lavas of which have partly obliterated a northwards branch of the Rhine rift valley. Deep weathering has been observed in some of the basaltic flows which have been eroded in a radial pattern around the mountain. Glacial ice only penetrated as far south as the 'Westphalian Gate' near Kassel, so the area was subjected to periglacial activity during the Pleistocene. The Teutobergerwald consists of two parallel escarpments lying between the Ems and Weser rivers; the northern ridge is formed from Triassic and the southern from Jurassic strata.

*The Harz massif*

The Harz massif lies on the border between the Federal and Democratic Republics of Germany. It is bounded by fault scarps and in its highest part it rises to 1142 m in the Broken massif but its surface morphology is a plateau resulting from remnants of a former planation

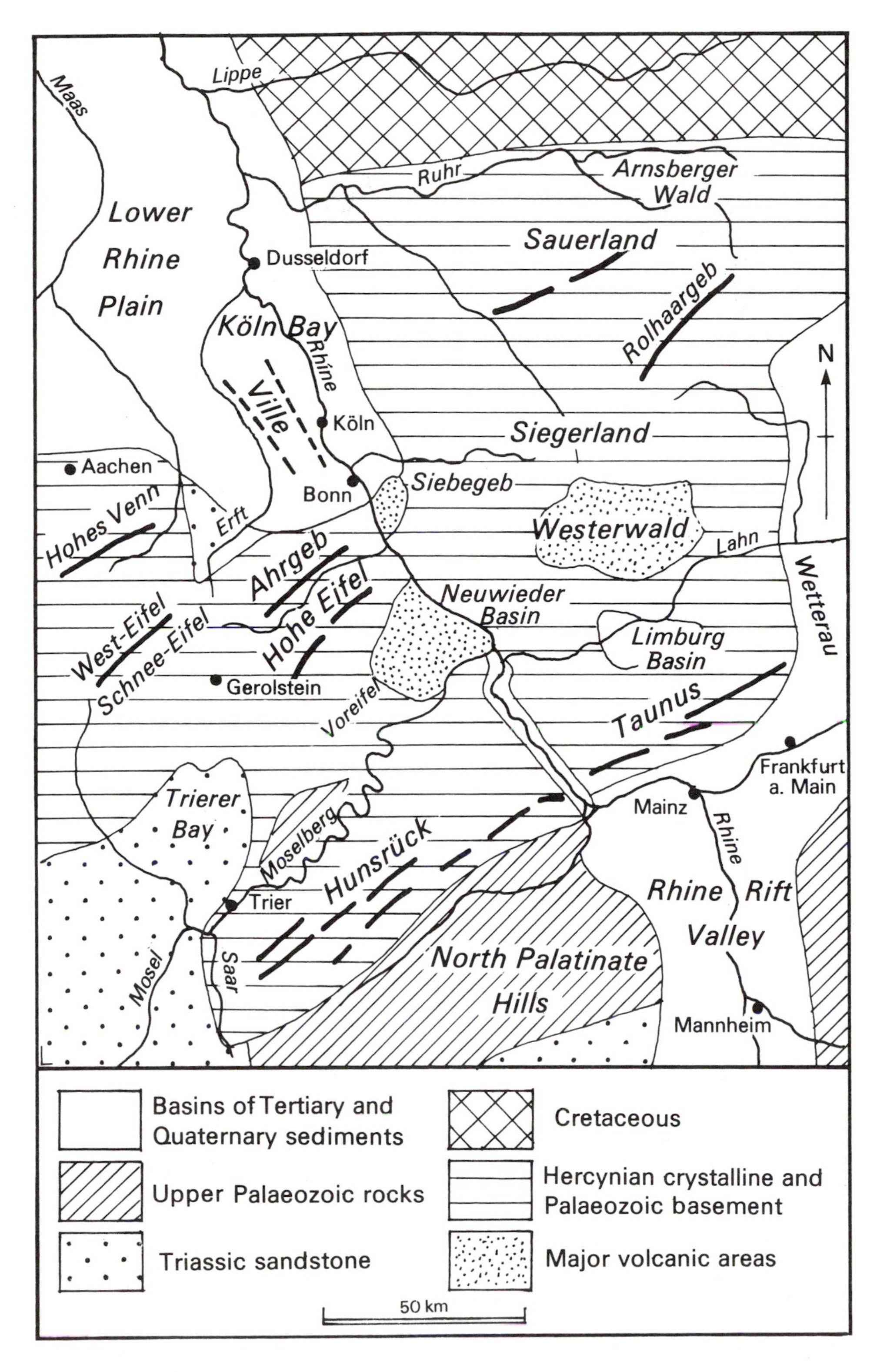


*Figure 8.21.* The Rhenish uplands.

surface. In common with the other areas described in this group of uplands, Lower Palaeozoic rocks were folded in a mid-Palaeozoic orogeny and consist of thick sandstones and shales which have been over-thrust towards the north-west and metamorphosed. Intrusion by granite also occurred but no Tertiary volcanic activity took place. The area experienced partial glaciation during the Pleistocene and the northern slopes are mantled with loess. An interesting salt karst occurs in the outcrop of the Zechstein beds.

### *The Thuringian basin*

South of the city of Leipzig is the Thuringian basin, drained by the headwaters of the Saale, Elster and Elbe. The basin is surrounded by scarps formed by the Bunter and Muschelkalk but its centre is an undulating plain at 200–400 m, developed on Keuper marl. Pleistocene ice just penetrated the basin from the north-west, so fluvioglacial deposits occur on the valley floors.

The Thuringian forest is an upland area which lies to the south of the Thuringian basin. It is bounded by fault scarps and consists of Pre-Cambrian and Lower Palaeozoic rocks intruded by granites. Permian sandstones crop out in the north-western part of the upland. As has been observed with the other uplands, a planation surface has been developed with pockets of deep weathering suggestive of a warmer climate than at present. Subsequent Pleistocene cold conditions produced angular frost debris and solifluction deposits.

### *The Polish uplands*

At the southern margin of the glaciated north European plain in Poland are the uplands of Silesia and the Gory Swietokrzyskie. These uplands consist of a core of Palaeozoic rocks, partially covered by Mesozoic sediments. After Hercynian folding, planation has taken place and, following uplift, river erosion has cut deeply into the upland plateau surface to reveal the Lower Palaeozoic strata beneath. The Lysogory upland rises to just over 600 m on a quartzite ridge strewn with a blockfield after periglacial disruption. North-west of Krakow a Mesozoic escarpment overlooks the exposed Palaeozoic foundations of the region. The Silesian uplands are separated from the Gory Swietokrzyskie by the valley of the Nida river which flows in a synclinal structure of Cretaceous and Tertiary sediments overlain by glacial and fluvioglacial deposits.

## The central group

### *The Amorican massif*

The peninsula of Brittany, which lies to the west of the Paris basin, is an upland of undulating plateaux interrupted by broad valleys. Its eastern margin bisects the Cotentin peninsula and passes east of Angers, reaching the Atlantic coast 75 km south of Nantes. Rocks from the Pre-Cambrian and Palaeozoic are folded and strongly metamorphosed to produce a rigid massif, the structures of the northern part trend east–west (Variscan trend) but those of the southern part of the massif diverge towards the south-east (Amorican trend).

The abrupt northern ridge of the Montagnes d'Arrée (384 m) is formed by Devonian quartzites but elsewhere schists form low plateau surfaces. The Montagnes Noires in the south are mainly granite. The Amorican massif was reduced to a peneplain by erosion during the Tertiary, and some Tertiary deposits, similar to the 'crags' of eastern England, occur on its surface in the central region. Generally the Lower Palaeozoic rocks have been preserved in synclines and intrusive granites form rounded uplands, rising to 400 m above sea level. This upland surface of Brittany tends to decrease in elevation westwards, finally ending in the abruptly cliffed coast. The western part of the Cotentin peninsula has a planation surface at 100 m with granite hills rising above it to 200 m above sea level.

Except for the Aune river, which flows between the two upland areas, the drainage pays little attention to the underlying complex structures, flowing either to the north or south coasts. The coasts are for the most part rocky with sandy bays and the Channel Islands are emergent parts of a submerged portion of the massif. The extreme western coast is characterised by rias such as the Baie de Douarnenez and at Quiberon a tombolo feature extends 7 km in length.

### *The Aquitaine basin*

The basin of Aquitaine lies to the west of the Massif Central and is drained by the Garonne river and its tributaries which reach the Bay of Biscay through the Gironde estuary (Figure 8.22). In south-west France the Pre-Cambrian and Lower Palaeozoic rocks dip southwards beneath an increasing thickness of Mesozoic and Tertiary rocks. At Bordeaux there are 7000 m

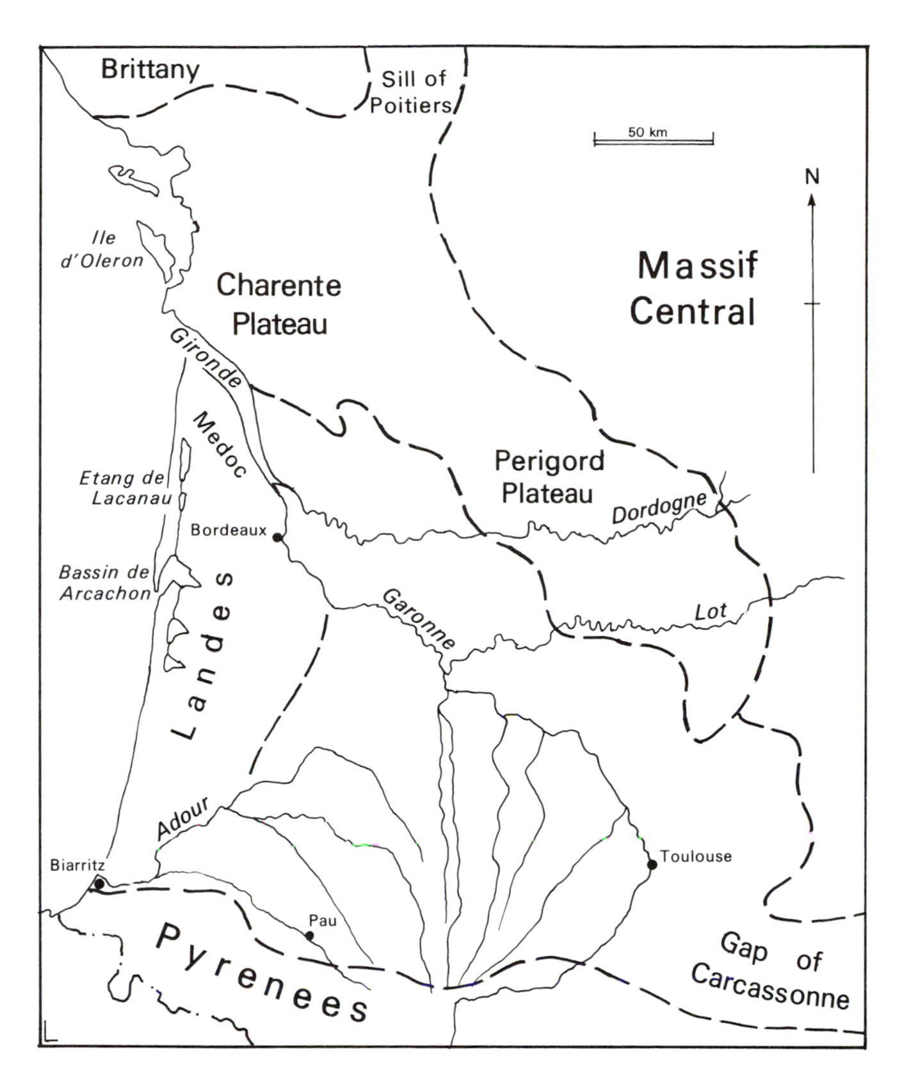


*Figure 8.22.* The Garonne valley.

of these rocks which have accumulated in a downwarp to the north of the Pyrenees. Erosion of the mountains has led to large quantities of sediment reaching the Aquitaine basin. Between Pau and Toulouse a large fan-like pattern of rivers draining from the Pyrenees can be observed on most atlas maps of France. Extending from about 600 m down to 200 m these valleys have cut into Pliocene pebbly deposits which overlie molasse also derived from the Pyrenees. At Toulouse the Garonne has a floodplain at 130–140 m with river terraces at 150-160 m and 200 m. The lower Garonne and Dordogne flow over alluvial sediments covering molasse. Their floodplain is bounded on the north by the Côtes de la Dordogne and on the south by the Medoc peninsula which are formed by Tertiary sands and gravel (Figure 8.22).

From Pointe de Grave at the mouth of the Gironde to Biarritz there is a zone of coastal dunes 8–10 km wide with sand bars and lagoons like the Etang de Lacanau which drain parallel to the coast into the estuary of the Leyre river, the Bassin de Arcachon, which is almost closed by a sandspit extending southwards across its mouth. Older sand dunes and plains extend inland as far as the Garonne in the north and up to 60 km south where they formerly constituted an extensive waste but now are planted with maritime pines.

In the north of the basin of Aquitaine, Jurassic limestones extend from the Paris basin through the sill

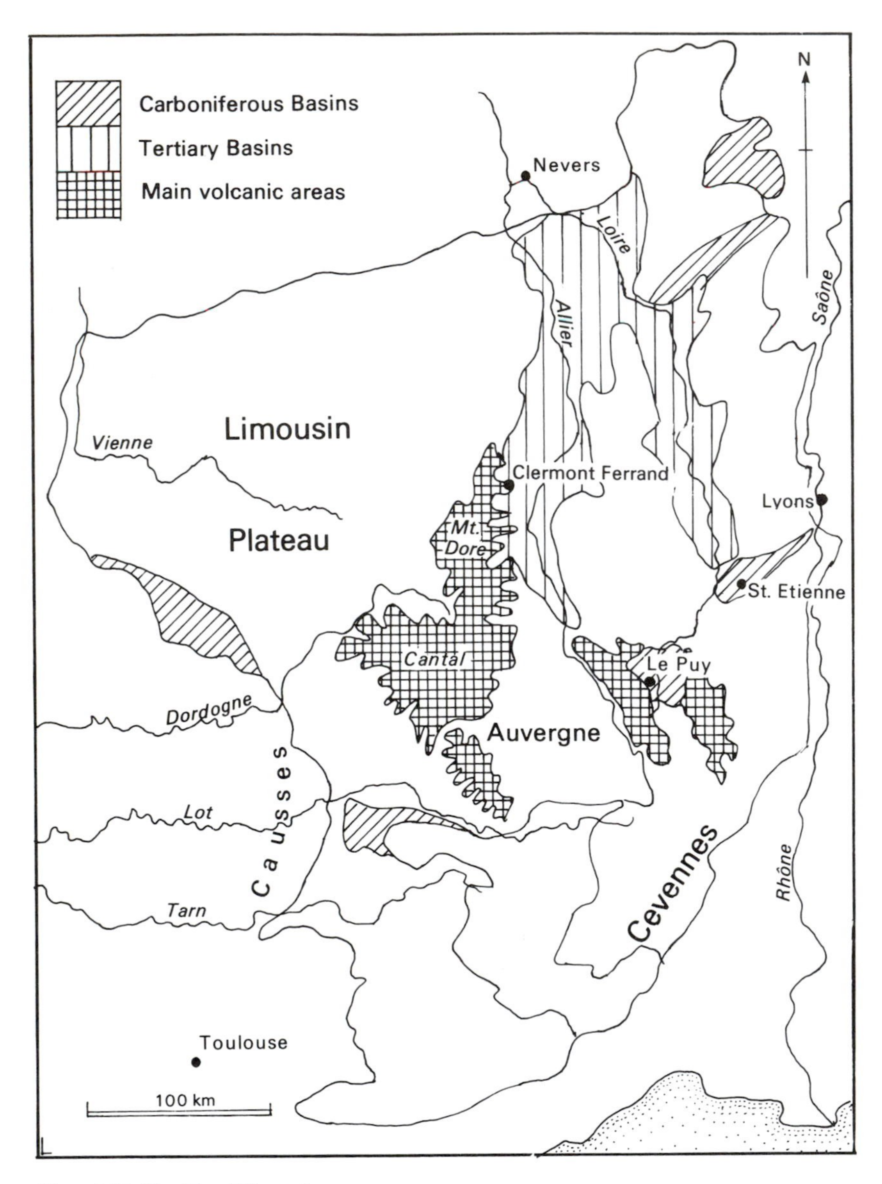


*Figure 8.23.* The Massif Central.

of Poitiers to reach the coast at Ile d'Oleron. In Charente the limestone forms a plateau, 250 km by 100 km, most of which is no more than 300 m above sea level. Inland the limestones extend along the western flank of the Massif Central where rivers such as the Dordogne and Lot have cut deep gorges across the outcrop, and the interfluves form plateaux with precipitous sides. These river cliffs and the caves in them provided shelter for Palaeolithic man.

### *The Massif Central*

The Massif Central rises abruptly from the Rhône–Saône trough where its south-eastern edge is known as the Cevennes (Figure 8.23). Its surface is inclined northwards to the Paris basin and south-west to the Aquitane basin. The plateau of Limousin in the north-west is not fractured by faulting and is the most stable area of the Massif. East of a tear fault which runs from

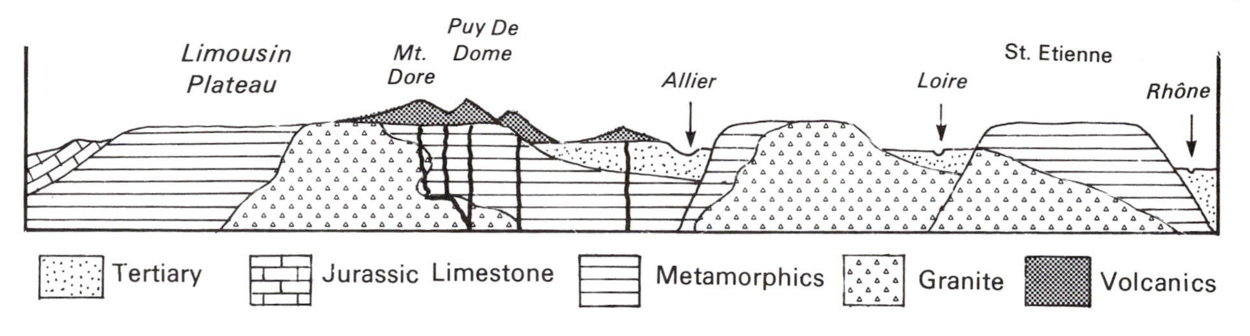


*Figure 8.24.* Section of the Massif Central.

Toulouse to Nevers the Massif is broken by other faults and rift valleys which have influenced the drainage pattern. The geological core of the massif lies in the Auvergne district around Clermont-Ferrand. It comprises strongly metamorphosed pre-Devonian rocks into which granites have been intruded. Coal measures and Permian strata rest unconformably on these older rocks which were folded in the Hercynian orogeny and subsequently uplifted. As a result, Mesozoic rocks only accumulate around the edge of the Massif, and especially in the west where they form the Causses (Figure 8.24).

Peneplanation had taken place by the early Tertiary resulting in a now uplifted surface of about 1000 m. At the time of the formation of the Pyrenees in the early Tertiary, faulting also occurred in the Massif Central producing graben and initiating the present drainage of the Allier and Loire. Volcanic activity began in the Oligocene and continued until the Pliocene. The earlier eruptions were of viscous trachyte lavas which form the steep-sided domes of Le Puy. The now eroded Cantal (1858 m) and Mont Dore (1886 m) are the composite cones of volcanoes which erupted in the Pliocene and fluid basalts accumulated to a depth of 1000 m. The Châine des Puys, a 35 km line of vents, represents the final eruptive activity when small composite cones were built up and lava flowed down adjacent valleys. During the Pleistocene, glaciation occurred on the higher peaks and fluvioglacial streams cut through the lavas to the crystalline rocks beneath, leaving the lavas on the crests of hills. On the western slopes of the Massif Central, the rivers Dordogne, Lot, Viaur and Tarn deeply dissected the Mesozoic limestones to produce the Causse plateau with elevations between 200 and 450 m.

### *The Rhône–Saône valley*

Although the underlying rocks were laid down before the Hercynian orogeny, the present lowland only came into being in the Miocene when the Massif Central was uplifted. Throughout the Tertiary and Quaternary sediment was brought into the valley by rivers from the Alps to the east and the fault scarp to the west. Low-lying areas have been infilled to make plains, but narrow gorge sections remain between Lyon and Avignon at Vienne, St Vallier, Donzere, Mondragon and Roquemaure.

### *The Vosges, Black Forest and the Rhine valley*

The Vosges and the Black Forest lie on either side of the graben which forms the middle part of the Rhine valley along the borders of France and Germany (Figure 8.25). In common with the other massifs of the central European uplands, these are also composed of pre-Devonian rocks which have been strongly metamorphosed and intruded by granites. During the late Cretaceous the area of the Vosges and Black Forest began to experience a broad anticlinal uplift and during the Eocene a planation surface was developed. Rifting of the Rhine valley began in mid-Tertiary and much of the Mesozoic sedimentary cover was stripped, exposing the crystalline core of granite surrounded by metamorphic rocks. The central part of the Rhine valley was rifted along a section 300 km long and 30-40 km wide from the vicinity of Basel to the Rhine highlands.

The crystalline rocks of the Vosges and the Black Forest reach about 1500 m above sea level but further north elevations are lower and the overlying Triassic

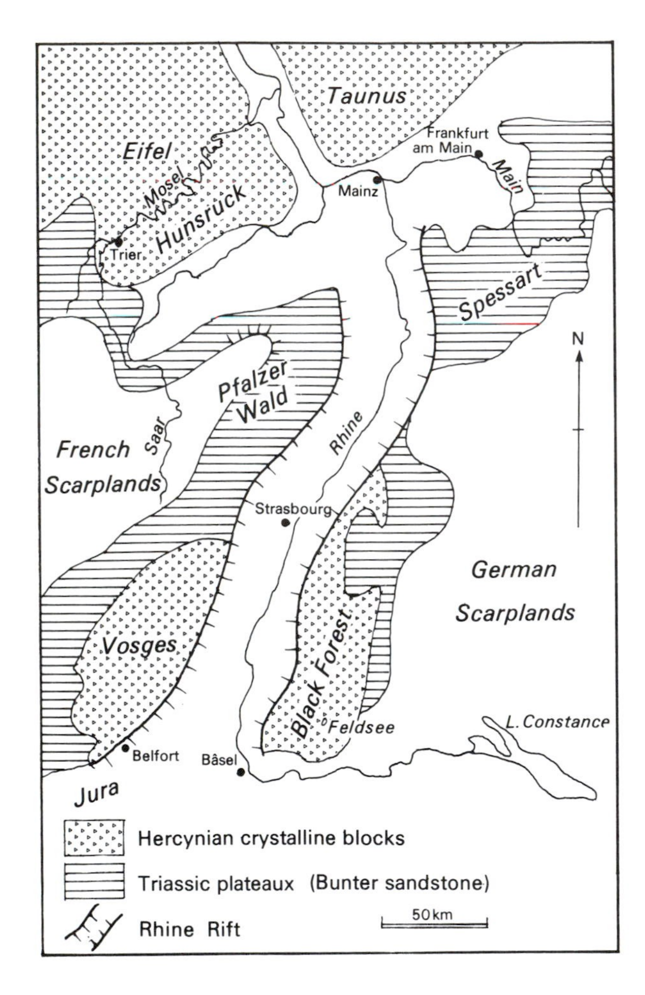


*Figure 8.25.* The Vosges and the Black Forest.

sandstones form lower plateau-like areas in the Pfalzerwald and Spessart districts. Above 900 m glacial landforms occur with corries on east-facing slopes, now containing lakes. Angular scree debris and solifluction features are common. On the Black Forest, the Feldsee Mummelsee are typical corrie lakes but the Titsee and the Schluchsee are moraine-dammed lakes.

In the rift valley, subsidence continued with accumulation of Tertiary sediment including salt deposits. During the Miocene, volcanic activity commenced and the basalts of the Vogelsberg were thrown out across the developing rift valley. It appears this arrested further rifting in this direction and another line of weakness further west became active to form the present Rhine valley across the Rhine highlands. Some volcanic activity continued until the Quarternary and thermal activity is associated with the fault scarps to the present day.

The floor of the rift valley is covered with Pleistocene and Holocene sediments which occur in fluvioglacial terraces and on the floodplain. High (10 m), middle and low terraces occur which decline in elevation from 250 m at Basel to 85 m in the north. However, the inclination of the terraces does not follow that of the floodplain as neotectonism in the Rhine highlands has affected the terrace levels. Despite this uplift, the Rhine appears to have maintained its course, cutting the gorge between the plateau of the Hunsrück and Eifel in the west and the Taunus and Westerwald in the east.

### *The Bohemian massif*

A similar geological history can be described for the Bohemian massif which forms the western part of Czechoslovakia, centred on the city of Prague. The core of this massif is the Pre-Cambrian, Archean and Proterozoic metamorphic rocks which were reduced to a peneplain before Lower Palaeozoic sediments were laid down. After the deposition of Devonian limestones the area was disrupted by earth movements. Subsequently fluvial processes developed the Moravian karst. The Bohemian massif remained as an upstanding area throughout most of the Mesozoic until it was partially submerged during the Cenomanian transgression. Sedimentation also took place in the Tertiary and Miocene vulcanism erupted along the north-western edge of the massif on the eastern flank of the Erzberge mountains. These mountains have a general elevation of 800 m but rise to just over 1200 m on the highest peaks. They are steepest on the Czech side where fault scarps cause the surface to drop rapidly into the Bohemian basin; the north-western slopes are more gentle. The Bohmerwald is the uplifted scarp edge of the ancient rocks of the massif on the south-western side. Glaciation affected the mountains, particularly in the Elster and Saale stages of the Pleistocene and periglacial phenomena occur including patterned ground, solifluction terraces and asymmetric valleys. Loess is also present.

## The southern group

The rectangular Iberian peninsula has an area of 596 718 km² and extends 750 km from north to south and the same distance from east to west (Figure 8.26).

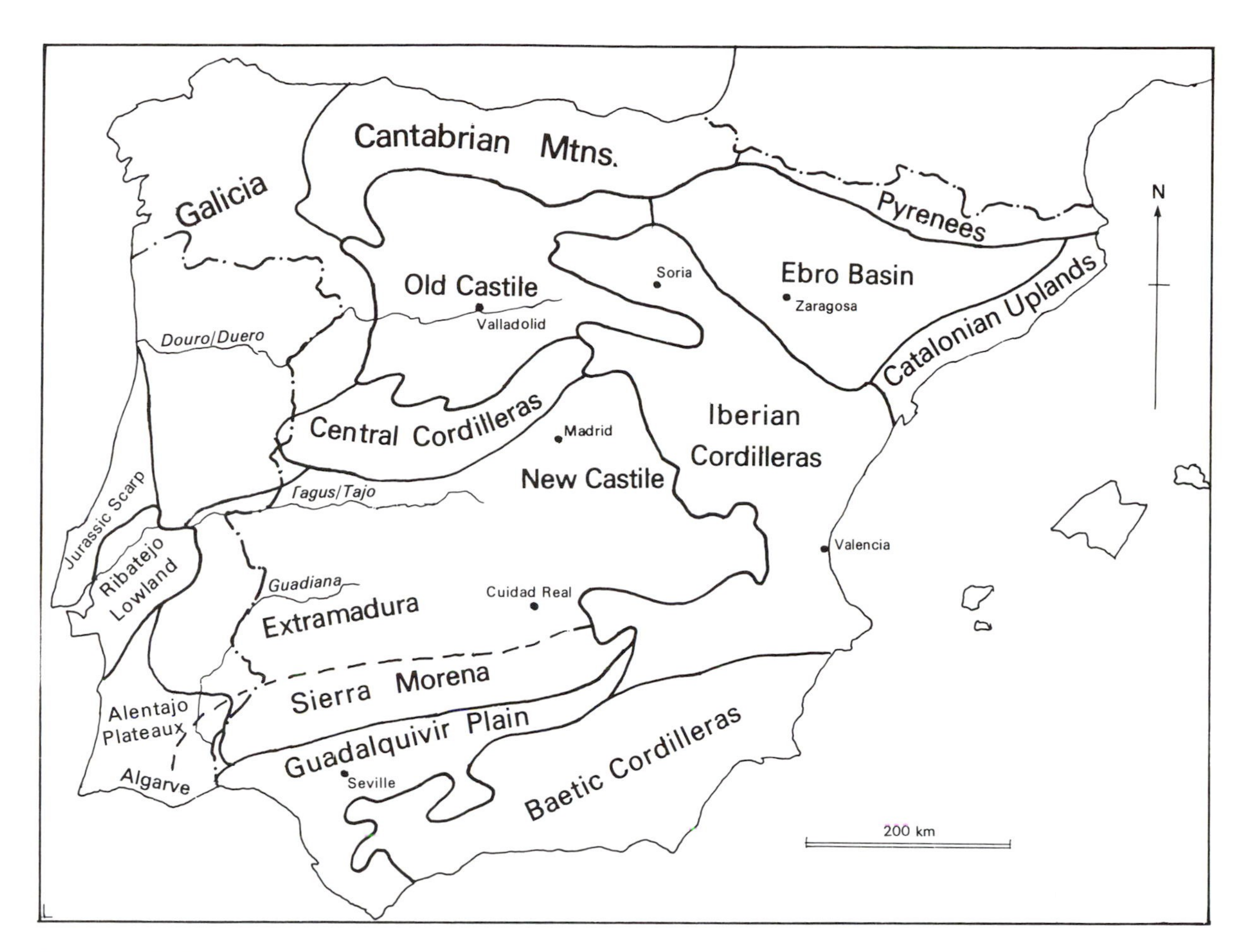


*Figure 8.26.* The Iberian peninsula.

Within the peninsula, the largest geomorphological unit is the Meseta, one of the Hercynian horsts which provide the basis of the landforms of the central European uplands. The geological foundation of the Meseta is to be found in the Cambrian slates and Silurian conglomerates of Galicia. This area was peneplained in the Carboniferous, but the trend of the Caledonian structures can still be seen. Extensive intrusion of granites also took place during the Caledonian orogeny. After block-faulting during the Hercynian orogeny the Meseta remained above sea level during the Mesozoic but the mid-Tertiary orogeny provided the basis of the present relief. The downfaulted basins were infilled with sediments to form the extensive plains of Old and New Castile. Although much of the surface of the Meseta is a plateau which lies between 500 m and 1000 m above sea level, ranges of mountains rise above it to give peaks of over 2000 m in the Sierra de Gredos, the Sierra de Guadarrama and the Cordillera Cantabrica. The basins of the Ebro and the Guadalquiver are downfaulted regions which may be identified as separate geomorphological regions. The Pyrenees and the Baetic cordillera (Sierra Nevada) are discussed in the Alpine highlands province (p. 000).

## *The Meseta*

In the north-west of the Iberian peininsula, north of the Duero river, is the plateau of Galicia with an average height of 500 m above sea level. It is dissected by rivers with narrow valleys often strongly influenced by the Caledonian structural trend and the faulting pattern. Typical granite landforms are not well developed; where granite does occur it forms broad domed interfluves. There is a summit plain which rises to about 2000 m on the highest mountains which is attributed to an oligocene planation. Below this there are distinct erosion surfaces at 800–900 m, 600–700 m and 500 m on the interfluves of the Miño valley. In some valleys a late Tertiary deposit overlain by coarse gravels in a red clay forms a broad plain at about 200 m. It is thought to have originated as a pediment deposit of early Pleistocene age. The valley floor of the Miño lies at 200 m near the coast rising to 500 m inland. The coast is one of cliffs with rias.

The Cantabrian mountains or Asturian massif of northern Spain were formed by the Hercynian folding and block-faulting of Palaeozoic sediments. Two ridges parallel to the coast with an elongated trough between them provide the northern edge of the Meseta at 300–500 m, rising in the mountains to 2042 m in the limestone horst of the Picos de Europa. Where limestone rocks crop out karstic features occur; the Comerza polje is 2.5 km long by 1 km wide. Inland the older rocks are progressively covered by Tertiary sediments which form the plateau landscape around Valladolid in Old Castile. As these mountains are near the coast they are strongly dissected by rivers which descend steeply to the sea. The coast is formed of quartzite headlands and bays occur where shales crop out.

Glacial features occur down to 1000 m above sea level and features are fresh which suggests they are of the last major glaciation. On the Picos de Europa, glaciers were up to 7 km in length, leaving spreads of boulder clay and moraine dammed lakes.

### *The Guadalquivir plain*

This lowland in the southern part of Spain is triangular in shape, 300 km from east to west and 200 km across at its western end. Its structure is interpreted as an alpine foredeep with a sharply plunging northern side and an irregular southern side where nappes from the Baetic cordillera rise out of the alluvium. The foredeep is infilled with sediments of Oligocene to Pliocene age which become thicker towards the west. The plain may be sub-divided into two parts: an upper valley east of Seville and the lower valley.

In the upper valley there is an extensive early Pleistocene pediment around the foot of the surrounding mountains. Terraces are present along the rivers, an upper one has gravels and lies at 80–170 m above sea level. There are two lower levels cut into this upper terrace which is always decalcified. The middle terrace lies between 50 and 80 m and has a red soil with a travertine crust. The lowest terrace at 20–50 m has a black, hydromorphic soil.

The lower valley is also characterised by pediments extending from the surrounding hills. However, much of the valley downstream of Seville is an infilled estuary covering 200 km$^2$. The actual coast is formed by sandbars with salt marsh accumulating on the landward side.

### *The basin of old Castile*

Old Castile is the name given to the plateau at 850 m above sea level which is drained by the Duero river. In the western part of the basin Palaeozoic rocks crop out and form a plain at 700-900 m with red clayey soils. In the centre of the basin the soils are developed from a horizontally bedded, white gypsiferous marl. There are two major levels in this plateau landscape; a higher surface, the Paramo, developed between 850 and 1000 m and a lower surface, the Campina, which occurs as wide, flat-floored valleys. Three pediment levels are present on the Paramo surface but only one of the lower Campina surface.

As the Duero (Douro) passes into Portugal it drops rapidly in height through a gorge 300–500 m deep where the plateau surface is being incised. In its lower course the Duoro has terraces at 80–130 m, 60–65 m and 12–40 m.

### *The central cordillera*

North of Madrid the two plateaux of the Meseta are separated by an east–west aligned range of mountains, 50–60 km wide and reaching 2430 m on the Pico de Peñalara and 2592 m on Pico de Almanzor. The mountains have gentle slopes on their northern side but to the south there is an abrupt descent to the plains of New Castile (Figure 8.27). In structure block-faulted, these mountains are formed from metamorphosed sediments of Lower Palaeozoic age. Mesozoic rocks covered the whole range at an earlier geological period as a central rift valley in the Sierre de Guadarrama contains Cretaceous rocks. After uplift in the Miocene, erosion has dominated the development of landforms. Extensive pediments occur between 900 and 1100 m and planation surfaces occur at 1500–2200 m on the summits. An extensive 'Meseta' surface occurs at 1150–1250 m. North of Madrid a valley floor surface is developed at 850-950 m.

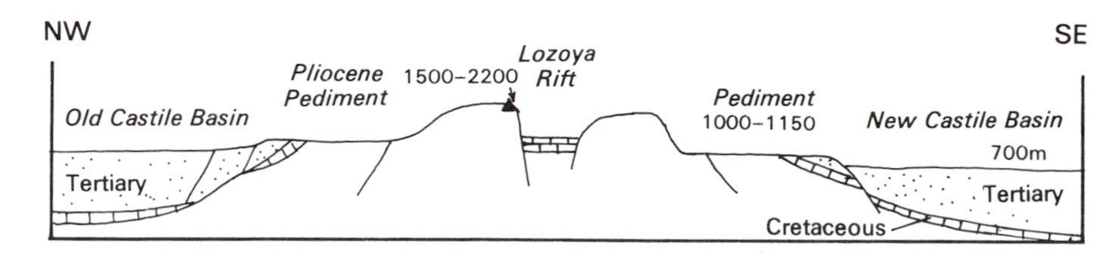


*Figure 8.27.* Section through the central cordilleras of Spain.

Glacial features of local significance occur in these mountains. In the western Sierra de Estrella they extend down to 1620 m but in the eastern, drier areas, glacial features only extend down to 1900 m. The glacial features include small corries, two sets of terminal moraines and fluvioglacial deposits.

### *The plain of New Castile*

The southern part of the Meseta rises in elevation from 200–300 m above sea level in Alentejo in Portugal and Estramadura in western Spain to an average height of 1200 m in the upper Tajo and Guardiana basins. The plain of New Castile is a peneplained area which is subdivided into two by the Monters de Toledo. The southern margin of the Meseta is marked by the Sierra Morena from which relief drops abruptly into the Guadalquiver basin.

The lower western plateau of Estramadura is separated from the plain of La Mancha by the volcanic district of the Campo de Calatrava, near Cuidad Real. Thermal springs are still active here, but from Upper Miocene to Pliocene times volcanic activity took place leaving domes, cones, crater lakes, and lava flows.

The plain of La Mancha is a flat plain with level-bedded strata and virtually no incision has taken place, but downstream the valley of the Tajo is similar to that of the Duero in that it is incised into a plain with surfaces similar to the Paramo and Campina of Old Castile. There are four or five river terraces along the Tajo developed on 4–5 m of alluvial infilling. Some karstic features occur in the Upper Tajo valley. Arid conditions become apparent in the Campo de Criptana, a tributary valley of the Guardiana where wadis, salt lakes and areas of internal drainage occur.

### *The Iberian cordillera*

The eastern margin of the Meseta occurs south of the Ebro river and the town of Zaragoza in the elevated lands known as the Iberian cordillera. Although reactivated by Alpine earth movements, these uplands are part of Hercynian Europe. Rocks up to the Carboniferous are incorporated in folding and subsequently were covered by only slightly disturbed Mesozoic and Tertiary sediments. The older rocks protrude through the younger Mesozoic and Tertiary rocks which form the lower ground. Where the older rocks crop out, gently rounded summits occur at 1900–2100 m with an erosion surface cut at the lower level of 1300–1500 m. Near Soria Mesozoic rocks are thrown into a sequence of regular anticlines and synclines, but south of the town tectonic disturbance is not so great. Pedimented slopes occur around the mountains including several levels aged from early Pleistocene and younger. The northernmost of these sierras had glaciers especially where relief was over 200 m and there are numerous corries and U-shaped valleys with moraines as evidence.

### *The Catalonian cordillera*

Similar comments may be made about the Catalonian cordilleras, parallel to the coast and north of Valencia. These uplands separate the Ebro basin from the western Mediterranean and the present course of the Ebro through the mountains only came into being following the Miocene earth movements when the formerly enclosed basin became linked to the Mediterranean.

These mountains are a mixture of block-faulting and folding in which Eocene sediments were folded and thrust westwards in the Oligocene. Tilted planation surfaces occur at 100 m, and 500–700 m above sea level and five partial erosion cycles have been identified on the Montsony massif. Typical granite landforms occur in the northern part of these mountains and karstic features are present in areas of limestones. Nivation and other periglacial processes occurred during the Pleistocene on the higher mountains.

*The Ebro basin*

The Ebro basin is a triangular-shaped trough infilled with Tertiary sediments, mainly Oligocene, resulting from the erosion of the Pyrenees which were uplifted in the Eocene. The surface features of the Ebro basin are the extensive Quaternary terraces and current flood-plain deposits. The Ebro has constructed a small delta (28 000 ha) at the coast after its passage through the Catalonian uplands to the Mediterranean.

*The Portuguese lowlands*

The Portuguese lowlands lie to the west of the Meseta, separated from the plateau by a major fault scarp (Figure 8.28). The lowlands are formed from Mesozoic strata which have been folded slightly to produce a scarped landscape of up to 700 m, but mostly below 500 m above sea level. Dissected Tertiary rocks form the Ribautejo lowland south of the Tagus but the Alentejo plateau, a lower segment of the Meseta plateau, extends to Cape St Vincent. The Algarve coastal fringe is formed from Tertiary sandstones which give interesting cliffs with stacks; Mesozoic limestones crop out further inland. From Lisbon, north-west to the Berlengas islands, Tertiary volcanic activity occurred on the limestone escarpment north of the Tagus. Parts of the western coast of Portugal are low-lying with sandbars and spits enclosing lagoons. However, where solid rocks approach the coast, cliffs occur.

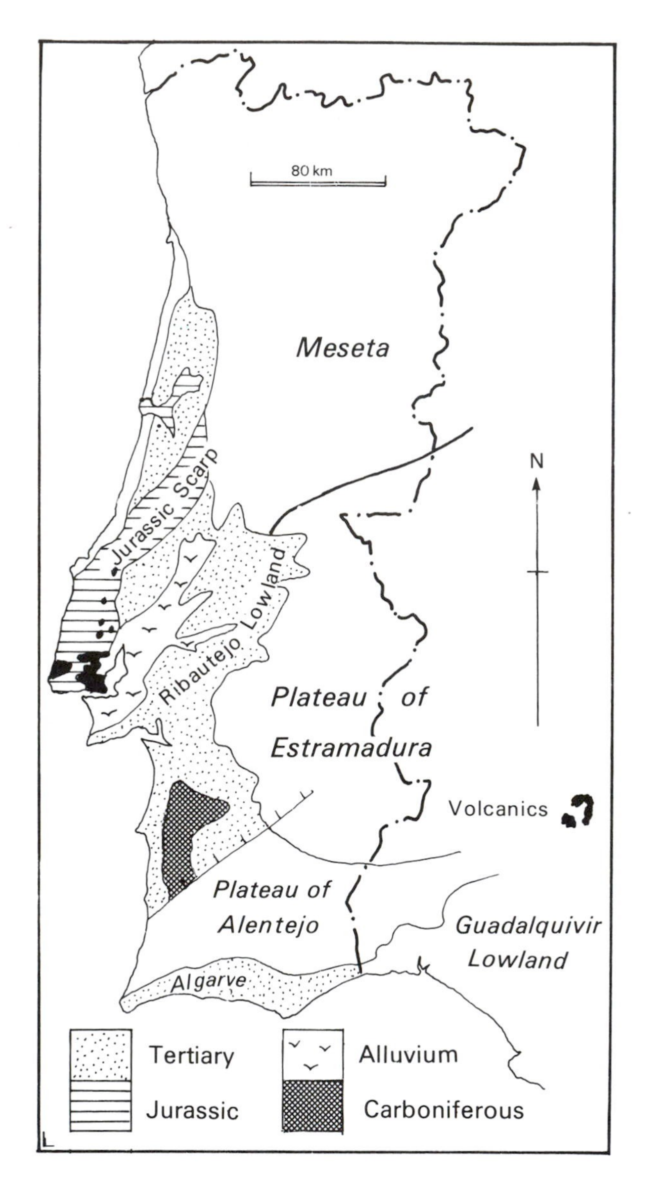


*Figure 8.28.* The Portuguese lowlands.

*The Maures and Esterel massifs*

Between Toulon and Cannes on the Riviera coast of France the attractive rocky coast with small bays is formed by the massifs of Maures and Esterel. The smaller Esterel lies east of the Argens river, but neither massif reaches much over 500 m above sea level. These two massifs are largely composed of gneisses and mica-schists with intrusions of granite cropping out north of St Tropez. Strong faulting was introduced during the Hercynian orogeny which gives a north–south grain to the country. Volcanic activity during the Tertiary was greater in the Esterel and most of the coast between Frejus and Cannes is composed of lavas. These lavas apparently came from a source lying to the south, possibly Corsica before it was moved southwards and turned through 90°.

*Corsica*

The island of Corsica may be sub-divided into three sections with distinct landforms. The western two-thirds is composed of granites, deeply dissected by streams which have exploited the north-east to south-west faulting which also gives the rocky headlands and bays of the west coast. The highest mountains, Cinto (2710 m) and Rolondo (2625 m), occur in this part of the island, the former consisting of volcanic rocks of similar age and composition to those in the Esterel massif of the French Riviera. Frost action is common above 1000 m and evidence of Pleistocene glaciation occurs above 2500 m in the form of corries, and glaciated valleys with moraines extending down to 1000 m.

In the north-east, Mesozoic sediments were metamorphed into schists during the Alpine orogeny; these rocks give rounded mountain scenery rising to a maximum of 1767 m in Monte San Pedrone. Downfaulted limestones give the karstic plateau of Bastia.

The third section of Corsica is the lowland of Miocene sediments along the east coast which is covered by alluvial fans developed during the Quaternary. The east coast, unlike the rest of the Corsican coast, is sandy and lagoons enclosed by sandbars occur north of the mouths of the Galo and Tagnone rivers.

### *Sardinia*

Sardinia is a mountainous, rectangular island the highest peaks of which do not exceed 2000 m and are mostly less than 1000 m in height. The greater part of the island is composed of Hercynian granite intruded into Lower Palaeozoic schists. Where the granite crops out tors occur but the schists form the highest land culminating in Monti del Gennargentu (1834 m). These older rocks are partly covered by Mesozoic limestones which crop out around the Gulf of Orosei on the east coast where they form strong escarpments, and in the south-west where they have mineralised. An early Tertiary fracture cuts across the south-west corner forming a fault trough (the Cixerri) which has partly been infilled with erosion debris. Tertiary volcanic activity has resulted in volcanic cones and a lava plateau north of the Gulf of Oristan on the west coast. The Campidano is the largest lowland plain of Sardinia; it is formed by a graben infilled with Pliocene sediments and covered by Pleistocene alluvial deposits.

Both Corsica and Sardinia appear from palaeomagnetic evidence to have migrated southwards from the Esterel and turned in an anti-clockwise direction. If this is so, the Tertiary fracture across the south-west part of the island originally might have been part of the faulting associated with the Rhone-Saône valley.

### *The Po plain*

The Po lowlands are a synclinorium which lies in the concave side of the Alps. Its complex structure can be seen in the anticlinal Monteferrato hills east of Turin. These structures gradually disappear beneath the Tertiary and Quaternary sediments of the Plain of Lombardy. Along the northern side of this lowland, beyond the Alpine front and the limit of glaciation, extensive outwash terraces coalesce to form the more elevated parts of the plain upon which Milan is situated. Between higher and lower terraces a spring line occurs (fontanelli).

Ancient deltaic deposits of the Po lie inland of the present delta and seaward of Lake Comacchio. In medieval times the delta moved north to accumulate between Comacchio and the present delta. The coast is of low elevation with sandbars enclosing lagoons, the most famous of which is the Venice Lagoon, 50 km north of the mouth of the Po river.

The structure of the Po valley is that of a foredeep in front of the Appennine folding. It has been a subsiding part of the Earth's crust since the mid-Tertiary so that in the vicinity of Rimini the base of the Pliocene is already 7000 m below sea level. The great downwarp has been infilled with the sediment eroded from the Alps as they were uplifted. This foredeep continues along the eastern side of Italy, but crosses the 'heel' of Italy along the Bradano valley.

### *The Hungarian plains*

The Pannonian basin is surrounded by mountains, the Alps to the west, the Carpathians to the north and east and the Dinaric Alps to the south. The basin is a feature which has developed since the Alpine orogeny, subsiding gradually as the surrounding mountains were uplifted. More than 4000 m of sediment were laid down in the basin between the Miocene and the Quaternary. The basin is sub-divided into the Danubian and great Hungarian plains (Figure 8.29).

The Danubian or little Hungarian plain is 100 m above sea level and extends for about 100 km east–west and 150 km in a north-east–south-west direction. The Danube breaks through the Little Carpathian mountains at Bratislava and has laid down an alluvial fan, extending 100 km downstream of Komarom, diverting the course of the Vah river eastwards. The Raba, a south bank tributary of the Danube also has laid down an extensive fan as it enters the southern part of the Danubian plain.

The great Hungarian plain is 230 km from east to

*Figure 8.29.* The Hungarian plains.

west and 450 km from north to south and the elevation of most of the plain is less than 100 m; at the Iron Gate gorge where the Danube leaves the Pannonian basin it is at 36 m above sea level. The geomorphology of the great Hungarian basin is dominated by the Danube and Tisza rivers. The interfluve between them is only 50 m above their floodplains and is dune-covered. The Danube has a floodplain 20–30 km wide south of Budapest and the lack of gradient on the Tisza results in many meanders. Between Budapest and Miskolc several alluvial fans border the plain and in the northeast of Debrecen another large fan has been laid down where the Tisza enters the plain. In the south of the basin the Drava and Sava rivers are important tributaries of the Danube.

*The Trans-Danubian uplands*

The Trans-Danubian uplands extend across the Hungarian plains in a north-east–south-west direction, dividing it into the Danubian and great Hungarian plains (Figure 8.29). These uplands, which cross the Danube north of Budapest, are block-faulted Palaeozoic rocks with upper Cretaceous limestones and remnants of slight volcanic activity. Hot springs occur, some of which are in the city of Budapest where they are used as thermal baths. The general elevation of the Trans-Danubian uplands is 500 m, rising to 704 m in Koris Hegy. A summit level is apparent and below it remnants of former pediment surfaces and valley benches occur. Present-day pediments occur around the block-faulted uplands and periglacial features occur on their summits. Lake Balaton (106 m) occupies a graben between the Trans-Danubian uplands and the Samogy hills, north of Pecs.

*The Apuseni mountains*

To the east of the great Hungarian plain, and south-west of the Romanian town of Cluj, the basement complex crops out in the Apuseni mountains of Romania. These mountains lie within the great bend of the Carpathians and represent an uplifted fragment of the Hercynian landmass. They reach 1848 m elevation in the western Bihor mountains, but mostly the land lies between 1000 and 1500 m above sea level. These mountains were affected by an orogeny during the latter part of the Cretaceous when nappes were thrust from the south. They have since largely been eroded to reveal metamorphic rocks and granite intrusions. The nappes were from a eugeosynclinal situation where flysch deposits were wedged up from the ocean floor at the site of a trench. Tertiary volcanic activity and associated faulting also affected this area, the faults forming the present river valleys. The Transylvanian plain has a similar geological history to the Hungarian plain but it is more elevated, having an average height of 200–450 m above sea level.

### *The Rhodope massif*

Most of the rocks of this massif are thought to be of Pre-Cambrian age and there is much metamorphism of the original sediments. It became an emergent feature in the Hercynian earth movements, and has been successively affected by orogenies including the Alpine when it received its last major uplift. It has been considered to be an ideal example of a zwischengebirge as it lies between the outward-thrust Balkan ranges and the Hellenide ranges. The Rhodope massif rises to 2925 m in Musalla Peak, 60 km south of Sofia and is the highest peak of the Balkans, but little information is to hand on its detailed geomorphology.

## The Alpine highlands

The mountain ranges formed during the Alpine orogeny extend across southern Europe from Spain to the Caucasus in the Soviet Union and beyond into Asia. Two lines of major folding may be discerned with Hercynian horsts and graben lying between. The northern line of Alpine folding can be traced from the Sierra Nevada of southern Spain (the Baetic Alps), extending through the Balearic Islands to the Alpes Maritimes of Provence and the great arc of the Alps in Switzerland and Austria. This line of fold mountains continues eastwards in the Carpathians around the Pannonian (Hungarian) basin, crossing the Danube at The Iron Gate and doubling back across Bulgaria to the shore of the Black Sea. This structural alignment appears to continue in the Pontic range of northern Turkey, but it may be related to the folding in the southern Crimea and the Caucasus in Russia (Figure 8.30).

A southern line of fold mountains may be traced along the coast of North Africa in the Atlas mountains (p. 34) which crosses into Sicily and Italy to form the Appennines, returning southwards through Yugoslavia and Greece as the Dinaric and Hellenic mountains respectively. This structural feature then curves round through Crete to link with the Taurus folding of Turkey (p. 153).

Between these folded lineaments lie several horsts and graben of older rocks, blocks of Hercynian Europe which became caught up with the Alpine orogeny. The graben have become infilled with sediment from erosion of the Alpine mountains and the upthrust blocks, which lie between the folded sedimentary rocks of the Alpine folding. An example of a horst is the Rhodope massif of the Balkans and of a downfaulted graben, the Hungarian plain.

A large gulf, the Tethys Sea, had been present between Eurasia and Gondwana since the end of the Palaeozoic. However, gradual rotation of the Eurasian landmass towards Africa almost closed this gulf but a second Tethys ocean developed by sea-floor spreading. Towards the end of Triassic times, this neo-Tethys sea had opened and accumulation of sediments had begun on the trailing margins of the bordering continents. Foundering then occurred to give deep water conditions and the accumulation of shales in the early Jurassic followed by calcareous sediments.

Marine sediment from the surrounding continents (flysch) became interleaved with fragments of ocean-floor material (ophiolites) at plate-consuming margins during the Cretaceous, accompanied by vulcanism similar to that of the Pacific island arcs today. In the period between the Eocene and the end of the Oligocene, interaction of Africa and Eurasia caused uplift and produced the nappe structure of the Alps. The approximate age of the main Alpine earth movements is given in Figure 8.30. Erosion of the rising landmass gave rise to continental clastic deposits (molasse) which were deposited in lowland areas north and south of the mountains, eventually reducing the mountains to a peneplain. Miocene uplift of the Alps was accompanied by down-faulting of the graben of the western Mediterranean which became desiccated, leading to the deposition of thick salt beds. The connection with the Atlantic only became re-established in the early Pliocene.

The impact of the Alpine orogeny was felt over a

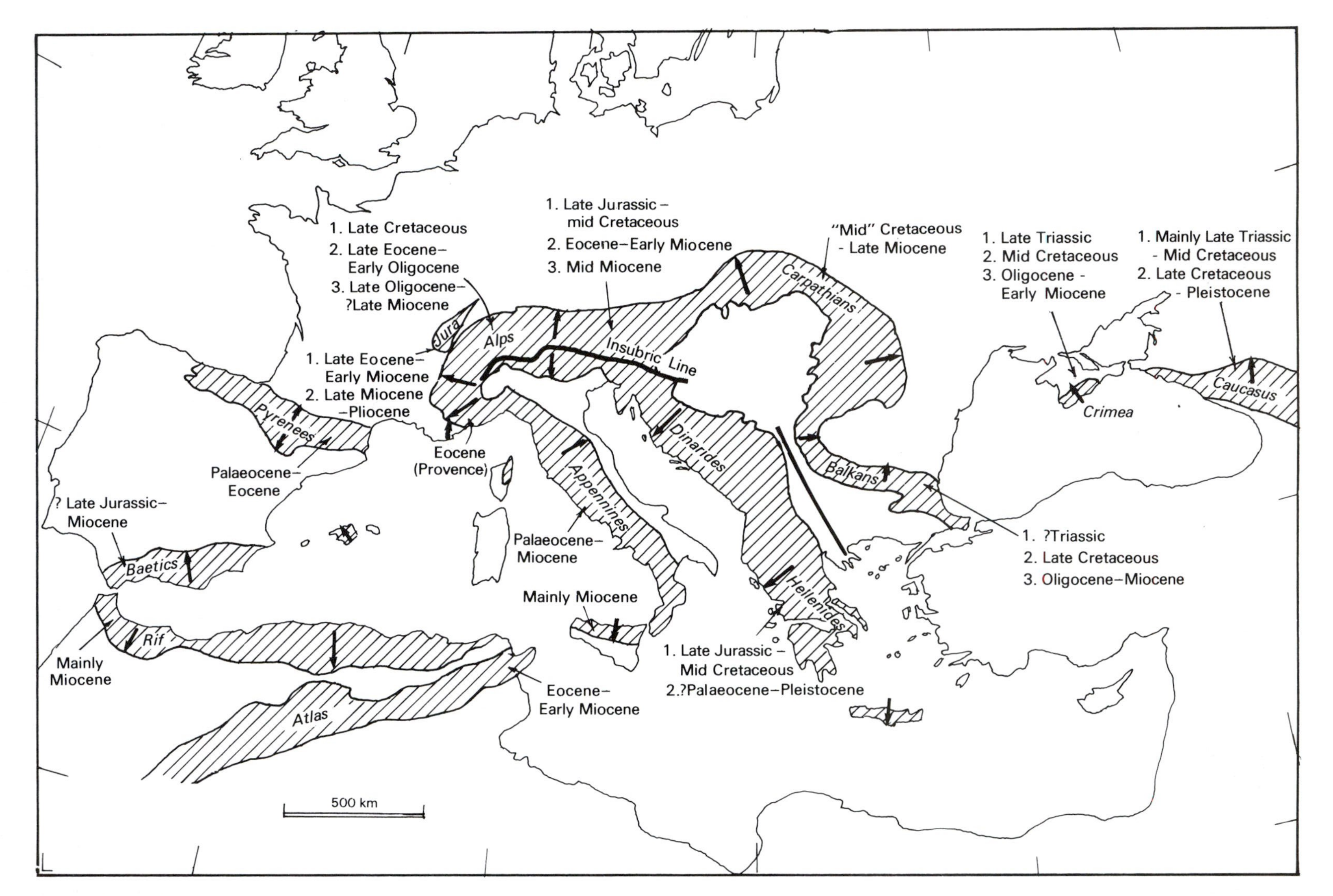


*Figure 8.30.* The Alpine mountains of Europe.

large area of Europe. Former Hercynian tectonics were reactivated in the Pyrenees and ripples extended well away from the mountains themselves. Folding of the Cretaceous strata on the Isle of Wight into a sharp fold with a vertical limb took place at this time and the more gentle synclines of the London, Hampshire and Paris basins and the anticline of the Weald were brought into being.

In the eastern Mediterranean basin, subduction of the Mediterranean floor continues to take place beneath the Aegean plate as it is forced southwards by the westward motion of the Turkish plate. This activity explains the vulcanicity present in the Aegean area where an explosive eruption of the island of Santorini took place in about 1450 BC. The ocean trench lies off the southern shore of Crete and further south a ridge on the floor of the abyssal plain suggests an active spreading centre may be present, creating new sea floor. In this respect, the eastern Mediterranean basin is very different from the western basins which have a thick sedimentary cover and until Miocene times were above sea level.

Today, the Alps are rugged with sharp glaciated peaks and deep valleys, but despite their complex structure, with faulting and over-thrusting, the area had been reduced to a peneplain by the end of the Pliocene and was then uplifted again. The present rugged relief was produced by erosion following this uplift. Originally the rivers of the Alpine region were thought to be antecedent, maintaining their courses as the mountains were uplifted. However, recent studies have questioned this assumption. Observations indicate that rivers such as the Rhone and Reuss may have been fixed in their courses by structural/lithological pattern and not as a result of superimposition or antecedence.

Glaciation had a profound effect on the landforms of the Alpine ranges which supported ice caps during the Pleistocene. Several of the high mountain valleys of Switzerland still contain valley glaciers nurtured by small ice caps and glacial features are typical of the summits. The highest peaks are Mont Blanc (4807 m), Monte Rosa (4634 m), Dent Blanche (4505 m), Matterhorn (4477 m), and Jungfrau (4158 m), all of which occur in the Swiss Alps. The Austrian Alps have slightly lower summits in the Grosser Glockner (3798 m) and the Wildspitze Alpen (3774 m).

The geomorphology of the Alpine highlands province will be considered under the following headings:

The Pyrenees
The Baetic cordillera
The Balearic Islands
The Alps Maritimes
The Alps
The Jura
The Carpathians
The Appennines
Sicily
The Dinaric Alps
The Hellenic mountains
The Cretan arc
The Balkan highlands
The Crimea
The Caucasus

### *The Pyrenees*

The Pyrenean mountains lie between the Bay of Biscay and the western Mediterranean basin forming the boundary between France and Spain. The easternmost limit is at Cape Creuse and although many parts of the mountains are over 2000 m in elevation, the highest peak is Pico de Aneto (3404 m). The core of the Pyrenees had already been folded in the Hercynian orogeny and included Lower Palaeozoic sedimentary rocks with granitic intrusions. Sedimentation took place during the Mesozoic at the western end of the Tethys Sea and the present fold mountains were brought into being during the early part of the Tertiary (Palaeocene–Eocene) and further folding took place in the Miocene. The Pyrenees have a symmetrical structure with thrusting to north and south but extensive nappes, similar to those of the Alps, are not present. Erosion has revealed older rocks and their structures in the centre of the range where they have been intruded by granite batholiths. Folded Mesozoic strata lie to north and south of this axial zone and beyond them lie the basins of Aquitaine in France and Ebro in Spain.

High level erosion surfaces have been identified on the Pyrenees at about 2400 m and 2000 m. These are thought to have developed in the middle or late Tertiary. A small ice cap accumulated during the Pleistocene and features of glacial erosion occur at high elevations. At lower elevations moraines and fluvioglacial gravels form terraces along the rivers. The moraines descend to lower elevations on the cooler northern slopes than on the warmer slopes of the Spanish side. Glacial and post-glacial erosion cut deep gorges as the rivers flow away from the central core of the mountains.

### *Baetic cordillera*

The westernmost alpine mountains, the Baetic cordillera, occur in the south-east of Spain from Cadiz to Cabo La Nao and the structures continue below the Mediterranean Sea as the Balearic islands. Also known as the Sierra Nevada, these mountains overlook the Guadalquivir trough to the north and rise to 3487 m in Mount Mulhacen.

The core of these mountains consists of Palaeozoic phyllites and marbles into which granites have been intruded. The succeeding Mesozoic strata have been thrust-faulted towards the north and the Palaeozoic rocks now appear through 'windows' where erosion has removed the superincumbent nappes. The northern part of the Beatic cordillera is characterised by a large masses of Jurassic and Cretaceous rock which have slid northwards on Triassic gypsum. A foothill zone consists of unmetamorphosed Mesozoic sediments, also subject to gravity sliding, and with diapirs of salt pushing up through the limestones. At the southern end of the range the 'Pillars of Hercules' on either side of the Strait of Gibraltar, are formed of Jurassic limestone, with the flysch of early Tertiary forming the lower-lying land of La Linea.

An uplifted, warped high level surface is described upon the highest parts of the Sierra Nevada attributed to pre-Miocene erosion but widespread erosion surfaces also exist between 1000 and 1100 m and below these, Pliocene strand formations occur at 450 m.

### *The Balearic Islands*

The line of the Baetic cordillera is continued in a submarine ridge which extends 400 km north east from Cabo La Nao. The islands of Ibiza, Mallorca and Menorca are the visible parts of this submerged ridge where it rises above the sea. The oldest rocks crop out in Menorca where Palaeozoic strata are exposed. However, on all three islands Jurassic and Cretaceous limestones, folded in the early Tertiary, form the major features of the scenery. As with the Baetic cordillera on mainland Spain, the main thrust of the folding was to the north and this is well expressed in the northern sierras of Mallorca. These are composed of thick limestones into which deep gorges have been eroded, e.g. Torrent de Pareis. Polje occur on the uplands between ridges of limestone and Soller is situated on a drowned doline. Some of the most spectacular scenery of Mallorca occurs on the Formentor peninsula where the steeply dipping limestones run out to sea. What happens to the structures east of Menorca is not known. Some authorities would link them with the Alpes Maritimes of Provence, but movements of the Corsican and Sardinian micro-plates make such hypothetical linkages difficult to prove.

South-eastern Mallorca has lower hills, also formed of folded limestones and containing the karstic features of the caves of Drach and Arta. Between the upland regions of Mallorca lies an undulating central plain composed of Tertiary deposits including Oligocene lignites. The geological sequence is completed by Pleistocene slope deposits, alluvium and fossil sand dunes of Tyrrhenian age. Late Tertiary or Quaternary faulting disrupted the ridge extending from Spain, parts of which foundered to isolate the Balearic islands.

The north-west coast of Mallorca has steep cliffs which descend into deep water but the lower south-east coast has many small bays, locally called calles. The cliffs of the rocky coast indicate different levels of the sea and the present sea level is marked by a pronounced 'nip'. Along the coasts of the central lowland both to the north-east and south-west, there are sandy beaches with older Pleistocene dune formations lying inland.

### *The Alpes Maritimes*

In southern France, alpine structures trend east–west through Provence in the Alpes Maritimes. Mesozoic sediments are crushed against the northern flank of the Hercynian block of the Maures–Esterel which lies along the coast south-west of Nice. Folding and faulting was most intense at the western end in Provence between the Maures–Esterel massif and the Massif Central. Further faulting took place in the Miocene in a north–south direction and uplift probably occurred in the late Tertiary or early Pleistocene. Between the rivers karstic plateaux occur upon which remnants of a strongly-weathered bauxite remain in pockets less affected by erosion. The rivers which appear to be superimposed, are deeply incised, as in the Gorges de Verdun, the walls of which are almost 500 m high. Many facets of former

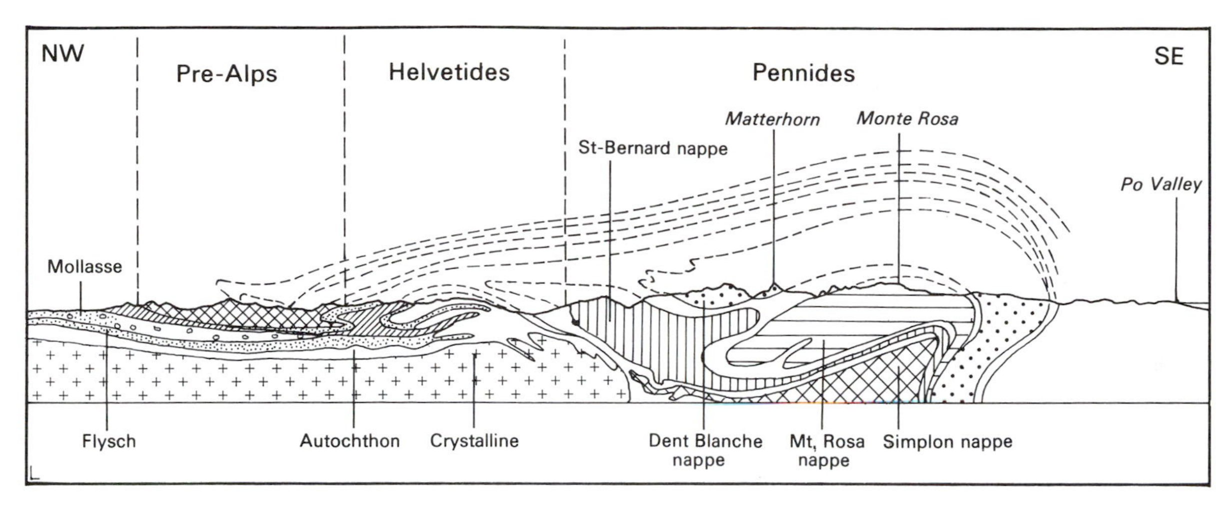


*Figure 8.31.* Section through the Alpine folding in Europe.

surfaces are preserved on the limestones; these occur at 400 m, 300 m, 200 m and about 80 m in western Provence., In the foothills soliflucted Pleistocene deposits form pediment features (called glacis by French geomorphologists). River terraces are developed at 25 m and 16 m along the Herfault and Aube rivers. Along the coast sand dunes and lagoons (étangs) characterise the lowland coast of western Provence but rocky headlands are typical of the coast near Nice. The Rhone delta has two major parts, an older feature, the Crau and the present delta, known as the Camargue.

### *The Alps*

The alpine ranges occupy an arcuate area 800 km long and 150 km wide beginning in the French Riviera and extending to Austria. Considerable doubt exists about the early development of this region, some protagonists finding evidence for Cambrian as well as Variscan folding along the line of the Alps. Hercynian earth movements certainly produced a number of large horsts and graben, some of which became involved in the later alpine folding. Large-scale thrust-faulting occurred (the Pennides and Helvetides) with extensive development of north and north-westward running nappes, characteristic of the French, Swiss and Austrian Alps (Figure 8.31). Major folding episodes began in the late Jurassic in the Alps and continued in the Cretaceous, late Eocene, Oligocene and late Miocene.

However, complex though the structure may be, it is the effects of subsequent erosion which have produced the dramatic scenery of this region. After each of the orogenic episodes, the landscape was reduced to a peneplain, only to be uplifted again. The initial arched surface of the alpine area has almost been completely eroded away but sufficient accordance of summits occurs for geomorphologists who have studied this area to recognise what has been called a 'gipfelflur' surface. Below this summit surface a second accordance of levels is attributed to Pliocene planation, and on the Lower Jurassic and late Tertiary sediments a further bevelling occurs which passes beneath Pleistocene outwash gravels. Superimposition of rivers appears to have taken place as they are generally independent of the structures; river terraces are a feature of alpine rivers and as they pass out onto the plains of Bavaria these terraces have been related to the four main phases of glaciation, namely Gunz, Mindel Riss and Wurm. However, since these four major phases were identified earlier this century, the number of glacial episodes recognised in the Pleistocene has increased to more than twenty.

It was in the Alps that features of glacial erosion were first observed and their implications for other landscapes realised. All the classic glacial features are present including the 'horn' peak of the Matterhorn, the arrêtes, the valley glaciers, and below them the U-shaped glaciated valleys with ice-scoured lakes, moraines and outwash plains (Figure 8.32). Solifluction and post-glacial fluvial erosion has added to the range of geomorphological features present.

*Figure 8.32.* The Rhône glacier.

### *The Jura*

This section of the Alpine province lies in an arcuate zone from the vicinity of Basel in the north-east, around the western end of Lake Geneva and beyond into the western part of the French Alps. The greater part of it consists of Jurassic rocks but some Lower Cretaceous rocks crop out, preserved in synclinal structures. The Jura is the type site for folded sedimentary rocks with anticlinal hills and synclinal valleys. In all, approximately 160 more or less symmetrical anticlines and synclines were produced during the Alpine orogeny by folds occurring in the Jurassic rocks overlying salt beds in the Triassic Muschelkalk beneath (Figure 8.33).

The largest folds occur in the eastern part of the section in the folded Jura, rising to 1723 m in the Crêt de la Neige. The synclines of this area also contain molasse from the denudation of the higher Alps further east. The more extensive plateau Jura occur in a parallel arc to the folded Jura with the tabular Jura lying north-west of Besançon.

Drainage from the area appears to be superimposed; the Rhone emerges from Lake Geneva at right-angles to the folds before turning south at its confluence with the Saône. Other transverse valleys, known as 'cluses' cut through some of the ridges to give inward-facing scarps with vertical limestone cliffs. Smaller dry valleys are cut in the flanks of some of the anticlines.

### *The Carpathians*

In general, the Carpathians are described as rounded, forested mountains. They may be sub-divided into three sections: the northern Carpathians occur in eastern Czechoslovakia along the Polish border; secondly, the eastern Carpathians which have an arcuate form in the USSR and Romania, and thirdly the Transylvanian Alps which run from east to west across Romania. The

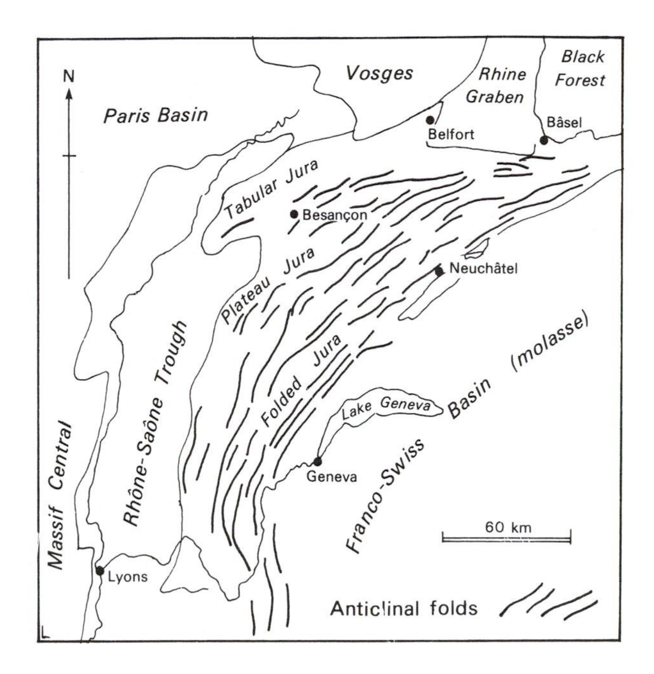


*Figure 8.33.* The Jura.

Carpathians curve around the upland area of the Apuseni mountains, which lie in western Romania and also enclose the Hungarian plains.

The northern Carpathians are situated between the north European plain and the Pannonian plain. The highest part, the High Tatra (2655 m) lies 100 km south of Krakow and is formed of Hercynian granite and metamorphic Lower Palaeozoic sediments which crop out in the Rudo Gory. Strong folding and thrusting took place in mid-Cretaceous times and again in the Miocene so that structural foreshortening of 40 km can be demonstrated in these mountains; movements of up to 100 km are claimed as a result of northward directed nappes. One completely over-rode the High Tatra and although mostly eroded, its remains can be seen plunging beneath later flysch deposits. The outer ranges, called the Beskid mountains, are formed from Mesozoic rocks where mineralisation has occurred. A deep structural trough, infilled with Tertiary molasse, lies north of the Beskid mountains on the edge of the north European plain.

The eastern Carpathians form an arc of highlands from the western Ukraine southwards into Romania. The average height of these mountains is over 2000 m culminating in Mount Ciucasul (3100 m). A crystalline core, the Rodna mountains, crops out on the western slopes but the eastern flank is composed of flysch rocks which have been strongly over-thrust. The molasse-filled trough continues around from the northern Carpathians. The eastern Carpathians are separated from the Transylvanian basin by the presence of the Harghita and Calimanuli mountains where cones and craters indicate Tertiary volcanic activity.

The Transylvanian Alps or southern Carpathians extend from south of Brasov westwards to the Iron Gate gorge where the Danube breaks through the Alpine ranges. A considerable part of the range is over 2000 m in elevation and the peaks reach 2500 m. Overfolding and thrusting occur to the south-east and south with a major uplift occurring in the Pliocene.

Despite this complex structure throughout the Carpathians, geomorphological studies which have been carried out suggest there is evidence for an upland surface formed from an uplifted peneplain with monadnocks. Below this surface three erosional terraces have been identified.

### *The Appennines*

The Appennines form the backbone of the Italian peninsula and although several sub-divisions can be distinguished, similar geomorphological features occur throughout this province. The Ligurian Appennines are 80 km wide in the north and 110 km in the south, their height is mainly between 1000 and 1500 m. Mesozoic limestones and marls enclose faulted basins containing Pliocene deposits. Extensive outcrops of clays (argille scagliose), have extensive landslip features.

The Etruscan Appennines have slightly higher peaks than the Ligurian Appennines (Monte Cimone, 2165 m), and their geology is more complex. Palaeozoic, Mesozoic and Lower Tertiary rocks crop out, exposed through windows of erosion in nappes which once covered the whole area. The thrusting in this section of the Appennines took place towards the north-east.

In the central Appennines, folded limestones are more extensive and are abruptly terminated on the eastern side by a large fault scarp. In general the landforms are plateau-like but higher peaks like Carno Grande (2914 m) rise above these plateaux and have glacial features. The Aquila plateau is a major karstic area with poljes and downfaulted tertiary basins containing Pliocene lake deposits. The western part of the central Appennines, both north and south of Rome, had late Pleistocene to Recent volcanic activity leaving some 52 craters, some of which now have lakes in them. Current vulcanicity is restricted to the Phlegrean Fields and Vesuvius. The Phlegrean Fields adjacent to the Gulf of Pozzuoli is part of a caldera with many subsidiary craters and ash deposits. The last major eruption of Vesuvius (1277 m) took place in 1944, but the volcano has been intermittently active since 79 AD when Pompeii suffered a nuée ardente followed by a blanket of 2.8 m of ash. Nearby Herculaneum appears to have been engulfed in an intensely hot, fluidised mass of pumice which resulted in a thick ignimbrite deposit which covered the settlement. As the city of Naples is built on a similar ignimbrite deposit, the implications for the city are clear.

In the southern Appenines, Cretaceous limestones and Triassic dolomites form the block-faulted landscapes south-east of Naples and coastal terraces have been observed at 80 m, 35 m, and 20 m above present sea level. The 'heel' of Italy is formed from almost level bedded limestones of the Apulian tablelands, 500 m above sea level. Lying between these two limestone areas is the Bradano valley, extending inland from the Gulf of Taranto. This is interpreted as a foredeep infilled with Tertiary sands and clays, lying between the folded zones of the Appennines and the Dinaric ranges of Yugoslavia. The 'toe' of Italy is formed by part of the Calabrian massif, the remainder of which forms the north-eastern corner of Sicily. This massif is composed of metamorphosed Devonian and Carboniferous rocks into which granite has been intruded.

*Sicily*

The island of Sicily has a distinctive triangular shape and is famous for Mount Etna and its connections with the Mafia. A fractured part of the crystalline massif of Calabria overlain with Mesozoic limestones and Tertiary flysch comprises the north-east uplands of the island. Along the north coast, Oligocene shales make up much of the Nebrodi mountains which are characterised by many landslides. The highest peak of this northern range is Mount Madonie (1979 m), formed from Triassic and Liassic dolomite in which karstic features have developed. The central part of Sicily from Catania westwards consists of Tertiary basins infilled with marls and beds of gypsum, the latter of Miocene age. Mount Etna (3323 m) is an active volcano which rests on sandstones and shales where it has built a composite cone. An eruption destroyed Catania in the seventeenth century and minor activity continues to the present day. In 1986, lava flows destroyed the upper part of the funicular access to the peak and a number of buildings.

The south-eastern part of Sicily is a stable foreland which structurally has more in common with north Africa than the rest of the island. It consists of Mesozoic limestones which accumulated on a gently subsiding platform, covered with Miocene carbonates upon which karstic features have developed.

*The Dinaric Alps*

The Dinaric Alps are mainly composed of carbonate rocks which have been folded parallel to the coastline of Yugoslavia. Strong folding and thrusting of the mid-Triassic to Cretaceous limestones during the Eocene and later in the Miocene was followed by uplift and erosion along major rivers which cut deep gorges such as along the Neretva river. It is possible to recognise a 'high karst zone' which extends the whole length of Yugoslavia where the 1000 to 1500 m plateau surfaces are pitted with karstic features. The Livro polje, inland from Split, is reputed to be the largest in the country, extending 70 km in length and 10 km in width. The 'Dalmatian karst zone' lies along the coast and has been partly flooded by the sea to provide the classic example of a 'Dalmatian type' of coast. It extends from the peninsula of Istria in the north as far as Albania in the south. The limestone of this region has been covered by flysch deposits, some of which are preserved in the relatively simple downfolds flooded by the sea between the islands. The Istrian plateau has been trimmed by erosion between 200 and 300 m above present sea level and spectacular cave systems exist at Postojna and Skocjanska further inland.

*The Hellenic mountains*

The Greek mountains are essentially a southerly continuation of the Dinaric mountains. They are most easily considered in relation to the Rhodope massif which has acted as a semi-stable block alongside which Mesozoic and later sediments have been crushed. Against the western flank of the Rhodope massif metamorphic rocks consisting of schists, possibly of late Pre-Cambrian age form the eastern and central prongs of the Chalkidiki

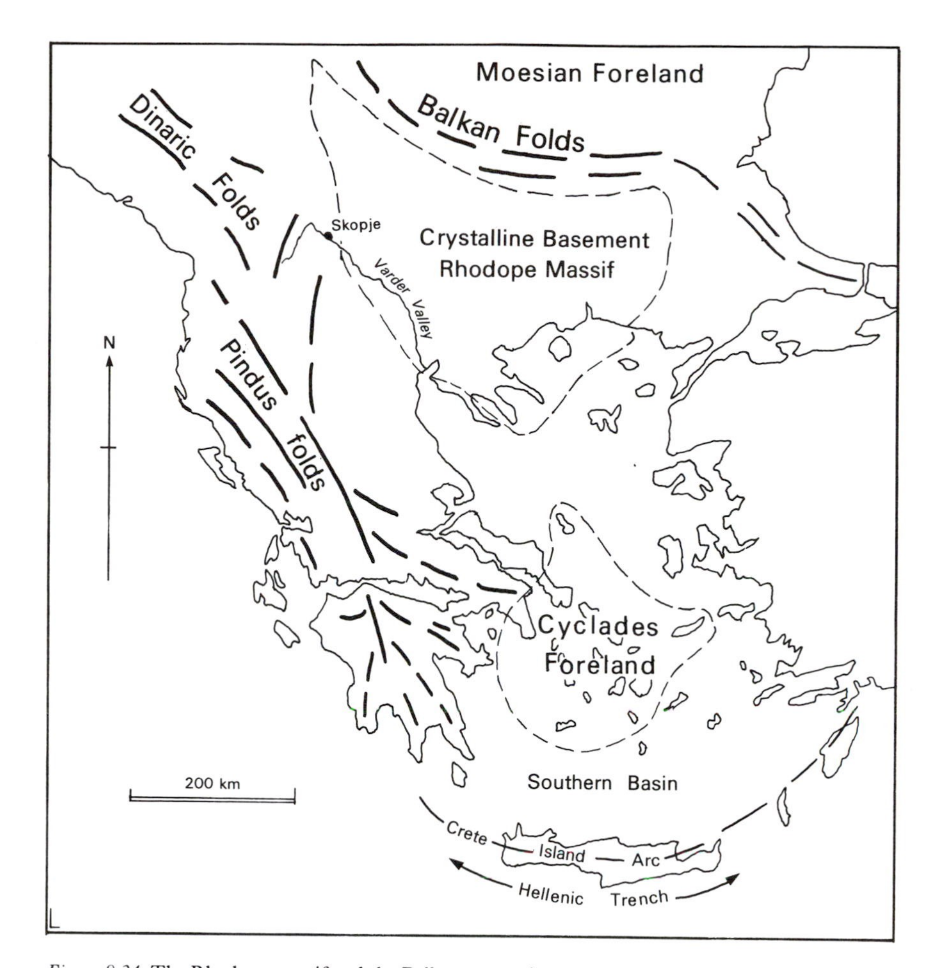


*Figure 8.34.* The Rhodope massif and the Balkan mountains.

peninsula. The western prong is of Mesozoic rocks which have been thrust westwards together with sea-floor material incorporated during orogenesis; there are extensive exposures of pillow lavas in the Vardar valley (Figure 8.34).

The eastern coast of Greece is formed from granites and gneisses covered by Mesozoic limestones and flysch. The limestones with the flysch deposits have been folded, faulted and pushed westwards to form the Pindus mountains which reach 2300 m above sea level. The west coast of Greece comprises a trough infilled by weakly folded Mesozoic limestones; if this zone extends northwards towards Yugoslavia, it lies beneath the Adriatic sea and the stable foreland against which folding took place may be the Apulian plateau of the 'heel' of Italy.

The Appennines and the Dinarides appear to have migrated northwards, as a separate micro-plate, being forced against the Alps, the suture being the 'insubric' line flying to the north of the Dolomites. If this is correct, then a large slip-fault must occur where displacements of up to 300 km may have occurred. This slip fault may run along the Vardar valley on the eastern flank of the Hellenic and Dinaric mountains; Skopje was the site of the severe earthquake on this fault. The doubling northwards of the Italian and Yugoslavian mountains accounts for the apparent sudden break in the alpine folding around the southern shore of the Mediterranean at the Gulf of Sirte, off the coast of Libya.

### *The Cretan arc*

The final part of the alpine folding in Europe concerns the arc of folding which runs through Crete. Currently, this is interpreted as an island arc situation. South of

Crete a deep sea trough occurs and back-arc vulcanicity dramatically destroyed most of the island of Santorini (Thira) in about 1450 BC. From Crete and Rhodes this line of folding links with that of the southern mountains of Turkey.

*The Balkan highlands*

The main mountain range of Bulgaria runs in a west to east direction parallel to, and south of, the course of the Danube river. These mountains have been formed by sediments accumulated between the Moesian platform (the lower Danube basin) and the Rhodope massif of southern Bulgaria and eastern Greece. When folding took place, open Jura-type folds were developed in the foothills between 50 and 100 km south of the Danube, but in the main range of the Stara Planina typically alpine structures with thrusting to the north occur. Subsequent erosion and uplift have resulted in a series of peaks including Yumrukchal (2376 m). South of the Stara Planina range, the Sredny Gora is composed of Palaeozoic rocks which have been eroded to a peneplain, covered with Quaternary deposits and uplifted. Although forming rounded hilly land in the west it becomes lower towards the Black Sea. The Maritza river which drains the area turns and flows to the Aegean around the eastern edge of the Rhodope massif.

*The Crimea*

The Crimean peninsula is clearly divided into a northern plain, the Crimean steppes, and a southern mountain range, 30 km wide, extending from Sevastopol and Balaklava to the Kerch peninsula. The geological deposits of the Crimea have been interpreted as a northwards-thrust fold mountain system composed of Triassic and later rocks with only minor igneous intrusions. The relief features of the highlands include the 1500 m main ridge, with the peak Roman-Kosh reaching 1543 m, which falls sharply to the Black Sea inland from Yalta. A lower, 700 m ridge lies along the northern side of the highlands and the foothills of 250 m merge gradually into the plain of the northern steppes. This plain is the infilled foredeep of the fold system.

The southern coast of the Crimea is the faulted, northern edge of the Black Sea graben which foundered in mid-Quaternary times allowing water to flow in from the Mediterranean. At first the sea extended further north than at present, producing the extremely flat plains along the southern USSR coast. As the sinking process continued beneath the Black Sea, the land to the north has re-emerged from the shallow sea which now only extends into the Sea of Azov at the mouth of the river Don.

*The Caucasus*

The Caucasus mountains extend 1000 km between the Black and the Caspian seas forming a natural boundary to the continent of Europe, but the political boundary between Russia and Turkey lies mainly along the Aracks river further south. The Caucasus mountains may be sub-divided into two ranges, the greater and lesser Caucasus, with a lowland trough between. The greater Caucasus tend to be higher in the west where Mount Elbrus (5642 m) is the highest mountain in Europe (Figure 8.35).

According to Soviet geologists, Pre-Cambrian schists and gneisses crop out on the southern side of the greater Caucasus which appear to have been disturbed first by the Variscan and subsequently by the Alpine orogeny. The former produced a series of upthrust horsts with downfaulted graben in which Carboniferous, Permian and Jurassic rocks occur. Cretaceous limestones unconformably rest upon these older rocks and result in karstic features in Dagestan. During the Alpine orogeny, folding took place beginning in the mid-Cretaceous and continued during the late-Cretaceous and Pleistocene with thrusting to the north. As uplift occurred the rising mountain range was eroded and partly buried in its own debris; extensive molasse deposits occur on both sides of the range. The final phase of geological activity in the Caucasus took place when a number of volcanos erupted during the Quaternary, producing the cones of Mounts Elbrus and Kazbek.

The Georgian–Azerbaijan lowland between the greater and lesser Caucasus mountains has three parts; the Kolkhida lowland in the west is separated from the Kura lowland of the east by the Suram granitic massif. These lowlands are infilled with Tertiary and Quaternary sediments, which form low plateaux and hills above the Kura alluvial lowlands at 200–600 m. The Kura river has built an extensive delta into the Caspian Sea. North of the Caucasus, the Stavropol plateau and Kuban–Azov lowland lie between the mountains and

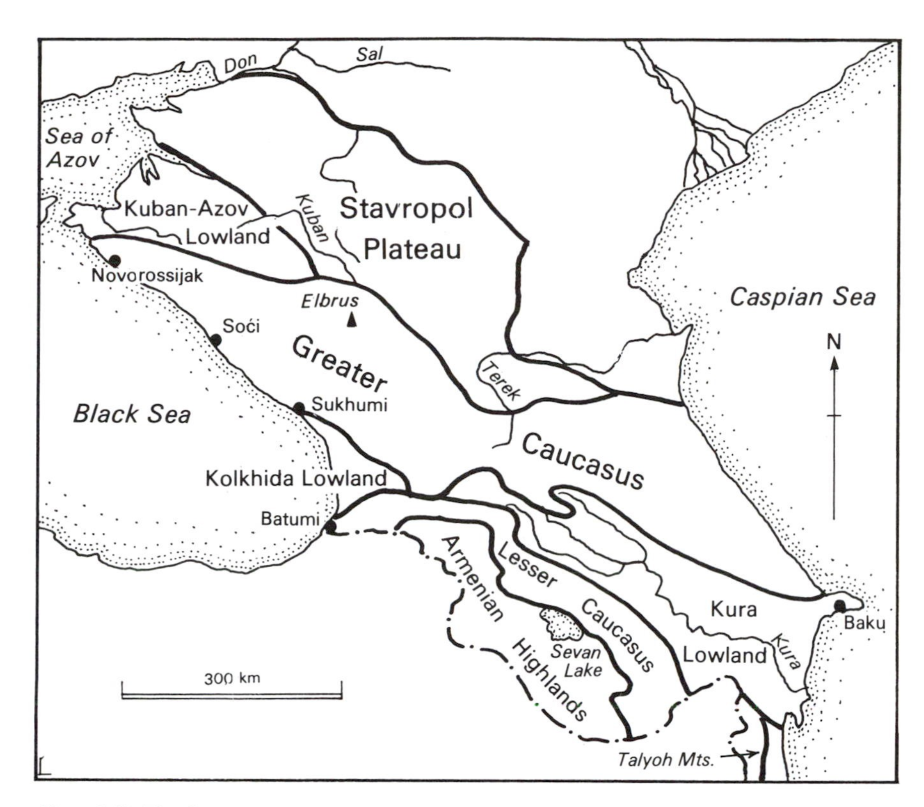


*Figure 8.35.* The Caucasus.

the south-west extremity of the European plain.

The lesser Caucasus are part of the folding associated with the Pontide mountains of northern Turkey. They include wedges of sea-floor material (ophiolites) produced at the site of a deep sea trench where subduction of a plate margin was occurring. Subsequent Tertiary volcanic activity resulted in extensive lava plateaux, making these mountains different from the greater Caucasus.

## The Ural mountains

The boundary of Europe and Asia is conventionally taken to be the Ural mountains which extend for 2500 km from the Arctic Sea almost to central Asia. The structural line is claimed by some to continue south to the Gulf of Oman. The Urals have been the scene of repeated orogeny from Pre-Cambrian to the Hercynian. The major relief features were formed by the end of the Mesozoic, but further tectonic disturbance occurred during the Tertiary. The Urals fit the model of a western miogeosyncline and an eastern eugeosyncline well and, in the process of mountain-building, many fragments of the sea floor were caught up in the folding and now occur in the eastern flank of the mountains. Transverse faults sub-divide the mountains into a number of sub-sections which are emphasized by differences in rock type and climate.

### *The northern Urals*

The Ural mountain range begins on the island of Novaya Zemlya and proceeds to make an S-bend before continuing due south along the line of 60°E as far south as the city of Perm. The highest peak in this section is the Gora Narodnaya (1894 m). Fault scarps bound the Urals on both sides and there is strong structural control of the landforms, in the Polyarny Urals an 800 m high fault-scarp extends for 100 km.

The northern Urals were strongly glaciated during the Pleistocene and ice still covers most of Novaya Zemlya. There are over 140 glaciers present on the northern Urals, they extend further downslope on the western side where there is more precipitation.

*The southern Urals*

The central and southern Urals are wider than the northern and have a general relief of 700–800 m with the peaks reaching 1500-2000 m above sea level. The central Urals are composed of a sequence of anticlinoria and synclinoria which have been eroded to reveal Pre-Cambrian quartzites and metamorphosed rocks. North of the Ural river (which drains to the Caspian Sea) a tectonic foredeep lies west of the Urals and has been infilled with molasse deposits eroded from the Urals; it forms a plain 200-300 m above sea level (see p. 225). On the eastern flanks, Chelyabinsk is situated in a graben. In the southern Urals, south of the Ural river, the mountains decrease in elevation and the climate becomes increasingly arid. The rocks which make up these mountains are Permian and Devonian limestones, dolomites, sandstones and shales with ultrabasic intrusives cropping out on the eastern slopes.

## Oceanic landforms of the European plate

The opening of the North Atlantic has resulted in the development of a broad area of abyssal plain west of Spain, France and the British Isles, referred to as the north-east Atlantic basin. The British Isles rest on a broad expanse of continental shelf but north of Britain the sea floor is interrupted by a ridge between Europe and Iceland which separates the Atlantic from the Arctic ocean. The European part of the Arctic ocean consists of a broad continental shelf and a restricted area of abyssal ocean floor.

*The north-east Atlantic basin*

The western margin of the European plate runs along the mid-Atlantic ridge from the Azores to Iceland. From the crest of the ridge, the depth of the sea increases to the abyssal plain which is between 5000 and 6000 m below the sea surface. Extensions of the abyssal plain reach into the Bay of Biscay and northwards between Britain and Rockall and further west, towards Iceland. A narrow continental shelf surrounds the Iberian peninsula, but from the northern part of the Bay of Biscay the shelf extends 800 km westwards, curving around Ireland, Scotland and Shetland. The North Sea is a shallow sea of the continental shelf which has been affected by rifting linking with the rifts which mark the edge of the Rhine valley further south. The northern margin of the north-east Atlantic basin is indicated by the submarine ridge between Scotland and Iceland upon which the Faeroe islands rest.

*Iceland*

The island of Iceland (102 828 km$^2$) lies astride the mid-Atlantic spreading centre and is unique as it is constructed entirely of oceanic crustal material. It is the only major area where a mid-oceanic ridge can be inspected on land. The rift-faulting on the crest of the ridge is about 40 km wide and results in the island being divided into an eastern third on the European plate and a western two-thirds on the North American plate.

Where the rift crosses Iceland is the site of many earthquakes, it also shows signs of tensional stress with many faults. It is also the location of many volcanoes including Heckla, Torfajokul, Skjaldbreidhur and Askja. Surtsey and Eldfjell have erupted during the last decade. Rocks at the eastern and western extremities of Iceland have ages of about 20 million years which indicates that the average rate of spreading is about 10 mm per year.

Eruptions of basalt have formed the plateaux upon which ice caps have accumulated; Vatnajokull and Myrdalsjokull are the largest and there are several smaller accumulations of ice.

*The Norwegian basin*

The mid-Atlantic ridge continues northwards from Iceland through Jan Mayen to pass into the Arctic ocean between Greenland and Spitsbergen. This basin is not as deep as the Atlantic basin discussed previously with the undersea Voring plateau extending well out from the Norwegian coast, almost dividing the deeper basin into two parts. A deeper channel, about 1500 m, extends around the southern coast of Norway to within 50 km of the Swedish coast.

*The northern continental shelf*

The Barents Sea covers an extensive area of continental shelf and slightly deeper waters enclosed by the islands of Spitsbergen, Franz Joseph and Severnaya Zemlya. North of these islands lies the continental slope which descends to the deep waters of the Arctic Angara basin.

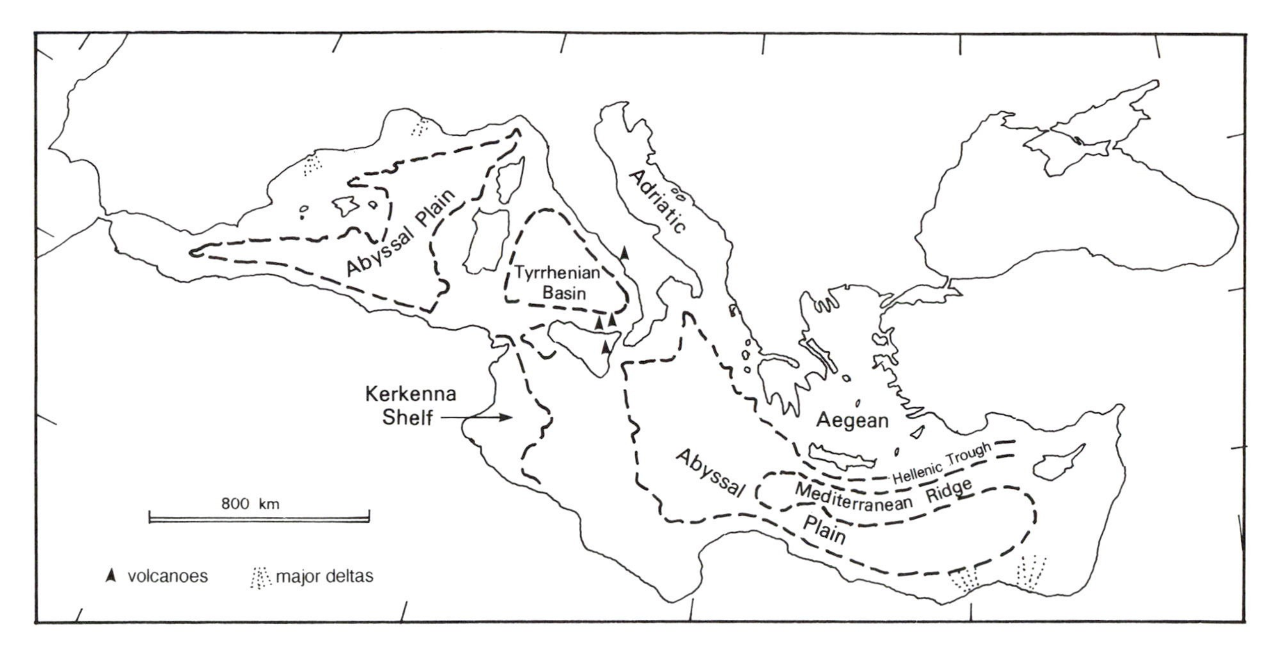


*Figure 8.36.* The Mediterranean Sea.

*The Angara basin*

After passing between Greenland and Spitsbergen, the mid-Atlantic ridge continues across the Arctic sea towards the New Siberian islands where it ends. The islands appear to be a pivotal point about which the continental plates have moved, the Arctic ocean becoming wider and compression resulting in the occurrence of folding and mountain building in the Verkhoyansk mountains. Sea-floor spreading has created an area of abyssal plain at depths of over 5000 m below sea level.

## The Mediterranean basin

As the Mediterranean Sea lies between Europe and Africa, between the African shield and the Alpine fold mountains, it has a complex structure. However, it may be simply divided into the western and eastern Mediterranean for the purpose of the present analysis (Figure 8.36).

*The western Mediterranean*

The triangular-shaped western Mediterranean is roughly split into two by the islands of Corsica and Sardinia. Between Italy and Sardinia is the Tyrrhenian basin and between Sardinia and the Balearic Islands is the western Mediterranean basin. Both these basins have depths of more than 2000 m and in places the sea floor is 3000 m below sea level.

The rocks of the western Mediterranean basin floor are continental and the basin has probably formed as a result of foundering of Hercynian blocks during the Alpine orogeny. Pliocene and Quaternary sediments overlie Miocene evaporites, the latter a remnant of the former Tethys Sea which ceased to exist as the Alpine orogeny developed. The Ebro valley continues across the depositional shelf in an undersea canyon cut in a deltaic fan, and a large deltaic fan brought down by the Rhone river lies in the Golfe du Lion. In contrast, the North African coast has a narrow coastal shelf.

The Tyrrhenian basin between Italy, Sicily and Sardinia is also a graben structure. It is a fragment of crustal material which has been depressed in excess of 3000 m, accompanied by volcanic activity. On the eastern margin the vulcanicity of the Italian mainland has already been mentioned, but the Lipari islands including Stromboli and Vulcano rise from the sea and others remain submerged. That this was once an area of dry land is demonstrated by the submerged canyons which have been eroded down into the basin.

### *The Sicilian–Tunisian plateau*

The south-eastern part of Sicily is a stable foreland which structurally has more in common with Africa than the rest of the island. It consists of Mesozoic limestones which accumulated on a gently subsiding platform, covered with Miocene carbonates upon which karstic features have developed. Most of the structural features of Sicily may be traced across the 150 km of the Sicilian Channel into Tunisia. The sea here is less than 1000 m deep and several islands, including Malta, Gozo, Pantelleria and Lampedusa rest on the shallow sea floor between the western and eastern Mediterranean basins.

Malta and Gozo are composed of block-faulted Tertiary limestones underlain with Jurassic limestones. Clays and marl partings between the limestones retain precipitation and maintain a groundwater table. The western part of Malta is a series of fault-blocks with scarps running east–west and with cliffs of over 200 m along the coast. An abrupt east-facing scarp overlooks the lower plateau of eastern Malta formed upon the Tertiary globigerina limestone.

### *The eastern Mediterranean*

The eastern Mediterranean may be sub-divided into three sections, the deep sea area of the Mediterranean itself and the shallow seas of the Adriatic and Aegean. The most extensive part of the abyssal plain of the eastern Mediterranean lies south-east of Italy and Sicily but it narrows eastwards parallel to the Libyan and Egyptian coasts. The depth of this part of the sea floor is between 2000 and 3000 m, reaching 4000 m in its deepest parts. South and east of Cyprus there is a shallow shelf and off the delta of the Nile there are two underwater deltaic fans, associated with the Rosetta and Damietta branches of the river.

The northern part of the eastern Mediterranean basin appears to be similar to an island arc situation. Off the southern coast of Crete there is a trench, 5100 m deep, and south of that a submarine ridge extends from Cyprus, curving around Crete towards the Ionian islands off the west coast of Greece. It has been suggested that this ridge is a spreading centre, but no magnetic strips have been identified on the sea floor. It is possible that the island of Crete is part of an outer island arc and that the inner volcanic arc is represented by the islands of Santorini and Nisiros. The volcanic activity occurring on these islands is the result of the subduction of sea-floor material beneath the Turkish micro-plate.

The Aegean north of the Cyclades islands is less than 1500 m deep and could be interpreted either as a subsided part of the southern foreland of the Rhodope massif or as a subsided part of the Turkish micro-plate. It has been block-faulted and the upfaulted areas protrude above the sea as islands. A fault-scarp bounds the northern side of the Dardanelles and the Sea of Marmora which is another graben.

The Adriatic has a shallow northern basin which is less than 250 m deep but the depth gradually increases south-eastwards until it joins the eastern Mediterranean basin. There is some speculation about its structure as it is in part a foredeep associated with the Appennine folding. Large quantities of sediment have accumulated in it, for example, as in the Po lowlands.

### *The Black Sea*

The area covered by the Black Sea is 450 000 km$^2$, and like the Mediterranean, there is speculation about its origin as it does not appear to have a spreading centre from which new sea-floor material has been extruded. Rather, it is interpreted as a graben, continuous with the lowland between the greater and lesser Caucasus. The northern side of the Black Sea has an extensive shelf, a continuation of the extremely flat plains of the southern Ukraine. The Sea of Azov does not exceed 14 m in depth and this southern coast is characterised by sandy shores with long sandbars such as the Tongue of Arabat and the Tendra peninsula. However, the southern coast of the Crimea has been faulted.

There is a large submarine fan on the western side where the Danube flows into the Black Sea, but the Turkish shores have precipitous slopes and a narrow coastal shelf with many submarine canyons. The deeper parts of the Black Sea are 2000 to 2200 m deep and these occur in two basins, separated by a ridge in the centre which has features of mud diapirism. Salinity is low and connection with the Mediterranean was only established during the Quaternary.

9

# The Pacific Ocean basin

The Pacific Ocean is the largest single feature on earth, covering 165 384 000 km$^2$, or almost half of the Earth's surface. From north to south it is 15 216 km and from east to west 16 885 km. The average depth is 4300 m, but in the trenches, which lie around the periphery, depths of 11 000 m occur. The eastern margin of the Pacific Ocean lies in the oceanic trenches of Central and South America and there are few islands other than the Galapagos islands and Easter Island in the eastern Pacific. In contrast the western Pacific has the many islands of Melanesia, Micronesia and Polynesia. Its western margins are complicated by the island arcs ranging from the Aleutians in the north to the Bismarck–Solomon islands in the south together with the marginal seas which they enclose.

## Geological history

That the floor of the Pacific basin was different from the surrounding areas was known from the early years of the present century; the 'andesite line' enclosed rocks of basaltic or sea-floor origin as distinct from the sialic rocks of continental origin (Figure 9.1). It was argued that the present Pacific is the remnant of the ocean which encircled Pangea, but this has since been proved incorrect because it appears that no part of the Pacific floor is older than 200 million years. However, tectonic features of the Andean mountains indicate that the margin of the Pacific may have been formed about 2000 million years ago. In contrast, the Indian, Atlantic and Southern ocean basins are younger and are still increasing in area, whereas the Pacific floor is being consumed

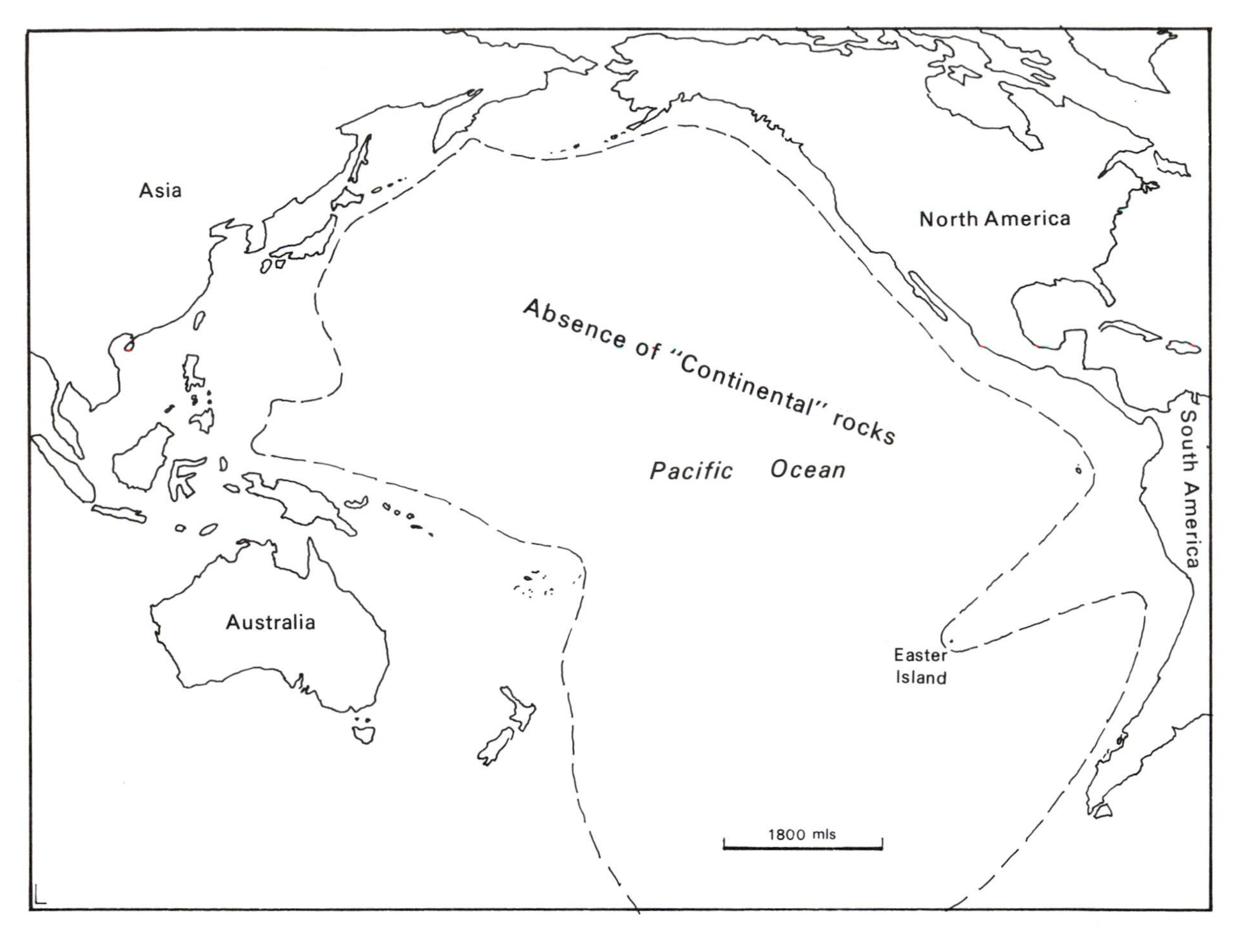


*Figure 9.1.* The absence of continental rocks in the Pacific basin is marked by the andesite line.

along the lines of the trenches which surround it. The marked difference between the eastern and western Pacific margins is caused by the relative movement of the lithospheric plates relative to the underlying mantle rocks.

The constructive spreading centre of the Pacific Ocean can be traced from south of New Zealand in a broad curve to Easter Island and then northwards to the Gulf of California. The line of the spreading centre is offset by many transform faults, one of which is the San Andreas fault (see p. 98) where the spreading centre has been overrun by the North American plate. It emerges again onto the Pacific floor west of the coast of Oregon and Washington finally disappearing off the northern tip of Vancouver Island.

Passage of the Pacific plate over the 'hot spot' of the Hawaiian islands gives an indication of its direction of motion. As the older islands lie north and west of Hawaii, the motion of the plate must be north-westwards. These volcanic islands and the others associated with transform faults are younger than the ocean floor upon which they rest. The older and lower parts of these oceanic islands are tholeitic and the exposed shield volcanoes are composed of basaltic lavas. The only exception to this is Easter Island which appears to have some continental material in its geological composition.

The geomorphological features of the Pacific basin may be described in the two major physiographic divisions of the Pacific Ocean basins and the island arcs; each with several provinces.

## The Pacific Ocean basins

### *The north-east Pacific basin*

This large basin, lying west of North America, south of the Aleutian islands and east of the Emperor–Hawaii range of seamounts is the western flank of the east Pacific spreading centre. Off the coast of North America several major crustal dislocations were discovered in the 1950s and are now interpreted as major transform faults. These are the Mendocino, Murray, Molokai, Clarion and Clipperton fracture zones which extend in roughly an east–west direction. The eastern flank of the spreading centre has mostly been subducted beneath the North American plate, but two small areas remain, the Cocos plate off the coast of Mexico, and the Juan de Fuca plate off the coast of Canada. The north-eastern Pacific basin becomes deeper further west and north until depths of over 5000 m are reached east of the Hawaiian islands and on the Aleutian abyssal plain. Glacial deposits lie on the ocean floor in the Gulf of Alaska, and the remaining part of the basin floor is covered by deep sea siliceous ooze and pelagic clay.

### *The south-east Pacific basin*

This basin lies between the east Pacific rise and the Peru–Chile trench along the South American coast. It may be sub-divided into the Peru and Chile basins, separated by the Nazca–Sala Gomez ridge of seamounts the tops of which lie only 300–400 m below the sea surface. This ridge extends from Easter Island, renowned for its huge statues carved from grey trachytic lava, to the central Chilean coast. The basin floors on either side of this ridge lie at more than 5000 m depth. Both Peru and Chile basins lie over the small Nazca plate which is being subducted into the Peru trench below South America. Its northern boundary is marked by the Cocos ridge which extends from the east Pacific rise towards the isthmus of Panama. On this ridge lie the Galapagos islands, with their large tortoises and the interesting pattern of evolution of finches as described by Darwin. The Galapagos islands are volcanic, composed of basalts and tuffs, and like many volcanic islands, have a tendency towards a circular shape. A further ridge between southern Chile and the east Pacific rise separates the Chile basin from the sea floor which surrounds the Antarctic plate. The ocean floor is mainly covered with pelagic clays.

### *The south-western Pacific basin*

The south-west Pacific basin lies between the Pacific Antarctic ridge, the plateau upon which New Zealand is situated, the Kermadec–Tonga trench and the islands of Polynesia. Although the deepest parts of the ocean occur in the peripheral trench (10 047 m), the abyssal plain is extensive and between 5000 and 5200 m deep. The northern boundary of this basin is a broad submarine ridge at about 15°S on which the Polynesian Cook islands, Society islands and Tuamotu group stand. The southern parts of the basin floor are covered by pelagic clays but around the equatorial regions the deep sea deposits are calcareous.

### *Polynesia*

The many islands of Polynesia rise from the sea floor as volcanic masses and in addition there are many seamounts which do not reach the sea surface. A feature of these Pacific islands is their coral reefs. After visiting Tahiti and Keeling Atoll, Darwin proposed his theory about the formation of coral islands in 1842. This explains the relationship of corals with the islands and their fringing reefs, barrier reefs and atolls. Darwin suggested that a volcanic island with a narrow fringing reef gradually subsides until a barrier reef is formed and then as subsidence continues further the island disappears leaving only an atoll. Other scientists proposed a development sequence controlled by glacial events during the Pleistocene, when lower sea levels and colder conditions may have limited coral growth. The post-glacial rise in sea level is thought to have been about 10 mm $yr^{-1}$ and coral can grow at about 15 mm $year^{-1}$, so coral growth could have kept pace with the post-glacial eustatic rise in sea level. However the base of the coral has been found in deep borings at depths of 600-1000 m which is well below eustatic sea level variations, so Darwin's hypothesis must be correct in its broad outline, even though individual sites may vary in detail.

### *The western Pacific basin*

The islands of Micronesia, the Caroline and Marshall islands and the Kiribati group (Gilbert Islands) lie in

*Figure 9.2.* Oahu, Hawaii. A costal view of old, vegetated lava flows.

the southern part of this basin which extends north to the Marcus–Necker rise on which the island of Wake is situated. The western margin of the basin is the Marianas trench where the Challenger deep reaches 11 033 m, the greatest ocean depth known.

*The north-western Pacific basin*

Between Japan, the Kuril islands and the line of the Emperor seamounts lie some of the deepest parts of the Pacific abyssal plain. Although more than 6000 m deep, this plain is bounded on its western side by the Kuril and Japan trenches where depths of over 10 000 m occur.

The Hawaiian islands lie between the north-western and north-eastern Pacific basins. From Hawaii the islands form a chain which may be traced 3000 km north-westwards across the Pacific until they link with the north–south line of the Emperor seamounts. The ocean crust upon which the Hawaiian islands rest is about 70 million years old and the age of the volcanoes ranges from 6 million years old in the north-west to Mauna Loa and Mauna Kea which are still active. Although the islands appear small on maps of the Pacific, these volcanoes are in fact the largest mountains on earth. Mauna Loa (4170 m) rises 6000 m from the sea floor to give a total elevation of over 10 000 m and its volume is estimated to be 40 000 $km^3$. The lava poured out from these intra-plate volcanoes is basaltic and they are typically shield volcanoes with fluid eruptions, usually from fissures on the flanks of the volcano rather than from the crater. These lava flows may be rough and cindery, when they are known as Aa, or ropy and flowing in form when they are called Pahoehoe. Explosive activity rarely occurs with these volcanoes (Figures 9.2 and 9.3).

*Figure 9.3.* Oahu, Hawaii. A dissected volcanic core seen from the University of Hawaii.

## The Pacific island arcs and marginal seas

The landforms of the eastern margin of Asia may be related to the interaction of the tectonic plates of the Pacific Ocean floor and the Asian landmass. Along a sequence of arcuate lines from the Aleutian islands in the north through Japan and the Philippines to the Solomon islands in the south, the Pacific plate is being subducted beneath the landmass of Asia, producing many volcanic islands.

Island arcs are zones of the earth's surface where earthquakes are frequent and vulcanicity a common feature. It has been observed that there are differences in the amount of heat flowing from the interior of the earth and that anomalies in the attraction of gravity occur. With the development of the theory of plate tectonics many of these features began to make sense. The ocean trench, 50–100 km wide and up to 11 km deep, occurs where the ocean floor begins to pass down into the asthenosphere. Some oceanic or terrigenous material may be caught up in a wedge of thrusted material to form the outer arc ridge. Between this and the arc itself there may be a forearc basin. The island arc is composed of volcanic and intrusive rocks together with any sedimentary materials present. Although not fully explained, the potassium content of the lavas increases and lavas change from tholeiitic to alkaline towards the back of the arc. Shallow-sited earthquakes are located towards the front of the arc and deep-seated ones to the rear of the arc demonstrating a relationship with the descending slab of subducted material.

A marginal sea sometimes occurs to the rear of an island arc; this may or may not be associated with a minor spreading centre. Some marginal sea basins have a high heat flow, but others have a normal heat flow and are older, formed during the Tertiary. Sedimentation in these basins is rapid, with debris derived from erosion of the volcanic rocks of the arc.

### *The Aleutian islands*

These islands extend the line of the Alaskan peninsula across 2400 km of the North Pacific in an arc of volcanic islands which were developed during the Tertiary. The Aleutian trench lies south of the island chain to a depth of 7822 m. The outer arc ridge is represented by Kodiak island and the forearc basin by Shelikof strait, further west both these features are drowned by the sea. The island arc is composed of about 80 major volcanoes, 36 of which have been active during historical time. Several calderas are also present, including that of Mount Katmai which collapsed in 1912 as a neighbouring volcano discharged an estimated 7 cubic miles of pyrolastic material and the Valley of Ten Thousand Smokes came into being. Great Sitkin volcano is a composite cone which rests upon a basaltic shield volcano. The Bering Sea is a backarc basin infilled with a thick accumulation of terrigenous sediments. It has a normal heat flow.

### *The Kuril arc*

The Kuril islands link the peninsula of Kamchatka with the Japanese island of Hokkaido. Three phases of vulcanicity have been identified on Kamchatka, which took place during the Tertiary and continue through to the present day. The pattern of development of the arc is similar to the model presented with the Kuril trench offshore (maximum depth 10 542 m), the island arc and the southern part of the Sea of Okhotsk forming the backarc basin.

### *The Japanese arc*

The islands which comprise Japan lie at the junction of three tectonic plates: the Pacific, Philippine and Eurasian. Along the east coast of Honshu, north of Tokyo, the Pacific plate is being consumed in the Japan trench where it dips beneath the Eurasian and Philippine plates. The Philippine plate in turn is being consumed beneath the Eurasian plate in the Nankai trench south of Shikoku and Kyushu. Many shallow focus earthquakes occur as the Pacific plate is moving down at an estimated rate of 7.5 cm/year. In Tokyo, there are an average of three noticeable earthquakes each month. Occasional stronger tremors occur as in 1923 when an earthquake registering 8.3 on the Richter scale collapsed buildings and started fires, resulting in 99 000 deaths.

Volcanic eruption is also a hazard on the Japanese islands with more than 200 volcanoes, 60 of which have been active in historic time. Vulcanicity reached a peak in the Pleistocene and large quantities of volcanic material erupted including the ash from a caldera in Kagoshima Bay, Kyushu, which left a layer of pumice more than 10 m thick.

The effect of tectonic uplift is evident in the landforms of Japan. During the last interglacial (120–130 000 years ago) the Shimosueyoshi coastal terrace was formed on Shikuko which is now at 190 m above present sea level. Subsequently, Holocene terraces have been cut at 18 m and 12 m. Inland, rivers have aggraded and then dissected to form terraces as the land and sea relationship changed.

Above the tree-line on the higher mountains from central Honshu northwards, glacial features are present. Glaciation was most extensive on Hokkaido where evidence for at least two glacial stages can be seen. Relict periglacial features indicate that lowland Hokkaido and the higher parts of Honshu were tundra during the glacial stages of the Pleistocene.

The plains of the Japanese islands are restricted to narrow river floodplains, deltas and coastal swamps. The mountains rise directly from the sea in many parts of Japan, creating a rocky indented coast, particularly on the Pacific coast. The most extensive coastal plains occur around Nagoya, Toyama and Tokyo. The mountain ridges which extend throughout the length of the Japanese islands were uplifted at the time of the Alpine orogeny, together with the mountains of the Kuril, Ryukyu islands and the Kamkatchka peninsula.

The sediments from which these mountains are formed accumulated as a trench–subduction complex which was crumpled against the mainland of Asia to form an island arc as the Pacific tectonic plate moved against, and was subducted beneath, the landmass. All the Japanese islands tend to be tectonically unstable, much affected by earthquakes and vulcanism. Folding has occurred parallel to the line of the island arcs and in some regions has been sufficiently intense to produce nappes with folds of over 3000 m. In central Honshu, the folded sedimentary rocks have been eroded to a rugged terrain and volcanic peaks have been built upon them including peaks such as Asamayama (3300 m). Intrusion of granodiorite occur with porphyritic tuffs.

The island of Honshu is transversely divided by a depression, referred to as the 'Great Ditch'. This has largely been infilled by successive lava flows and other ejecta. However, its boundaries can be seen along the valley of the Fuji river, in the Hima rift valley and Siwa lake depression as a sequence of graben features. Mount Fuji (Fuji San, 3776 m) on the eastern side of the depression is the highest mountain in Japan and an excellent example of a composite volcanic cone. West of this depression the high ground is formed by the Hida and Akaishi mountains rising to over 3000 m in the peaks of Ontake San and Dai-Tenjo. Between Nagoya and Toyama the higher land is narrower and plateau-like, forming the Hida plateau. Lake Biwa, the surface of which is at 100 m above sea level, lies in a structural depression and a smaller depression parallel to the Great Ditch runs inland from Nagoya.

### *The Philippine plate and Marianas arcs*

The boundaries of the Philippine plate, comprising a small area of Pacific sea floor, are marked by deep sea trenches and subduction zones. On the west side, the Ryukyu trench and its arc of volcanic islands extends from Japan towards Taiwan and the Philippines trench lies off the eastern coast of those islands. The Marianas trench which extends along the eastern side of the Phillippine plate contains the world's greatest ocean depth, the Challenger deep (11 033 m).The main Pacific plate is being subducted beneath the eastern edge of the Philippine plate; both are composed of sea-floor material. Consequently there is no forearc ridge of siallic sediment scraped from the surface of the descending plate, but the forearc basin may contain level-bedded sediments derived from the volcanic arc. The marginal basin at the rear of the Marianas, the Parece Vela basin, may be distinguished from the west Philippine basin as it is the site of active sea-floor spreading and has a high heat flow. By contrast the west Philippine basin has a normal heat flow and has a 100 m covering of pelagic mud, indicating a Tertiary sea floor area in which sea-floor spreading is no longer active.

Opinion about the structure of the Philippine islands is divided, one interpretation being a symmetrical double arc system and the other a complex subduction situation at the edges of three lithospheric plates. The morphology of the islands is a central trough, partly infilled with sediments derived from the highly deformed Miocene rocks which form the east and west coast mountains. Volcanic activity is confined mainly to the margins of this central trough; altogether there are over 20 active craters in the Philippines. West of the Philippines lie the marginal basins of the South China Sea, Sulu Sea and Celebes Sea all of which have high rates of heat flow from the interior of the earth and are thought to be youthful, active basins.

### *The Bismarck archipelago and Solomon islands*

This is another area of the west Pacific where there is great complexity and differing opinions about the origin of the landforms. The simple model invoked to describe the principle of island arcs is difficult to apply to this complex of islands and deep sea trenches. The northern limits of the Australian plate are marked by an inactive trench in the northern part of the Coral Sea. Immediately north of this trench is the island arc of the Louisade archipelago which continues the line of the 'tail' of New Guinea where there are several volcanic peaks. Between the Louisade islands and the Solomon islands two inactive basins with high heat flow occur, the Woodlark basin and the Solomon basin. A similar structure occurs with the islands surrounding the Bismarck basin and these aree interpreted as horst-like uprisings alongside of which material is sinking into the asthenosphere. All these smaller systems are enclosed by the west Melanesian trench which lies north of the Solomon islands, marking the southern edge of the Pacific plate.

### *The Tonga–Kermadec arc*

The Tonga–Kermadec arc is the remaining major island arc to be considered on the Pacific margin. It is associated with an oceanic trench, maximum depth 10 047 m, lying east of these islands. Unlike the other island arcs of the Pacific, the Kermadec Islands and their accompanying trench lie in a straight line. The trench is interpreted as the site of a large shear zone between the Pacific and Australian plates, and evidence from western New Zealand indicates there has been an offset movement 480 km northwards.

# 10

# Geographical implications of major geomorphological features

The landforms of the Earth have been described within the framework of the major physiographic divisions and provinces of each lithospheric plate. Although the boundaries of the plates are well established and most people would agree about the major physiographic divisions, there is less agreement about the boundaries of provinces, and different authors may take slightly different areas. An aim of geomorphological analysis is to sub-divide the Earth's surface into areas with similar geological and geomorphological development, to provide a meaningful basis for discussion and explanation of present landforms.

This theme is but one of the many facets of interest in geomorphology, as is pointed out by Gregory (1987). Essentially, the foundation of the modern study of landforms begins with the writings of W.M. Davis whose evolutionary approach dominated the first half of the 20th century. Problems with dating the various stages in a denudation chronology and over-enthusiastic correlation of surfaces over large distances led to disillusionment with this approach. This coincided with a growth of interest in the natural processes involved and this approach has tended to dominate geomorphology work during the last 40 years.

The emergence of the concept of plate tectonics and its potential for providing an explanation of the distribution of the continents has given renewed encouragement to the study of the large-scale features of the Earth's surface; the features which formed the framework of the geomorphological analysis of the previous chapters. Some implications for broader geographical studies are summarised briefly in this chapter.

### *A framework for landscape description*

Geomorphologists wish to provide a scientific description and explanation of the landforms of the Earth's surface. In a very small area it is possible to make an objective map of every facet of the landsurface showing each break and change of slope. The resulting map shows the morphology, indicating as simply as possible the geometry of the landscape, but it would not necessarily provide an interpretation of the origin of the landforms. An interpretive element is introduced as soon as the landforms are given technical names, and at this stage speculation about their origin is inevitable. The search for a geomorphological common denominator which may be used to indicate a homogeneity or common origin of an area of the Earth's surface is best carried out upon the units discussed in this volume, such as the province or section.

The scientific description of landscape is something physical geographers have striven for since the earliest times, and many geographical terms for landforms originate from Classical times. In geographical analysis a good description is the basis for an interpretation of the origin of the landforms present. As has been observed throughout this book, it is often necessary to examine the geological history, particularly since the Tertiary to explain landform development.

Geographers require the analysis of geomorphology for their regional synthesis. Landform is used extensively by geologists when mapping the distribution of outcrops, and soil scientists find a geomorphological analysis of the landsurface invaluable for mapping the distribution of different types of soil and parent material, different sites tending to have different combinations of soil-forming factors. Ecologists and hydrologists also find the scientific description of landform important in their studies. Land use planners and civil engineers have found knowledge of the origin, character and distribution of landforms useful in avoiding unnecessary development expenses.

### *Earthquake location and volcanic activity*

The location of earthquakes in specific regions was one of the features which led to the identification of the margins of the tectonic plates. Creation of material at spreading centres and its destruction in subduction zones both result in crustal movement and earthquakes. Along the mid-oceanic spreading centres numerous earthquakes occur, but they are shallow (less than 70 km deep) and not usually severe as the nature of the rocks at shallow depth favours plastic deformation rather than sudden fracture. The most active earthquake zones are those associated with the subduction zones. As the plate being subducted descends into the asthenosphere, friction with the material on either side produces earthquakes from shallow depth down to beyond 300 km depth. The epicentres of these earthquakes can be plotted to indicate the angle at which the descending slab of cool crustal material is entering the asthenosphere; dips range from 45° to 70°. Thus, shallow earthquakes take place near the trench and the deeper ones further away from it on the continental side. Many major earthquakes also occur on transform faults where one part of a tectonic plate moves against another. Displacement obviously occurs laterally, as can be seen by the position of the spreading centre, but also vertically as the newly formed sea floor sinks away from the mid-oceanic ridge. Movement between plates can be observed in the San Andreas fault where the Pacific plate is moving past the North American plate at an average rate of about 4 cm per year. At intervals of 50 to 200 years the accumulated strains are relieved by a major earthquake such as occurred at San Francisco in 1906 and 1989. Smaller events occurred in 1899, 1907 and 1916; unfortunately it is difficult to forecast when the next large event may occur. Some major earthquakes occur in China, away from obvious plate margins. It is considered that these relate to major intraplate faults, rather than the edges of tectonic plates.

The location of volcanic activity is related closely to the same areas. As subduction takes place melting occurs and the magma migrates up to the surface where it erupts in volcanoes. Thus the island arcs of the Pacific are the best examples of this type of activity. However, volcanic activity also occurs where there has been crustal extension and block-faulting. In this case partial thinning of the crust by extension and the provision of conduits to the surface by faulting combine to give volcanic activity, such as seen at present in parts of the basin and range province of the western USA. The vast outpouring of basaltic lavas which occurred in the Deccan, Ethiopia, Australia and other parts of the world preceded rifting and separation of the tectonic plates.

The disposition of the continents and the mid-oceanic ridges has exercised great influence on sea levels. When ocean-floor spreading is active and the volume of the mid-oceanic ridges is increased, water is displaced onto the continental shelves. (At present it has been estimated that the volume of the mid-oceanic ridges is sufficient to affect the sea level by 0.5 km.) Equally, when a spreading centre becomes inoperative the mid-oceanic ridge subsides and the greater volume of the ocean basin leads to lower sea levels.

The distribution of land and sea at different times in the geological past has greatly influenced the pattern of oceanic circulation and the climates of the continents. As the continents have moved from polar to equatorial locations and beyond into the opposite hemisphere, so their climates have changed. Where mountains have been raised as a result of tectonic activity, the climatic patterns also have been altered. These changes have influenced the plant and animal life of the world.

### *Mineralisation and plate boundaries*

It has long been known that certain minerals occur in relatively restricted areas (metalliferous zones) of the Earth's surface. The relationship of these zones to the broad geomorphology of the continents was not fully appreciated until recent times. Most of the mineral wealth of Europe was emplaced during the Hercynian orogeny which was responsible for the uplands of the central European area. Copper, tin and molybdenum are deposited from hydrothermal solutions associated with granitic magmas. These are intruded into the core of mountain ranges and so if it is possible to trace the line of former mountain chains it should be possible to predict the occurrence of these minerals. Many metallic elements amount to extremely small percentages of the earth's crust, and only where there has been natural enrichment or accumulation are they worth mining. The process of creation and destruction of crustal material at plate margins leads to a natural concentration of metallic elements.

Along the mid-oceanic spreading centres new crustal material is created as the lithospheric plates are pushed apart. Water circulating through this new crustal material leaches out metals. The best known example is the Red Sea rift which is the source of some of the richest sulphide ores, 2000 m below sea level in 20–100 m thick beds. They are deposited from metal-charged brines which bring iron, zinc, copper, lead, silver and gold from the underlying rocks of the spreading centre. Thus if a search is made for the spreading centres of previous oceans where similar conditions prevailed this should reveal their associated metalliferous zones. Samples taken from the sea floor show that the mid-oceanic ridge rocks are enriched in iron, manganese, copper, cobalt, chromium, uranium and mercury. The rocks associated with the destructive margins of tectonic plates are also enriched with metallic minerals. As one plate is subducted beneath another, it passes down into the asthenosphere where it melts and its metallic elements are released to migrate up into the rocks above. This has occurred in the Andes, the source of many metalliferous ores and the site of the mythical El Dorado. The goldfields of Africa and Western Australia occur on the sites of former subduction zones. Enrichment also can be seen in the Troodos mountains of Cyprus where copper, iron and chromium are found and iron and manganese-rich layers are interbedded in an ophiolite rock sequence.

There is also evidence that metallic elements are released in sequence from the melting slab of crustal material as it sinks. At shallow depth, close to the oceanic trench, iron and manganese ore bodies occur, gold and copper occur in the rocks overlying the zone where the descending slab is 150–200 km below the surface and further away from the trench, where the Benioff zone is deepest, lead, zinc and silver are given off. The amount of potassium in erupted materials also increases with distance from the trench.

When the location of these past and present plate margins are known, the search for economic minerals is made easier. Conversely, the knowledge of the presence of ore bodies helps to establish the former position of the constructive and destructive margins of lithospheric plates and to add to our knowledge of paleogeography.

### *Fossil fuels and plate boundaries*

The conditions of sedimentation and lithification associated with the trailing margins of lithospheric plates provide a suitable environment for the preservation of hydrocarbons, the raw materials for oil and natural gas.

Around Britain, associated with rifting before the opening of the Atlantic in the Jurassic, graben structures developed in the North Sea where organically rich sediments accumulated. Covered by later sediments and acted upon by limited pressure and heat, the hydrocarbons migrated from their organic sources to accumulate in fold and fault structures or against salt diapirs rising from the underlying Permian salt deposits.

Somewhat similar, but not identical, geological conditions exist in other major oil and gas fields of the world. The most common site for these is on continental shelves where the seaward slope of the rocks encourages the oil and gas to migrate up dip towards the continent. The American Gulf coast and the Arabian Gulf oil fields conform to this structural picture. The foreland basin, which develops on the continental side of a subduction system, is another suitable place where there exists the fortunate coincidence of an organic source, an appropriate thermal history for oil and gas production and suitable rock reservoirs in which the oil and gas can be trapped. By contrast, oil and gas have not been found in outer arc basins, which have similar structures but the thermal regime at depth is apparently not right for oil formation.

The accumulation of organic matter in the geological past to form the deposits of coal which have been the basis of economic wealth for many countries can also be linked to the large-scale geomorphology of the lithospheric plates. At the time of accumulation of the deposits of the Carboniferous coal measures, the European and North American continental areas lay near the Equator. It is thought the coal swamps accumulated in conditions similar to a foreland basin which was repeatedly flooded by fresh water so that luxuriant plant growth could take place. A total depth of up to 3000 m of sediments was deposited, with individual seams of coal between 1 and 3 m thick, the thickest reaching 6 m. Occasional inundation by sea waters gave thin strata with marine fossils, but most of the sequence is composed of sandstones, shales and coals. In South Wales, the coals were laid down behind barrier beaches in deltas produced by northwards-flowing rivers which brought the sandy and clayey sediments.

Although the repeated changes of sea level which led to the cyclic accumulation of the sandstones, shales, seat-earths and coals could have resulted from the activity of mid-oceanic spreading centres, the southern continental area of Gondwana had drifted over the South Pole and was subjected to glaciation. As ice sheets increased in thickness, sea level was reduced and as the ice melted it was raised again. Which of these two mechanisms was responsible for the cyclic changes in sea level and indirectly the accumulation of the coal deposits is not clear.

### *Biogeographical implications*

The consequences of the changes in geographical position which have affected the continental masses since the end of the Mesozoic have been profound. The climate, phytogeography, zoogeography and soil geography have all been affected and the known movement of the continents helps to explain some of the anomalous distributions of flora and fauna revealed by geographical investigations. Some of the most important factors are:

1. The presence in the Mesozoic of the Gondwana and Laurasian land masses, but not completely separated by the Tethys Sea.
2. The rifting of the North American plate from Africa and Europe and its movement westwards in the Lower Cretaceous and the beginnings of the splitting of the Indian and Antarctic plates from Africa.
3. The separation of the South American plate from Africa during the Upper Cretaceous and the beginning of the separation of Australia from Antarctica.
4. The northern drift of the Indian plate, eventually closing the Tethys Sea and, with Africa, raising the Alpine–Himalayan mountain chain in a continent to continent collision.
5. The development of the Rockies and Andes as the separate parts of America drifted westwards.
6. The linkage of North and South America in the Pliocene.
7. Changes in climate as a result of continental position at different periods of geological history.
8. The effects of multiple glaciation during the Pleistocene.

To explain the distribution of many plants and animals raised problems for biogeographers who wished to account for their presence or absence in different parts of the World. Distributions of plant and animal species in the fossil record are different from those of the present day and this helped to support the ideas of continental drift as originally advanced by Wegener.

In the nineteenth century land bridges were proposed as an alternative to continental drift to enable the movement of animals and plants between continental masses. This certainly was a possibility in some cases, as in the Indonesian islands, but it did not account for the presence of similar species on either side of the Atlantic, or for the distribution of the southern beech in South America, Australia and New Zealand. One of the distributions Wegener drew attention to was the occurrence of fossils of a small Permian reptile which occupied a restricted area in South Africa and Brazil. Undoubtedly, the splitting of the continental masses also resulted in the isolation of different groups of animals, allowing the independent development of the marsupials in Australia for example.

During the Mesozoic, the two large continental areas, Gondwana and Laurasia, were separated by the Tethys Sea. This was the age of the reptiles and the fossil record shows that the crocodiles and two orders of dinosaurs originated in Gondwana. Turtles, lizards and snakes originated in Laurasia and the evidence indicates that there was free movement between the two continents until the middle of the Cretaceous. The climate was warm and equable, so it has been suggested that there was little incentive for the reptilian fauna to evolve further, each animal having its own niche in the ecology of those times.

By the late Cretaceous, conditions had changed; the continental fragments had begun to drift apart and high sea levels flooded the lower continental shelves to create isolated land areas, each with its own fauna and flora. On these core areas evolution was stimulated to take place more rapidly, particularly by mammals. Many present-day orders can be traced back to this important phase of evolution. In Laurasia, the early primates, bats, carnivores, horses, cattle, rabbits, and hares can all be traced in the fossil record. The southern continents were also isolated: South America had six orders of mammals, three of which have since become extinct, and in Africa the higher primates evolved concurrently with the elephants and rhinoceros. In this way the basic pattern of distribution of many animal species was laid down.

# *Further reading*

Information about the major sub-divisions of the continents and oceans may be obtained from many scattered specialist and general sources. Regional geographical texts should provide an assessment of the physical background of the area studied but often do not, depending upon the outlook of the author. The study of geography can offer the student a unique blend of the significance of the physical environment and the effect of mankind's occupation and utilisation of it. Understanding how the continents and oceans have developed, the geography of their major physiographic regions and provinces, and the landforms they contain, is an important basis for all geographical studies.

In order to place the study of world geomorphology into an appropriate framework the reader is encouraged to refer to K.J. Gregory's (1985) *The nature of physical geography* (Edward Arnold). For further historical details about the development of geomorphology, the reader should consult R.J. Chorley, A.J. Dunn and R.P. Beckinsale's (1964) *The history of the study of landforms* (Methuen). A good general introduction to the earth sciences is given in D.G. Smith's (1982) *The Cambridge encyclopaedia of earth sciences* (Cambridge University Press) and the problem of identifying geomorphological units in the landscape is discussed by D.L. Linton (1950) in 'The delimitation of morphological regions' in *London essays in geography* (London School of Economics), edited by L.D. Stamp and S.W. Wooldridge, pp. 199–217. The Russian approach to the analysis of landforms is discussed in G. Rikhter, V. Preobrazhenskiy and V. Nefed'yeva's article in *The Geographical Magazine*, **48** (5), 267–72.

The early theory of continental drift was expounded

in A.L. du Toit's (1937) *Our wandering continents* (Oliver and Boyd), and J.F. Umbgrove's (1947) *The pulse of the Earth* (Martinus Nijhoff, The Hague), and the first proposal for a suitable driving mechanism came in A. Holmes' (1939) *Principles of physical geology* (Nelson). Titles which provide the reader with current information on plate tectonics include T.H. van Andel's (1985) *New views on an old planet* (Cambridge University Press), and P. Cattermole and P. Moore's (1985) *The story of earth* (Cambridge University Press). A series of scientific papers on plate tectonics edited by J.M. Bird and B. Isachs (1972) from the *Journal of Geophysical Research* was published by the American Geophysical Union. A more popular approach was adopted in J.T. Wilson's (1971) edited papers from *The Scientific American* entitled *Continents adrift* (W.H. Freeman). A more specialised geological treatment is given in B.F. Windley's (1984) *The evolving continents* (John Wiley). The Open University geologists, I.G. Gass, P.J. Smith and R.C.L. Wilson, have produced *Understanding the Earth* (Artemis Press) as a background text for the science foundation course.

The major physiographic features of the Earth have received scant attention until recent years, but R. Gardner and H. Scogging's (1983) *Mega-geomorphology* (Oxford University Press) deserves attention from the reader. C.D. Ollier's (1981) *Tectonics and landforms* (Oliver and Boyd) contains many examples of the importance of endogenous processes for landform development. Textbooks on geomorphology are numerous, but are almost all devoted to the problems of process, rather than the distribution of features. In the compilation of this book, the author has found K.W. Butzer's (1976) *Geomorphology from the Earth* (Harper and Row) most useful as well as R.J. Rice's (1977) *Fundamentals of geomorphology* (Longman). Changes over the past three million years are most critical for landform development and to this end A.S. Goudie's (1977) *Environmental change* (Oxford University Press) should be consulted.

No study of world geomorphology can afford to ignore the tremendous influence of L.C. King's *Morphology of the Earth* (Oliver and Boyd), a 699 page state of the art study published in 1962. No other book reaches or aspires to the scale of King's work, a study which is concerned with every continent.

*Africa*

In addition to King's *Morphology of the Earth*, and Goudie's *Environmental change*, the reader interested in Africa will find it useful to refer to C. Buckle (1978) *Landforms in Africa* (Longman) and J.M. Pritchard's (1979) *Landform and landscape in Africa* (Edward Arnold).

*The Americas*

The landforms of Canada have been well described by B.J. Bird (1980) in *The natural landscapes of Canada* (John Wiley). The regional geomorphology of the USA has several books devoted to its study. The first was the *Physiography of the western United States* by N.M. Fenneman (1931), followed by the *Physiography of the United States* (1938), both published by McGraw-Hill. Subsequent books by Thornbury (1965) *Regional geomorphology of the United States* (John Wiley) and C.B. Hunt (1967) *Physiography of the United States* (W.H. Freeman), provide an excellent overview of the North American landmass. Information on central and southern America is less easily found, other than an incidental comment in specialised reports, not usually specifically concerned with geomorphology. Articles on the development of the Appalachian mountains and on the San Andreas fault are included in *Continents adrift*.

*Asia*

Geomorphic knowledge of Asia is patchy, but the features of Siberia are discussed by S.P. Suslov (1961) in *The physical geography of Asiatic Russia* translated by N.D. Gershevsky (W.H. Freeman). India was well served by the Geological Survey in the early years of the present century and W.N. Wadia's (1926) *Geology of India* (Macmillan) provides a good background to a study of that country's landforms. Z. Songqiao's (1986) *Physical geography of China* (John Wiley) is cast in a regional geomorphological framework and was most helpful in the compilation of this chapter.

*Australia*

Geologists have been fortunate in the compilation of I.F. Clark and B.J. Cook's (1983) *Perspectives of Earth* (Australian Academy of Science) which deals specifically with Australasia and many of that continent's landforms. G.W. Leeper's edited volume (1960) *The Australian environment* (CSIRO) discusses the broad detail of the regional geomorphology of Australia. The reports of the CSIRO Land Research Division all contain detailed geomorphological reports, mainly of the centre and north; far more detail than can be included in this book. N. and A. Learmouth's (1971) *Regional landscapes of Australia* (Heinemann) is attractively produced but is not geomorphologically oriented. E. Loffler's (1977) *Geomorphology of Papua New Guinea* by contrast is concerned solely the landforms of the island.

*Europe*

Readers who wish to pursue the geomorphology of Europe further than has been possible in this book should turn to C.E. Embleton's (1984) *The geomorphology of Europe* (Macmillan) and D.V. Ager's (1980) *The geology of Europe* (McGraw Hill). In Britain the Methuen University Paperbacks series (edited by E.H. Brown and K.M. Clayton, various dates) covers the geomorphology of the British Isles in great detail.

Many of the European countries have books and journals devoted to geomorphology or specific aspects of the subject. Those immediately available in Britain include *Geography*, *The Geographical Journal*, and the *Proceedings of the Institute of British Geographers*, which have occasional articles strictly concerned with geomorphology. Illustrations of many of the features discussed in this book may be seen in *The Geographical Magazine* during the past decade.

This bibliographical commentary does not attempt to be exhaustive; the information contained in the preceeding chapters has been gleaned from a wide range of sources. To avoid interruption of the text with frequent and often obscure references, these have been omitted and grateful acknowledgement is made here of the many sources used.

# Index